T0306104

FORENSIC RADIO SURVEY TECHNIQUES FOR CELL SITE ANALYSIS

FORENSIC RADIO SURVEY TECHNIQUES FOR CELL SITE ANALYSIS

Joseph Hoy
Forensic Analytics Ltd
UK

This edition first published 2015
© 2015 John Wiley & Sons, Ltd

Registered Office
John Wiley & Sons, Ltd, The Atrium, Southern Gate, Chichester, West Sussex, PO19 8SQ, United Kingdom

For details of our global editorial offices, for customer services and for information about how to apply for permission to reuse the copyright material in this book please see our website at www.wiley.com.

The right of the author to be identified as the author of this work has been asserted in accordance with the Copyright, Designs and Patents Act 1988.

Wiley also publishes its books in a variety of electronic formats. Some content that appears in print may not be available in electronic books.

Designations used by companies to distinguish their products are often claimed as trademarks. All brand names and product names used in this book are trade names, service marks, trademarks or registered trademarks of their respective owners. The publisher is not associated with any product or vendor mentioned in this book. This publication is designed to provide accurate and authoritative information in regard to the subject matter covered. It is sold on the understanding that the publisher is not engaged in rendering professional services. If professional advice or other expert assistance is required, the services of a competent professional should be sought.

Limit of Liability/Disclaimer of Warranty: While the publisher and author have used their best efforts in preparing this book, they make no representations or warranties with respect to the accuracy or completeness of the contents of this book and specifically disclaim any implied warranties of merchantability or fitness for a particular purpose. It is sold on the understanding that the publisher is not engaged in rendering professional services and neither the publisher nor the author shall be liable for damages arising herefrom. If professional advice or other expert assistance is required, the services of a competent professional should be sought.

Library of Congress Cataloging-in-Publication data applied for

ISBN 9781118925737

Set in 11/13pt Times by SPi Publisher Services, Pondicherry, India

1   2015

For Nicola, Ellie and Isabel, who always find much more interesting things for me to do when I should be writing.

Contents

6 Other Cellular Network Types

About the Author

Joseph Hoy has a background in telecoms engineering and training. Gaining experience initially as an IT and telecoms engineer with BT, NCR and AT&T, Joseph moved across to cellular telecoms and worked on a variety of engineering and training projects for Nokia around the world.

He has also worked as a cell site analyst and expert witness and has compiled forensic reports for a variety of police forces and agencies and has presented them in a range of courts, including the Old Bailey in London.

Joseph specialises in 4G LTE training and is Managing Director of Forensic Analytics, which has developed a suite of software applications that automate many of the processes involved in cell site analysis.

He is a member of the Institution of Engineering and Technology (IET) and is a member of the United Kingdom Forensic Regulator's cell site analysis working group.

Joseph lives in the United Kingdom with his wife and two young daughters.

Preface

This book is intended to serve two purposes: to provide a coherent explanation of the theories and procedures that underpin forensic radio surveying and of the network technologies being surveyed in a form that can be read cover to cover as a text book; but also to act as a reference resource that can be dipped into as needed.

Forensic radio surveying is undertaken in support of the digital forensics discipline of cell site analysis and is, on the face of it, a very simple process: 'go to a location, switch on the survey device, go back and process the results'. But without a proper understanding of the operation of the cellular networks being surveyed, of the issues related to different networks or technologies and without knowledge of the things that can go wrong with surveys (and their remedies), survey results will not be as accurate or useful as they could be.

The ability to demonstrate a full understanding of the cellular technologies and forensic radio surveying techniques is also of use if surveyors are called to court to explain their evidence. A lack of technical knowledge or understanding may be quickly discovered under cross-examination and will be used to undermine the credibility of any cell site evidence being presented.

In general, the aims of the this book are to provide a readily understandable introduction to the topic to those new to forensic radio surveying and to act as an aide memoire to remind more experienced forensic radio surveyors of information related to surveying that they have learned on training courses but may sometimes have trouble remembering.

The forensic disciplines of cell site analysis and radio surveys are dynamic and challenging.

New technologies, updated techniques and evolving networks ensure that the specific details of the topic change over time. We endeavour to keep up with these changes and will update the information in this book at regular intervals.

We recognise, however, that we will not always get everything right and may not always be quick enough to amend outdated material, so we welcome comments or criticism from readers. We will be happy to debate the topics and issues raised, provide further information and generally engage with the forensics community as required to ensure that this book is as accurate, comprehensive and up to date as possible.

Questions, comments and feedback can be sent to: forensic.radio.surveying@forensicanalytics.co.uk

Acknowledgements

A great number of people have helped with the development of this book, including:

My colleagues at Forensic Analytics, Martin Griffiths and Andrew Hausler.

Ian Clark, David Bell and Tom Hoy at Lynross Training. Much of the basic network overview content in Chapters 4 and 5 is based on material that we jointly developed for Lynross courses and they kindly gave me permission to adapt it for this book.

My colleagues at LGC Forensic – Ceri Walsh, Sue Carter, Sue Delahaye, Nick Chandler and Mick Shelley – who helped me to understand what cell site is all about.

David Bristowe and Professor Jan Stuart, who were instrumental in developing the discipline of cell site analysis.

Matthew Tart and Dr Iain Brodie of CCL Group Ltd (both formerly of the Forensic Science Service) for suggestions and ideas and also for their 2010 paper '*Historic cell site analysis – Overview of principles and survey methodologies*', which they co-authored with Nicholas Glead and James Matthews when they were all working for the Forensic Science Service.

Fellow cell site experts and practitioners who provided comments that were used in the precursor document to this book, including Dominic Kirsten, Ben Spencer, Thea Selby, Phil Gardiner, Vicki Meaton, Mark Johnson, Greg Smith, Nicky Haigh, Peter Brown and Duncan Brown.

Chris Cox, of Cox Communications, gave advice on presenting the mathematics of radio systems in Chapter 2. Professor Berthold K.P. Horn at MIT, Bruno Xaiver of CelPlan, Brazil and Don Hill of Proactive Technical Solutions, Inc who provided information about CDMA2000 in Chapter 6.

Lester Wilson of 3G Forensics, Brian Edwards of FMS Ltd, Steven James and John Morfit of Ascom and Shaun Desmond of Anite who provided assistance and information related to their respective survey devices.

Tom Hoy and Ian Church for proofreading and grammatical advice. Anna Smart, Sandra Grayson, Clarissa Lim, Alan Mill and Radjan Lourde Selvanadin at Wiley.

A number of figures and tables in this book were taken from various 3GPP Technical Specifications (TSs) or Technical Reports (TRs). In relation to this content: © 2013. 3GPP TM TSs and TRs are the property of ARIB, ATIS, CCSA, ETSI, TTA and TTC who jointly own the copyright in them. They are subject to further modifications and are therefore provided to you 'as is' for information purposes only. Further use is strictly prohibited.

A number of tables in the book were taken from various 3GPP2 Technical Specifications. In relation to this content: COPYRIGHTED MATERIAL reproduced and distributed by John Wiley & Sons under written permission of the Organizational Partners of the Third Generation Partnership Project 2 (3GPP2).

Glossary

0G	Pre-cellular radiotelephone networks
1G	First Generation mobile networks
2G	Second Generation mobile networks, e.g. GSM
2.5G	Enhanced 2G networks, e.g. GPRS
2.75G	Enhanced 2G networks, e.g. EDGE
3G	Third Generation mobile networks, e.g. UMTS
3.5G	Enhanced 3G networks, e.g. HSPA/HSPA+
3GPP	Third Generation Partnership Project – global standards body
3GPP2	3GPP mark 2 – United States standards body
4G	Fourth Generation mobile networks, e.g. LTE
5G	Fifth Generation of mobile networks
802	IEEE family of networking standards
802.11	IEEE WiFi standards family
802.16	IEEE WIMAX standards family

A

ACC	Access Control Class
Active	a 3G cell currently selected to serve a mobile device's Connected Mode connections
ADC	Analogue to Digital Conversion
AM	Amplitude Modulation
AMPS	Advanced Mobile Phone System
ANPR	Automatic Number Plate Recognition

ANSI	American National Standards Institute
ARFCN	Absolute Radio Frequency Channel Number in 2G
ARIB	Association of Radio Industries and Businesses
ATIS	Alliance for Telecommunications Industry Solutions
AuC	Authentication Centre

B

BA List	BCCH Allocation List – neighbour cell list in 2G
BCC	Base Station Colour Code (part of BSIC)
BCCH	Broadcast Control Channel
BER	Bit Error Rate
BLER	Block Error Rate
BSC	Base Station Controller (in 2G)
BSIC	Base Station Identity Code (in 2G)
BSID	Base Station ID (in CDMA2000)
BSS	Base Station Subsystem
BTS	Base Transceiver Station (in 2G)
BWA	Broadband Wireless Access

C

C1	Cell Selection algorithm (in 2G)
C2	Cell Reselection algorithm (in 2G)
Camp On	To select a cell as the serving cell in Idle Mode
CCCH	Common Control Channel
CCH	Control Channel
CCTV	Closed Circuit Television
CCSA	China Communications Standards Association
CDG	CDMA Development Group
CDR	Call Detail Record
CDMA	Code Division Multiple Access
CDMA2000	3G network type
cdmaOne	2G network type
CELL_DCH	Cell Dedicated Channel state (in 3G)
CELL_FACH	Cell Forward Access Channel state (in 3G)
CELL_PCH	Cell Paging Channel state (in 3G)
CGI	Cell Global ID
CI	Cell ID
CINR	Carrier to Interference and Noise Ratio
Connected Mode	The state a mobile device is in when a connection has been established to a base station and traffic flow is possible

CPICH	Common Pilot Channel (in 3G)
CRH	Cell Reselection Hysteresis
CRS	Cell-specific Reference Signal (in 4G)
CS	Circuit Switched, e.g. traditional voice telephony service
CSA	Cell Site Analysis
CSAS	Cell Site Analysis Suite
CSFB	Circuit Switched Fallback
CSG	Closed Subscriber Group (for 3G/4G femtocells)
CSP	Cellular Service Provider

D

DAC	Digital to Analogue Conversion
D-AMPS	Digital Advanced Mobile Telephone System
dB	decibels
dBm	decibel milliwatts
dBW	decibel watts
dBi	decibel isotropic
DC-HSPA	Dual Carrier HSPA
DECT	Digital Enhanced Cordless Telephone
Dedicated Mode	Original term for Connected Mode used in GSM
DCS	Digital Communications Service
DL	Downlink
DSA	Derived Service Area

E

E.164	ITU international phone number standard
E.212	ITU network numbering (MCC + IMSI) standard
EARFCN	Evolved Absolute Radio Frequency Channel Number (in 4G)
Ec/Io	Energy per chip/Interference – signal to noise ratio measurement (in 3G)
Ec/No	Energy per chip/noise – signal to noise ratio measure (in 3G)
ECGI	EUTRAN Cell Global Identifier (in 4G)
ECM	EPS Connection Management (in 4G)
EDGE	Enhanced Data Rates for Global Evolution, PS data for 2G networks
E-GSM	Extended GSM900 band
EIR	Equipment Identity Register
eNB	EUTRAN Node B (also Evolved Node B) – 4G base station
eNB ID	eNB Identifier
EPC	Evolved Packet Core (4G core network)

EPLMN	Equivalent PLMN
ESN	Electronic Serial Number
ETSI	European Telecoms Standards Institute
EUTRAN	Evolved Universal Terrestrial Radio Access Network (in 4G)
EV-DO	Evolution – Data Optimised (or Data Only)

F

FACCH	Fast Associated Control Channel
F-BCCH	Forward Broadcast Control Channel
FCH	Frequency Correction Channel
FDD	Frequency Division Duplex
FDMA	Frequency Division Multiple Access
Femtocell	A small-scale cell/base station designed to be deployed at a user's home or office, which provides a small bubble of network service
FM	Frequency Modulation
F-PCH	Forward Paging Channel
F-PICH	Forward Pilot Channel
F-SYNC	Forward Synchronisation Channel
FWA	Fixed Wireless Access

G

GERAN	GSM/EDGE Radio Access Network
GGSN	Gateway GPRS Support Node
GHz	Gigahertz (billions of cycles per second)
GPRS	General Packet Radio Service, PS data for 2G networks
GPS	Global Positioning System
GSM	Global System for Mobile, 2G network type
GSM-R	GSM for Railways

H

Handover	The process of passing the active connections for a mobile device in Connected Mode from one cell/base station to another
HF	High Frequency
HHO	Hard Handover
HLR	Home Location Register
HPLMN	Home PLMN
HSDPA	High Speed Downlink Packet Access
HSPA/HSPA+	High Speed Packet Access, fast PS data for 3G networks

HSS	Home Subscriber Server (evolved form of HLR)
HSUPA	High Speed Uplink Packet Access
Hysteresis	A process that attempts to prevent an Idle Mode mobile device reselecting to a new cell too quickly after a previous reselection
Hz	hertz (cycles per second)

I

iDEN	Integrated Digital Enhanced Network
Idle Mode	The state where a mobile device is powered on and attached to a network but has no active control or traffic connections
IEEE	International Electrical and Electronics Engineers
IFAST	International Forum on ANSI-41 Standards Technology
IMEI	International Mobile Equipment Identifier
IMEISV	IMEI and Software Version number
IMSI	International Mobile Subscriber Identifier
IP	Internet Protocol
IRAT	Inter-Radio Access Technology
IS	Interim Standard
IS54	D-AMPS/TDMA 2G system
IS95/A/B	cdmaOne 2G system
IS136	Enhanced D-AMPS/TDMA 2G system
IS2000	CDMA2000 system
ISDN	Integrated services Digital Network
ISHO	Inter System Handover
ITU	International Telecommunications Union

K

K	Subscription-specific secret security key
kHz	kilohertz (thousands of cycles per second)

L

LA	Location Area (in 2G and 3G)
LAC	Location Area Code
LAI	Location Area Identifier (LAC plus country code, network code)
LAU	Location Area Update
LEA	Law Enforcement Agency
LF	Low Frequency
LMDS	Local Multipoint Distribution Service

LOS   Line of Sight
LTE   Long Term Evolution, a 4G network type
LW   Long Wave

M

MAC   Medium Access Control
Mbps   Megabits per second
MCC   Mobile Country Code, e.g. 234 for the UK
MC-HSPA   Multi-Carrier HSPA
MDN   Mobile Directory Number
ME   Mobile Equipment
MEID   Mobile Equipment ID
MF   Medium Frequency
MFN   Multi Frequency Network
MGW   Media Gateway
MHz   Megahertz (millions of cycles per second)
MMDS   Multi-channel Multipoint Distribution Service
MME   Mobility Management Entity (in 4G)
MMS   Multimedia Messaging Service
MNC   Mobile Network Code, e.g. 10 for O2 UK
MNO   Mobile Network Operator
MS   Mobile Station, a 2G mobile device
MSC   Mobile Switching Centre (2G/3G CS core network node)
MS-ISDN   Mobile Station ISDN number – mobile phone number
MSIN   Mobile Subscriber Identification Number
MSS   MSC Server
MTPAS   Mobile Telephony Privileged Access Scheme
MuNST   Multi Network Survey Tool, as CSurv device
MVNA   Mobile Virtual Network Aggregator
MVNE   Mobile Virtual Network Enabler
MVNO   Mobile Virtual Network Operator
mW   milliwatts
MW   Medium Wave

N

NCC   Network Colour Code (part of BSIC)
NCL   Neighbour Cell List (in 3G and 4G)
NID   Network ID
NLOS   Non Line of Sight

NMT   Nordic Mobile Telephone
NR   National Roaming

O

OFDMA   Orthogonal Frequency Division Multiple Access (in 4G)
OTSR   Omni-directional Transmit, Sectorised Receive

P

PCCH   Paging Control Channel
PCH   Paging Channel
PCI   Physical-layer Cell ID (in 4G)
PCS   Personal Communications System
PDC   Personal Digital Cellular
PDN-GW   Packet Data Network Gateway (4G core network node)
PDSN   Packet Data Service Node
P-GSM   Primary GSM900 band
PHS   Personal Handyphone System
PLMN   Public Land Mobile Network
P_MAX   Maximum permitted uplink transmit power
PN   Pseudo Noise
POI   Period of Interest
PRL   Preferred Roaming List
PS   Packet Switched, e.g. the data transmission mechanism used by data
           networks like the Internet
PSC   Primary Scrambling Code (in 3G)
P-SCR   Primary Scrambling Code – alternative abbreviation (in 3G)
PSS   Primary Synchronisation Signal (in 4G)
P-TCH   Packet switched Traffic Channel (n 2G)
P-TMSI   Packet switched Temporary Mobile Subscriber Identifier (in 2G and
           3G)

R

R   Cell Reselection algorithm (in 3G and 4G)
RA   Routing Area (in 2G and 3G)
RAC   Routing Area Code
RACH   Random Access Channel
RAI   Routing Area Identifier
RAU   Routing Area Update

RAN	Radio Access Network
RAND	Random number used in authentication
RAT	Radio Access Technology
RB	Resource Block (in 4G)
RES	Response sent during authentication
Reselection	In Idle Mode, the process by which a mobile device selects the serving cell that it will camp on
RF	Radio Frequency
RFPS	Radio Frequency Propagation Survey
RIPA	Regulation of Investigatory Powers Act
RNC	Radio Network Controller (in 3G)
RNC ID	RNC Identifier
RNS	Radio Network Subsystem (in 3G)
RRC	Radio Resource Control
RRM	Radio Resource Management
RSCP	Received Signal Code Power (in 3G)
RSRP	Reference Signal Received Power (in 4G)
RSRQ	Reference Signal Received Quality (in 4G)
RSSI	Received Signal Strength Indicator
RTT	Radio Transmission Technology
RXLev	Received Signal Level (in 2G)
RXQUAL	Received Signal Quality (in 2G)

S

S	Cell selection algorithm (in 3G and 4G)
SAC	Service Area Code
SACCH	Slow Associated Control Channel
SC-FDMA	Single Carrier Frequency Division Multiple Access
SCH	Synchronisation Channel (in 2G)
SDCCH	Standalone Dedicated Control Channel
Serving	Term applied to the cell that an Idle Mode device is currently camped on or that a Connected Mode device is connected to
SF	Spreading Factor
SFN	Single Frequency Network
SGSN	Serving GPRS Support Node (2G/3G PS core network node)
S-GW	Serving Gateway (4G core network node)
SHO	Soft Handover (in 3G)
SID	System ID
SIM	Subscriber Identity Module
SINR	Signal to Interference and Noise Ratio
SMS	Short Message Service

SNR	Signal to Noise Ratio
SNR	Serial Number (part of an IMEI)
SPoC	Single Point of Contact
SR	Spreading Rate
SrHO	Softer Handover (in 3G)
SSID	Service Set ID
SSS	Secondary Synchronisation Signal (in 4G)
S-TMSI	Serving Temporary Mobile Subscriber Identifier (in 4G)
STSR	Sectorised Transmit Sectorised Receive
SVN	Software Version Number (part of an IMEI)

T

TA	Tracking Area (in 4G)
TAC	Tracking Area Code
TAC	Type Allocation Code (part of an IMEI)
TACS	Total Access Communications System
TAI	Tracking Area Identifier
TAU	Tracking Area Update
TCH	Traffic Channel (in 2G)
TDD	Time Division Duplex
TD-LTE	TDD version of LTE
TDMA	Time Division Multiple Access
TD-SCDMA	Time Division – Synchronous Code Division Multiple Access
TETRA	Terrestrial Trunked Radio
TIA/EIA	Telecoms Industry Association/Electronics Industries Alliance
TMSI	Temporary Mobile Subscriber Identifier
TO	Temporary Offset
TRX	Transceiver
TS	Technical Standard

U

UARFCN	UMTS Absolute Radio Frequency Channel Number
UE	User Equipment (in 3G and 4G)
UHF	Ultra High Frequency
UL	Uplink
UMTS	Universal Mobile Telecommunications System, a 3G network type
$UMTS_{HCR}$	High Chip rate version of UMTS (e.g. standard UMTS)
UMTS-FDD	FDD version of UMTS
$UMTS_{LCR}$	Low Chip rate version of UMTS (e.g. TD-SCDMA)
UMTS-TDD	TDD version of UMTS

URA_PCH   UTRAN Registration Area Paging Channel (in 3G)
USIM   UMTS SIM
UTRAN   Universal Terrestrial Radio Access Network (in 3G)
UWB   Ultra Wide Band

V

VHF   Very High Frequency
VLF   Very Low Frequency
VLR   Visitor Location Register
VoIP   Voice over IP
VoLTE   Voice over LTE

W

W   watts
WCDMA   Wideband Code Division Multiple Access
WiFi   Wireless Fidelity
WIMAX   Wireless Interoperability for Microwave Access
WLAN   Wireless Local Area Network
WLL   Wireless Local Loop
WRC   World Radio Conferences

X

XRES   Expected Response during authentication

1

Forensic Radio Surveys for Cell Site Analysis

1.1 Cell Site Analysis

Cell site analysis attempts to provide evidence of where a mobile phone may have been located when certain significant calls were made.

Mobile phone networks consist of a large number of radio 'cells', each of which covers a limited geographical area. Each cell is assigned a unique 'Cell ID', which is captured in the billing record (CDR or Call Detail Record) when calls are made.

Network operators are able, under tight regulatory guidelines, to provide details of the calls made by 'target' phones and can also provide details of the locations of the cells used by those phones.

Cell site analysis is designed to enable an investigator to determine whether calls made at or around the time of an incident or offence used cells that are located near the location of that offence.

1.2 Forensic Radio Surveying

Forensic radio surveys are designed to provide solid evidence to back up the assumptions made by investigators and cell site analysts.

Forensic radio survey equipment captures details of the cells that can be detected at a location and can indicate which cells would be selected for use by a phone being used at those locations.

Forensic Radio Survey Techniques for Cell Site Analysis, First Edition. Joseph Hoy.
© 2015 John Wiley & Sons, Ltd. Published 2015 by John Wiley & Sons, Ltd.

Forensic radio survey results can be used to prove that particular cells provide coverage at significant locations and can therefore indicate whether it is possible for a phone using those cells to have been at or near those locations when particular calls were made.

The only totally definite conclusion that can be drawn from cell site analysis is that the use of a particular cell by a target phone means that the phone must have been within the serving coverage area of that cell at the time.

Forensic radio surveys can set approximate limits to the area within which the target phone must have been located. This type of evidence can be very useful when attempting to prove or disprove an alibi or other statement.

Overall, forensic radio surveys add empirical rigour to an area of investigation that would otherwise fall prey to assumptions and wishful thinking.

Cell site analysis, based on a combination of CDR, cell location details and forensic radio survey results, can provide compelling evidence to support the allegations made by investigators.

2

Radio Theory

Cellular networks use communications methods based on basic RF (Radio Frequency) transmission principles.

2.1 RF Propagation

2.1.1 Radio Theory

Radio signals are created when an alternating electrical current is applied to an antenna.

Any electrical current applied to a conductor generates a magnetic field around the conductor. This field extends for only a short distance.

As shown in Figure 2.1, if the electrical current through an antenna is made to alternate – that is, to change its direction of flow from forwards to backwards, which causes the electrical current to move through a cycle of positive and then negative values – the entangled electrical and magnetic (or 'electromagnetic') field generated around the antenna begins to extend far beyond the antenna and turns into a radio signal.

As the current travelling through the conductor alternates, the electromagnetic field generated around the antenna expands to match each peak positive value and then collapses back towards the antenna, it then expands again to match the peak negative value and then collapses, and so on for each cycle of alternations.

Forensic Radio Survey Techniques for Cell Site Analysis, First Edition. Joseph Hoy.
© 2015 John Wiley & Sons, Ltd. Published 2015 by John Wiley & Sons, Ltd.

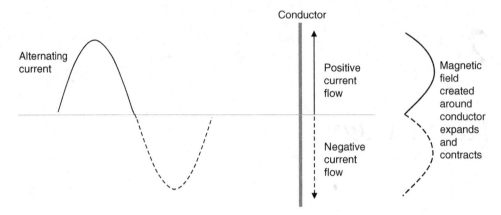

Figure 2.1   Alternating current

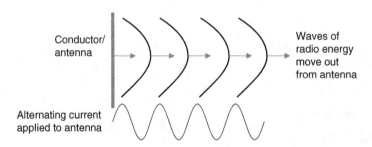

Figure 2.2   Generating a radio wave

If the rate of alternation (i.e. the number of cycles of changing positive to negative values per second) is sufficiently fast, each instance of the electromagnetic field that is generated does not have time to fully collapse before the instance generated by the next cycle of alternation begins to expand.

A conceptual way of imagining the effect of these alternating cycles could be as follows: A change in the electric current sets up a disturbance in the magnetic field close to the conductor. In turn, that disturbance causes the electric and magnetic fields further out from the conductor to change. Continuation of this process leads to a ripple of electric and magnetic fields travelling away from the conductor, which take the form of an electromagnetic wave. The whole process is very like the formation of a water wave when a stone is dropped into a pond.

As the source signal continues to cycle, wave after wave of electromagnetic fields are pushed out from the antenna as a phenomenon that we term 'radio waves'. This is demonstrated in Figure 2.2.

Table 2.1 SI units related to radio signal measurements.

Cycles/s	Scientific	Description	RF notation
1000	10^3	Kilohertz	kHz
1 000 000	10^6	Megahertz	MHz
1 000 000 000	10^9	Gigahertz	GHz
1 000 000 000 000	10^{12}	Terahertz	THz

Each alternation of the source electrical signal is termed a 'cycle' and the 'frequency' of a signal is calculated by counting the number of 'cycles per second'.

One cycle per second is known as 1 hertz (after Heinrich Hertz, the scientist who first demonstrated the existence of electromagnetic waves in the late nineteenth century) and is abbreviated as 1 Hz.

1000 cycles per second is 1 kilohertz (1 kHz), 1 million cycles per second is 1 Megahertz (1 MHz) and so on. More standard SI (International System of Units) descriptions of magnitude are shown in Table 2.1 [1].

Scientific notation is generally employed to represent large numbers or to standardise the way in which collections of numbers of both large and small magnitude are presented. This notation indicates the base value and a multiplier, which would usually be 10 raised to a power.

The value 1000 would be represented in scientific notation as 1×10^3, or a value such as 3 240 000 would be represented as 3.24×10^6.

The 'radio effect' can be created at any frequency, however low. There are, for example, systems that use very low frequencies (VLFs) of just a few tens or hundreds of hertz to send very long distance signals that can communicate with submarines on the other side of the world. VLF transmission is, however, quite difficult to achieve and very limited in the amount of information that can be transmitted. The most common forms of radio transmission use higher frequencies, with typical applications starting above around 3 kHz. The upper end of the range of frequencies that can be used to carry radio signals is generally accepted to be up at around 300 GHz, which is near the point where radio energy begins to be perceived as infra-red radiation and then light energy.

The range of frequencies that can be used to carry radio signals is therefore commonly classed as being between 3 kHz and 300 GHz. These frequencies are often collectively known as RF.

A radio receiver essentially consists of an antenna connected to a 'tuner' circuit that allows the user to specify the characteristics of the radio signal they wish to recover. The moving electromagnetic wave of the transmitted signal induces a current as it passes the receiving antenna, which can then be filtered and amplified to allow any information carried by the signal to be recovered.

Radio can, therefore, be thought of as 'induction over a distance'.

2.1.2   Basic Terminology

The basic terms employed to describe aspects of RF transmission are illustrated in Figure 2.3 and include:

- **Frequency:** The rate at which a source electrical signal alternates and therefore also the rate at which the generated electric and magnetic fields cycle from their peak positive values to their peak negative values, and back to their peak positive values again. Frequency is measured as 'cycles per second', with 1 cycle per second equal to 1 Hz. Frequency is usually represented using the symbol 'f'.
- **Wavelength:** The distance a radio signal travels during one cycle and hence the physical length of one cycle. Radio waves have a velocity, meaning the rate at which they move away from a transmitter, of the speed of light (300 000 km/s), so a 1 Hz signal (1 cycle/s) has a wavelength of 300 000 km for each cycle – it will have travelled 300 000 km during 1 s but will only have cycled once during that period. A 2 Hz signal has wavelength of 150 000 km for each cycle and so on. The speed of light is usually represented using the symbol 'c'. At cellular frequencies, a 900 MHz signal has a wavelength of approximately 30 cm and an 1800 MHz signal has a wavelength of around 15 cm. Wavelength is usually represented using the Greek lambda symbol 'λ'.
- The relationship between the velocity of a radio signal, its frequency and its wavelength can therefore be stated as $c = f\lambda$.
- **Amplitude:** Relates to the strength of the electrical and magnetic fields and is measured when the reach their peak positive and negative values.

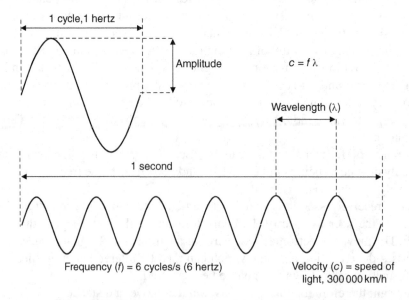

Figure 2.3   The frequency, wavelength and amplitude of a signal

- **Spectrum:** The range of frequencies that can be classed as being of RF are termed the 'radio spectrum'. This extends up to around 300 GHz at the highest. Electromagnetic frequencies above 300 GHz begin to be classed as 'infra-red' radiation and then 'light' rather than 'radio'.
- **Bandwidth:** A radio signal is typically centred on a 'carrier centre frequency' (or just 'carrier frequency') but extends to cover a range of frequencies either side of this centre point. The range of frequencies covered by a transmission is known as its 'bandwidth', that is the width of the radio band occupied by that transmission. A graphical representation of this is shown in Figure 2.4.

2.1.3 Propagation Modes

The frequency of a radio signal has an impact on the manner in which that signal propagates (i.e. the way in which the signal travels) as demonstrated in Figure 2.5.

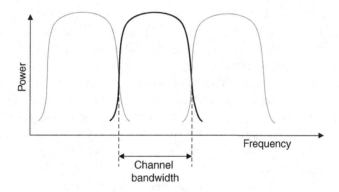

Figure 2.4 Bandwidth of a radio channel

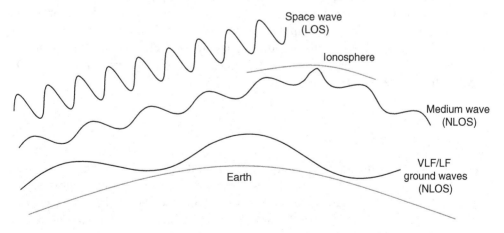

Figure 2.5 Radio propagation modes

Below 30 MHz, VLF, LF signals (which are also sometimes termed 'long wave' due to the long wavelength/LF) and Medium Frequency (also known as 'medium wave') signals are generally classed as 'ground wave' signals, as they tend to stay close to the ground and follow the curvature of the Earth following transmission.

This phenomenon is due to the properties of the ionosphere, a layer in the Earth's atmosphere that starts around 85 km above sea level and which reflects radio signals with a frequency below 30 MHz.

VLF and LF (also known as 'long wave') frequencies are useful for very long distance transmissions as these signals can hug the ground to travel beyond the transmitter's horizon. Signals with frequencies that are above the point where they can break free from the 'ground wave' effect but that are still below 30 MHz are able to reflect off the ionosphere and be carried beyond the natural horizon of the transmitter. This is the principle employed by MW (medium wave) radio stations.

Both of these are examples of 'non line of sight' (NLOS) transmission, in which a transmitter and receiver do not necessarily need to have a clear view of each other in order to exchange signals.

Higher frequency (and shorter wavelength) signals above 30 MHz tend to travel in straight lines and are also able to travel through the ionosphere, which generally makes them suitable only for 'line of sight' (LOS) transmission, which means that the transmitter and receiver do need a clear view of each other in order to exchange signals. This distance over which this type of transmission system can operate is limited by the curvature of the Earth. This means that signals from a terrestrial (ground based) transmitter can rarely extend past the transmitter's horizon to reach very distant ground-based receivers, although the range of this type of radio service can be increased by placing the transmitter and receiver as high as possible, for example on top of a tall building or a hill.

This type of high frequency wave is often called a 'space wave', due to the tendency of signals to pass through the ionosphere and travel out into space.

Cellular systems use frequencies in the UHF (Ultra High Frequency) band, which exists between 300 MHz and 3 GHz and are therefore limited to LOS transmission. However, the physical and geographical 'clutter' that exists in most areas where cellular services are deployed allows radio signals to be deflected and reflected over short distances in ways that allow them to reach places where there is no direct LOS between transmitter and receiver.

2.1.4   Multipath Transmission

Cellular systems deployed in very mountainous rural areas or heavily built-up urban areas often struggle to achieve LOS, as there is often some form of obstruction between the transmitter (base station) and receiver (mobile phone). In these scenarios a phenomenon known as 'multipath transmission', which is illustrated in Figure 2.6, becomes important.

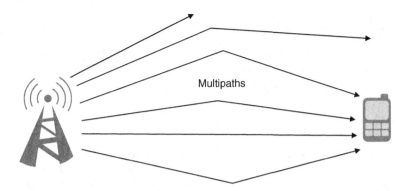

Figure 2.6   Multipath transmission

Radio waves propagate in much the same way as light waves; just like a beam of light, a radio signal can be blocked or attenuated by a large building or a hill, causing a 'radio shadow' to be created behind the obstruction. Also like light, however, radio signals can be diffracted (bent) as a result of travelling close to an object, or can reflect off smooth surfaces like windows or the sides of buildings, or scatter off rough surfaces; and each of these events can allow some of the signal's energy to travel along different propagation paths than would be possible using just LOS.

Some forms of interaction can cause a single beam of radio energy to be split into several different beams, each deflected along a different path.

This means that in a dense urban environment, signals from base stations can be received by mobile devices even if there is no direct LOS path between them, due to the signal bouncing off buildings or other objects and being reflected into areas that would not be reached by pure LOS transmission. The same is also true of the connection that travels in the reverse direction between a mobile phone and a base station.

Several duplicate elements of a signal may reach the mobile device having been reflected along different propagation paths to get there – each of these is known as a 'multipath'.

The signal being received by a mobile phone at any moment may consist of several multipaths combined together and will therefore be an aggregate of those separate 'copies' of the same signal. Multipaths can combine 'constructively', in which case the sum of their values creates a stronger signal, or they can combine 'destructively', in which case some or all of the multipaths cancel each other out and reduce the strength of the received signal. This is illustrated in Figure 2.7.

As multipaths are typically created by reflections, their paths can be altered by changes to the surface on which they are reflecting, so if a bus stops in front of a wall that had been causing a reflection the multipath created could be redirected along some other path. The same may happen if the phone moves. Multipath energy is added to and removed from the set being detected by a phone all the time, causing the signal strength measured by the phone (which is an aggregate or sum of all of the multipaths being received) to fluctuate quite markedly.

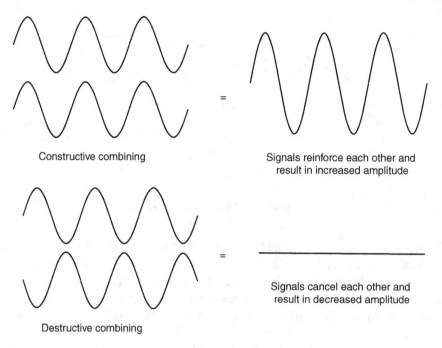

Figure 2.7   Multipath combining

LOS connectivity offers the best signal quality for a cellular service, but it is important to understand that a connection can still be maintained via diffracted, refracted or reflected signals even if no direct LOS exists.

2.2   Carrying Information on a Radio Signal

Radio is an analogue medium, in the sense that a radio signal is a continuously changing stream of energy that moves through an infinite number of values during each cycle.

All radio systems are therefore based on analogue transmission techniques. When the various types and generations of radio system are examined, however, some are described as 'analogue' systems and others as 'digital' – it is important to understand the differences between these concepts if the differences between the associated radio technologies are to be understood.

2.2.1   Analogue Transmission Systems

All early radio systems, including quite a significant number of systems that are still in use, relied on an analogue information transmission method.

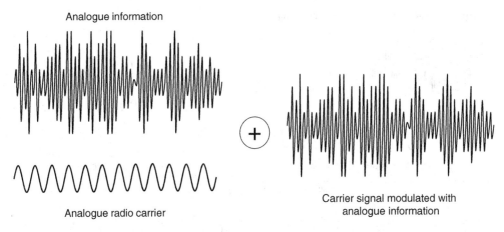

Figure 2.8 Analogue transmission

In an analogue system, a copy of the raw information to be transmitted – a person's voice or some music, for example – is simply overlaid onto a radio carrier frequency and the combined signal is then transmitted. This process is illustrated in Figure 2.8.

Sound is simply another form of analogue medium, so a voice, music and other forms of audio information exist as streams of analogue energy; and so the combination of an analogue sound stream and an analogue radio carrier creates a combined analogue radio signal.

The content of an analogue radio transmission is carried in the variety of 'modulations' or changes to the frequency and amplitude of the transmitted radio signal, which makes for a comparatively simple transmitter/receiver architecture but also creates a transmission medium that is easily disrupted.

All radio transmissions are susceptible to interference; sources of radio interference create 'noise' that combine with the radio signal. Too high a level of interference can impair a receiver's ability to understand the nature of the information being conveyed. Analogue transmission systems provide poor quality services in the presence of too much interference.

Analogue transmission also offers limited scope for security, as it can be difficult to apply encryption to analogue information streams.

Analogue transmission is still widely employed to carry services like broadcast radio – AM and FM radio stations transmit using comparatively basic analogue transmission techniques – but the majority of cellular systems migrated to digital techniques during the 1990s.

2.2.2 *Digital Transmission Systems*

Most modern radio systems are described as being 'digital radio' systems, which can be confusing.

As previously stated, all radio systems use analogue transmission techniques, as radio is an analogue medium. The distinction between analogue and digital

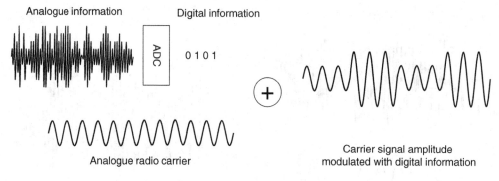

Figure 2.9 Digital transmission

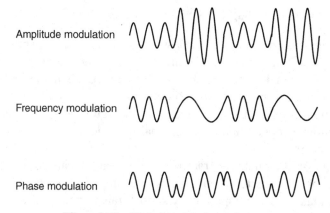

Figure 2.10 Digital modulation techniques

transmission is instead related to the format of the information that is conveyed via the radio connections.

An analogue transmission system modulates an analogue radio carrier with analogue information, such as an audio signal. A digital transmission system modulates an analogue radio carrier with a stream of digital ones and zeroes, as illustrated in Figure 2.9.

If the information to be transmitted is already in a digital format – computer data, Internet traffic and so on – then it can be conveyed directly to the transmitter. Information that starts in an analogue format, such as voice, must be converted from analogue to digital before being transmitted (and converted from digital back to analogue at the receiving end). Most digital transmission devices, such as a modern digital mobile phone, include the capability to perform ADC (Analogue to Digital Conversion) to allow audio 'traffic' to be transmitted over a digital radio service and DAC (Digital to Analogue Conversion) to convert it back to audio at the receiving end.

Digital systems encode binary data onto a radio carrier by modulating one or more of the basic properties of that radio carrier – this involves making changes to the frequency, amplitude or phase of the carrier.

Frequency modulation could, for example, involve increasing the frequency of the radio signal for a short period of time to represent a '1' in the transmitted information stream and decreasing the frequency to represent a '0'.

Amplitude modulation works in the same way but varies the power of the signal – higher power to represent a '1', lower power to represent a '0'.

Phase modulation is more complex and more difficult to visualise but involves rapidly jumping the transmitted radio signal from one part of its cycle to another without passing through the intervening parts – this manifests itself as a sharp change in the radio signal rather than the expected smooth progression through a cycle.

Simple examples of the various digital modulation schemes are outlined in Figure 2.10.

A simple digital modulation scheme would require one type of modulation to represent a '1' and a different type of modulation to represent a '0'; each modulation made to a radio carrier is known as a 'symbol' and the more modulations or symbols that can be encoded per second, the greater the data rate that can be carrier by a radio service.

With two modulations available, each symbol can carry one bit of data: 1 or 0.

Modern digital systems use advanced modulation schemes that use more than two modulation types; so if four different modulations (four different amplitude levels, for example) are supported then each change in the radio carrier can be used to carry two bits of data: 00, 01, 10 or 11.

With 16 modulation types (combinations of four amplitudes and four phases, for example) each symbol can carry four bits of data: 0000, 0001, 0010 and so on.

The fastest modern radio data systems can encode millions of symbols per second onto a radio carrier and each symbol can carry 2, 4, 16 or more bits of data.

Digital transmission techniques lie at the heart of the ability to access fast mobile broadband services. The fact that information is transmitted in a simple format, at least when compared to the infinite variety of properties that can be carried by an analogue transmission system, means that digital systems typically offer more consistent quality, especially in the presence of radio interference. The quality of a digital transmission can be further enhanced using complex 'error correction' techniques and the security of a radio link can be assured using sophisticated digital encryption schemes.

2.3 Radio Spectrum

2.3.1 Radio Bands and Channels

The range of possible radio frequencies is known as the radio spectrum. The usable range of frequencies available within the radio spectrum runs from around 3 kHz up to over 300 GHz. This spectrum may appear to be very wide but it is not infinite.

The radio spectrum in each country is controlled by that country's government, but governments cooperate to implement regional or global spectrum allocation plans. To ensure that interference between users is kept to a minimum, individual systems or networks are 'licensed' to operate within a particular range of radio frequencies – this

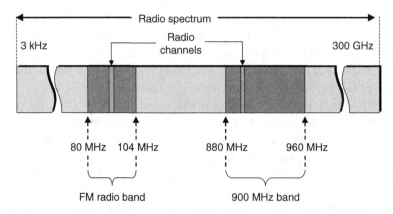

Figure 2.11   Radio bands and channels

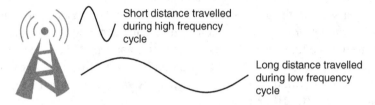

Figure 2.12   Frequency versus distance

is known as a frequency band. Depending upon the type of service being operated, these bands might cover just a few kilohertz or many Megahertz of bandwidth.

Radio bands are usually labelled using the main frequency that the band is based around – that is, 900-Band networks would use frequencies in a wide band based around 900 MHz and 1800-Band systems would be based around 1800 MHz. Within each band smaller allocations of frequencies are defined for individual users of the network – these are known as radio channels. The bandwidth of the radio channels used by a network is determined partly by the radio technology being used and partly by the amount of capacity the network assigns to each user.

Generic examples of spectrum, bands and channels are shown in Figure 2.11.

2.3.2   Effects of Frequency on Propagation

It is generally the case that, at a similar transmit power level, a LF (long wavelength) signal will be usable over longer distances than a high frequency (short wavelength) signal.

One way of visualising this is to imagine that there is only a finite amount of energy carried by each cycle of a signal; a long wavelength allows that energy to dissipate over a long distance, a short wavelength uses that energy up over a shorter distance. This concept is illustrated in Figure 2.12. This explanation is technically inaccurate, the relationship between frequency and propagation is based on a more complex set of principles, but it makes for a readily understandable mental image.[1]

In practice this means that it is more economical to use LF bands to send signals over longer distances rather than high frequency bands as LF transmission can be achieved using lower transmit power levels.

2.3.3   Cellular Bands

Modern cellular systems tend to be based on frequencies in the UHF band, between 300 MHz and 3 GHz (although some 4G networks are based on frequencies slightly above 3 GHz).

Systems based on frequencies at the lower end of this band (300–900 MHz) typically offer good long distance coverage, which is useful for creating large radio cells in rural areas. Other systems, based on frequencies in the upper end of this band (e.g. 1800–2600 MHz), tend to be used to generate small-sized radio cells to serve urban areas.

The set of radio bands employed to support cellular services in various regions around the world, as illustrated in Figure 2.13, is detailed in Table 2.2.

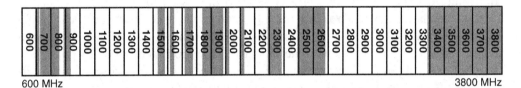

600 MHz   3800 MHz

Figure 2.13   Cellular radio bands

Table 2.2   Cellular radio bands.

Frequency band (MHz)	Network types	Characteristics
300	Public safety networks	Long distance, wide area cellular coverage
450	2G, 3G, 4G	
700	2G, 4G	
800	2G, 3G, 4G	
900	2G, 3G, 4G	
1500	2G, 3G	
1700	2G, 3G, 4G	Medium distance, medium area coverage
1800	2G, 3G, 4G	
1900	2G, 3G	
2000	3G	
2100	3G, 4G	
2300	3G, 4G	Short distance, local area coverage
2500	3G, 4G	
2600	4G	
3400	3G, 4G	
3500	3G, 4G	
3600	3G	

Details of the exact spectrum allocations currently in force in each country are published by the relevant national regulator. As an example, spectrum allocations in the United Kingdom are published by Ofcom (Office of the Communications Regulator) in the United Kingdom Frequency Allocation Table [2].

The Third Generation Partnership Project (3GPP) – the organisation responsible for coordinating development of most modern cellular systems – currently defines around 40 different radio bands for various cellular technologies, but all of these are in (or near) the UHF band [3].

The differences between 2G, 3G and 4G network types will be explained in a later section.

2.4   RF Measurements

Radio signal strength measurements form the foundation of forensic radio surveying.

The unit in which radio signal strengths are measured is the watt (W), although the milliwatt (mW) scale is also commonly used – 1 mW is 1/1000 watt.

It is often necessary when taking radio measurements to compare the strength of a signal when it leaves a transmitter to the strength of the signal when it arrives at a receiver. Radio is an enormously inefficient transmission medium and signals lose large amounts of power as they propagate. This means that a comparison of 'transmitted' versus 'received' signals is often a comparison of a large number versus a very small number. For example, a signal might be transmitted with a power level of 100 mW, but might be received with a power level of 0.000 001 mW.

To allow for simpler comparisons and calculations to be made when performing radio measurements, engineers generally use the decibel (dB) and decibel milliwatt (dBm) scales. By using decibels, the enormous variations encountered between transmitted and received signal strengths can be represented using simpler numbers.

2.4.1   Decibel Notation

The decibel uses a logarithmic scale to allow for simpler comparisons of large and small numbers.

A logarithm is a mathematical term that can be paraphrased as 'the power that number X must be raised by to get number Y'. An alternative way of writing this is:

$$X^a = Y$$

where 'a' is the logarithm of X that equates to Y (the inverse of which is $\text{Log}_x(Y) = a$).

A simple example of a logarithm is: $\text{Log}_{10}(100) = 2$.

A more mathematically rigorous term for 'power of' is 'exponent'. In 10^2, for example, a number (10) is raised to a power by an exponent (2).

The logarithm of 10 (or the base 10 logarithm) required to make 100 is 2 as the exponent that 10 must be raised by to get 100 is 2:

$$10^2 = 100 \text{ and conversely } \log_{10}(100) = 2.$$

Similarly, $\text{Log}_{10}(1000) = 3$ as $10^3 = 1000$.

The real purpose of logarithms is to simplify calculations involving very large and/ or very small numbers and this is due to the mathematical 'law of powers'. This states the following:

$$X^a \times X^b = X^{a+b} \text{ and also } X^a / X^b = X^{a-b}$$

So, to multiply two numbers together, it is only necessary to add their logarithms. For example:

$$10^3 \times 10^2 = (10 \times 10 \times 10) \times (10 \times 10) = 10^{3+2} = 10^5 = 100\ 000$$

Similarly, to divide two numbers it is necessary only to subtract their logarithms:

$$10^3 / 10^2 = (10 \times 10 \times 10) / (10 \times 10) = 10^{3-2} = 10^1 = 10$$

An example of a logarithmic system that makes use of these concepts is the decibel.

2.4.2 Decibels

The unit known as the decibel was designed to enable easier calculations of power gains and power losses in a system. If these gains were each expressed as a logarithm, then the total gain would be the sum of these values, following the law of powers. This logarithmic value is known as a 'Bel' (named after Alexander Graham Bell).

The logarithm of a radio signal's power gain or power loss – that is, $\text{Log}_{10}(\text{mW})$ – is expressed as a 'decibel' (or dB), the value of which is one-tenth of a Bel. The standard notation employed for dB values is therefore to multiply the Log value by 10 to make the outcome equivalent to a Bel value:

$$dB = 10 \times \text{Log}_{10}(\text{value})$$

The multiplication symbol is often omitted, making:

dB = $10\text{Log}_{10}(\text{value})$

Using the values mentioned earlier (power at transmitter = 100 mW, power at receiver = 0.000 001 mW), the benefit of using the dB scale becomes clearer:

$$100\,\mathrm{mW} = 10\mathrm{Log}_{10}(100\,\mathrm{mW}) = 10 \times 2 = 20$$

$$0.000\,001\,\mathrm{mW} = 10\mathrm{Log}_{10}(0.000\,001\,\mathrm{mW}) = 10 \times -6 = -60$$

The power loss experienced during transmission is therefore the ratio of the transmitted and received values:

$$\text{Power loss} = 100\,\mathrm{mW} / 0.000\,001\,\mathrm{mW} = 100\,000\,000$$

Using the law of powers with dB values (where exponential dB values are subtracted, as opposed to the division that would be performed on linear values):

$$20 - (-60) = 80\,\mathrm{dB}$$
$$\text{or Power loss (dB)} = 10\,\log_{10}(100{,}000{,}000) = 80\mathrm{dB}$$

This shows the received signal experienced a loss of 80 dB compared to the transmitted signal, which equates to it being 100 million times less powerful.

From Table 2.3 it can be seen that every time the power level doubles, 3 dB is added and each time a power level halves, 3 dB is subtracted.

This corresponds to a doubling or halving of signal strength for each change of ±3 dB.

A 10 dB gain/loss corresponds to a 10-fold increase/decrease in the signal level.

A 20 dB gain/loss corresponds to a 100-fold increase/decrease in signal level.

In other words, a device like a cable that has 20 dB loss through its length will lose 99% of its signal power by the time that signal is received at the other end. It can be seen, therefore, that by using the decibel scale, big variations in signal levels are easily handled with simple digits.

Table 2.3 Typical decibel values.

Ratio of transmitted to received signal power	Decibels (dB)
10 000	40
1000	30
100	20
10	10
2	3
1	0
1/2	−3
1/10	−10
1/100	−20
1/1000	−30
1/10 000	−40

The dB scale is known as a 'logarithmic' or 'non-linear' scale as the measurements represented by the dB values do not increase in a linear fashion. Each increase of 10 dB is not an increase of 10 units (as it would if linear), it is an increase of × 10 units.

So where the normal linear counting system would increase in steps of 10, 20, 30... the dB scale increases exponentially in steps of 10, 100, 1000...

2.4.3   Decibel Milliwatts

The dB scale provides a comparison of gain or loss between two values. A dB measurement itself is therefore not an 'absolute' value but a 'comparative' value.

Where dB will show the comparative difference between two values, the dBm scale will provide a result that can be mapped to a specific or 'absolute' milliwatt value. The dBm scale is therefore used to describe specific measurements, while the dB scale is used to compare the value of two different measurements.

dBm employs the same logarithmic scale as dB and is calibrated around the value 1 mW, which is equal to 0 dBm.

To convert an 'absolute' milliwatt value to dBm, use the following method: $dBm = 10Log_{10}(mW)$.

A signal measured with a strength of 100 mW will therefore equate to a value of 20 dBm.

The milliwatt value is known as a 'linear' value as the measured units progress in a linear fashion (e.g. 1 mW + 1 mW = 2 mW), this compares to the 'non-linear' progression of the logarithmic dBm values. A comparison of linear (mW) values and logarithmic (dBm) values is provided in Table 2.4.

Note: Linear and logarithmic values cannot be mixed in the same calculations, so if a calculation requires the use of a dBm value and a multiplying or dividing value,

Table 2.4   Linear mW values compared to exponential dBm values.

Linear power level (mW)	Decibel milliwatts (dBm)
100 000 (100 W)	50
10 000 (10 W)	40
1000 (1 W)	30
100	20
10	10
2	3
1	0
0.5	−3
0.1	−10
0.01	−20
0.001	−30
0.000 1	−40
0.000 01	−50

the dBm value must either be converted back to linear mW or the multiplier/divisor must be converted to its logarithmic equivalent.

To recap:

To convert mW to dBm: $dBm = 10Log_{10}(mW)$

To convert dBm to mW: $mW = 10^{(dBm/10)}$

A similar measurement scale, known as dBW (decibel watts), is also sometimes used, which is based on watts instead of milliwatts. 0 dBW = 1 W and dBW values are 1000× stronger than the same dBm values (because 1 W = 1000 mW).

There is also a scale known as dBi (decibel isotropic), which is used to measure the 'gain' of an antenna. Antenna gain is a way of measuring the power increase conferred on a signal due to the physical properties of a transmitting antenna that focuses or concentrates its output signal and is a comparison with the power of an 'isotropic' antenna that radiates in all directions. The output power of a base station antenna is often measured in dBi (whereas the RF signal output of the base station itself before being applied to an antenna is measured in dBm).

2.4.4   Cellular Measurements

Measurements taken by normal mobile phones and by forensic survey devices are usually expressed in dBm (or use reporting values that map to dBm values).

Radio is an extremely inefficient transmission medium, mainly due to the fact that a radio signal spreads out as it propagates, thereby diluting the transmitted power, and the power loss associated with sending radio signals can be significant.

Radio signals typically leave a base station with power levels of up to a few hundred milliwatts (20–50 dBm), but can be reported by GSM mobile devices at power levels of, at best, –48 dBm (0.000015 848 931 924 611 1 mW) and are usually much lower than that.

A mobile phone measuring signals from a nearby base station would typically report values of –48 to –80 dBm, whereas a phone being used some distance away from a suburban or rural base station will commonly report signal strengths of –80 to –100 dBm.

Table 2.5 presents a selection of common cellular dBm values and their linear equivalents (in mW). The mW values shown are unrealistically precise (in reality it is not likely that a mobile device would be capable of capturing measurements to 15 decimal places), the values have been shown in this format simply to emphasise how small they are.

The lowest usable received signal strength for a GSM phone is around –110 dBm.

To put this into some perspective:

If a GSM signal is transmitted with an output power of 100 W (100 000 mW or 50 dBm) and is received by a distant mobile device at or near the minimum value of

Table 2.5 Examples of common cellular dBm values.

dBm	Linear power level (mW)	
−45	0.000 031 622 776 602	3.16×10^{-5}
−50	0.000 010 000 000 000	1.0×10^{-5}
−55	0.000 003 162 277 660	3.16×10^{-6}
−60	0.000 001 000 000 000	1.00×10^{-6}
−65	0.000 000 316 227 766	3.16×10^{-7}
−70	0.000 000 100 000 000	1.00×10^{-7}
−75	0.000 000 031 622 777	3.16×10^{-8}
−80	0.000 000 010 000 000	1.00×10^{-8}
−85	0.000 000 003 162 278	3.16×10^{-9}
−90	0.000 000 001 000 000	1.00×10^{-9}
−95	0.000 000 000 316 228	3.16×10^{-10}
−100	0.000 000 000 100 000	1.00×10^{-10}
−105	0.000 000 000 031 623	3.16×10^{-11}
−110	0.000 000 000 010 000	1.00×10^{-11}

−110 dBm (1.00×10^{-11} mW or 0.000 000 000 01 mW), the power loss will be 160 dB, which means that the received signal would be:

- 1/10 000 000 000 000 000, or
- one-ten thousand billionth (or one-ten trillionth), or
- a factor of 10^{-16}

of its original power, which is a power loss level of 99.999 999 999 999 9%, but that radio signal should still be able to carry a reasonable quality phone call.

The performance requirements of 3G UMTS and 4G LTE systems can be even more spectacular, with the minimum receiver sensitivity in UMTS set at around −120 dBm and in LTE set at around −130 dBm.

2.4.5 Measurements Used by Different Cellular Generations

2G GSM networks employ mandatory frequency reuse techniques, which means that no neighbouring cells should be using the same radio channel as each other. GSM devices are therefore only required to take measurements of the strength of the 'wanted' cell's signal without needing to compare it to anything else.

The primary 2G GSM signal strength measurement is known as RXLev (received signal strength level) and is measured in dBm; this means that it provides an 'absolute' measurement of received signal strength and is not required to compare that signal against anything else.

3G and 4G technologies offer the opportunity for networks to operate as 'single frequency networks', in which neighbouring cells can all use the same radio channel. Measurements taken in these circumstances must be 'comparative' rather than 'absolute', meaning that they need to provide an indication of the strength of the 'wanted' cell's signal in comparison to the amount of 'unwanted' noise and interference produced by neighbouring cells.

3G and 4G systems capture a range of measurements, including:

- A measurement of the 'wanted' cell's signal, measured in dBm;
- A measurement of the total interference (also known as 'noise') received on the channel (known as RSSI – Received Signal Strength Indicator), measured in dBm;
- A 'signal to noise ratio' comparison of wanted signal versus channel noise, measured in dB.

The 'comparative' value is usually considered to offer the most useful signal strength measurement in 3G and 4G networks as it provides an indication of how 'usable' the cell is in relation to the current noise level.

This is important to know because a 3G/4G cell might have a very strong 'wanted' signal strength (–70 dBm, for example), but if it was being received in a cell that was currently suffering from a very high background noise level (e.g. –90 dBm), then the signal to noise ratio would be very high [(–90 dBm) – (–70 dBm) = –20 dB], meaning that the signal was significantly lower than the noise, making the traffic carried by that signal difficult to recover.

If a wanted signal were to be received at –70 dBm in a cell experiencing low levels of background noise (e.g. –75 dBm) then the resulting signal to noise ratio would be much lower [(–75 dBm) – (–70 dBm) = –5 dBm] and the 'wanted' signal would compare much more favourably to the noise level. This would offer a much better quality connection for users.

The calculations shown in the above examples are generic and are used for demonstration purposes only, the actual signal to noise ratio calculations performed in 3G and 4G networks can be more complex.

So, it can be seen from the above that knowledge of the 'wanted' received signal strength (in dBm) alone can potentially provide a misleading view of the quality of the cell being measured.

2.4.6 Describing Signal Strengths

One of the difficulties experienced by cell site analysts and expert witnesses is conveying the details of cellular operation to investigators, lawyers and jurors who have little understanding of the technologies or concepts involved.

One way of making cell site evidence more understandable for a lay audience is to put information such as signal strengths into more everyday language. Instead of talking about dB and dBm values it is often more effective to map these values to a set of simple labels. An example of a potential method of mapping signal strength values to simple descriptions is shown in Table 2.6.

Table 2.6 Example of mapping signal strength values into simple descriptions.

Description	2G (dBm)	3G (dB)	4G (dB)
Very strong	−45 to −85	−3 to −6	−3 to −10
Strong	−86 to −90	−7 to −10	−11 to −15
Moderate	−91 to −100	−11 to −18	−16 to −20
Poor	−101 to −110	−19 to −25	−21 to −30
Lowest reportable value	−110	−25	−30

Great care should be taken when using this approach however, as there is currently no agreed standard for mapping values to descriptions and any scale used must be regarded as subjective.

Different cell site experts use different scales of values and some refuse to map values to text labels at all.

Disagreements about the subjective labelling of signal strength values are often highlighted in defence cell site reports, so it is recommended that if a value to text mapping formula is employed in a cell site report there should be a paragraph of explanatory text somewhere in the report that gives the ranges of signal strengths that map to each text description. This should forestall at least some of the criticism that could be levelled by defence experts.

Note

[1] There are several more technically accurate reasons for why lower frequency signals tend to travel further than higher frequency ones. The first, and more important, comes from mathematical calculations of the reception of radio waves. Such calculations tell us that the effective collecting area of the receive antenna depends on the square of the wavelength of the radio waves. At lower frequencies (longer wavelengths), the receive antenna has a larger collecting area than it does at higher frequencies. It therefore acts as a larger bucket for the incoming radio waves, receives a stronger signal, and can detect the incoming radio waves more easily.

The second reason is that, at high radio frequencies (above a few GHz), radio waves are absorbed by atmospheric water vapour and oxygen as they travel. The higher the frequency, the greater the absorption, and the weaker the radio signal itself will be.

References

[1] Bureau International des Poids et Mesures (2014) *SI Prefixes*, http://www.bipm.org/en/si/si_brochure/chapter3/prefixes.html (accessed 30 May 2014).

[2] OFCOM (2014) *The United Kingdom Frequency Allocation Table*, http://stakeholders.ofcom.org.uk/spectrum/information/uk-fat/ (accessed 30 May 2014).

[3] 3GPP Technical Specification (2013) *Evolved Universal Terrestrial Radio Access (EUTRAN); Base Station (BS) radio transmission and reception*, TS 36.104 v12.2.0 Section 5.5, www.3gpp.org (accessed 24 July 2014).

3

Wireless Technologies and Deployments

3.1 Coordinating Cellular Development

Cellular communications networks span the globe. It is common nowadays for users to expect to be able to use their mobile devices in whichever country they happen to be visiting. The fact that this is usually possible is testament to the enormous efforts that have been made over the last 30 years to convince different countries, equipment manufacturers and network operators to cooperate with each other and coordinate the joint development of common standards.

To ensure that common standards are employed around the world, a number of 'standards bodies' have emerged to coordinate system development.

Overall responsibility for the development of telecoms standards of all kinds rests with the ITU (International Telecommunications Union), an agency of the United Nations tasked with ensuring that the various national and regional telecoms networks interconnect as a coherent global system [1].

An offshoot of the ITU, the WRC (World Radio Conference) is responsible for coordinating the use and allocation of radio spectrum globally. The WRC divides the world into three regions: Region 1 covers Europe, the Middle East, Africa and Russia; Region 2 covers the Americas; and Region 3 covers Asia Pacific. Countries within each region theoretically apply a consistent set of spectrum allocation rules and conventions, which is one of the main reasons that there are different allocations of spectrum in use for cellular systems in different parts of the world [2].

Forensic Radio Survey Techniques for Cell Site Analysis, First Edition. Joseph Hoy.
© 2015 John Wiley & Sons, Ltd. Published 2015 by John Wiley & Sons, Ltd.

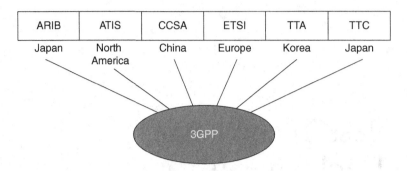

Figure 3.1   3GPP members

Many nations and some regions have their own telecoms standards bodies, respon-sible for coordinating the development and deployment of new telecoms systems. Examples of these bodies include ETSI (European Telecoms Standards Institute) in Europe [3], ARIB (Association of Radio Industry Bodies) in Japan [4], ATIS (Alliance for Telecommunications Industry Solutions) in the United States [5], CCSA (China Communications Standards Association) in China [6] and many others.

In the past, cellular systems were developed by individual regions or nations and some were developed by individual companies, which led to a lack of common stand-ards and fragmentation in the industry. In the late 1990s a number of separate standards bodies (as shown in Figure 3.1) grouped together to jointly develop new cellular systems; the first of these was 3GPP (the Third Generation Partnership Project) [7] which was later joined by a sister organisation known as 3GPP2 (Third Generation Partnership Project 2) [8].

3GPP is the organisation responsible for coordinating the development of the most widely deployed cellular technologies, which includes 2G GSM (the Global System for Mobile Communications), 3G UMTS (Universal Mobile Telecommunications System) and 4G LTE (Long-Term Evolution). 3GPP2 coordinates the development of the CDMA2000 and EV-DO standards.

3.2   Evolution from 0G to 4G

The successive waves of cellular technology can be broadly and loosely classified into a series of 'generations', each of which can be further subdivided into the sets of technologies that are employed in different regions.

3.2.1   0G – Pre-cellular Networks

The very first types of 'mobile' radio communications systems were developed to serve police forces and other types of emergency users. It is often stated that the first mobile, two-way, car-mounted 'radio telephone' system was developed by Bell

Laboratories (then part of AT&T) in 1924 and that the Detroit police department started to use a one-way broadcast radio system in 1928.

Various forms of radiotelephone system were developed over the following decades, partly driven by the need to provide battlefield communications systems during the Second World War and led to the commercial radiotelephone networks that began to be deployed in the late 1940s and early 1950s. These developments are sometimes grouped and classified as 0G (zero G) networks and were precursors to the cellular generations developed later.

3.2.2 1G – First Generation Networks

The cellular concept was developed in 1948, again by Bell Laboratories, but networks based on these principles were not developed until the 1970s.

The first cellular phone call is reputed to have been made in 1973 by Martin Cooper of Motorola, who used an experimental handset and a test network to call a rival developer at Bell Laboratories.

Commercial cellular networks began to be deployed from the late 1970s and have come to be known as the First Generation (1G) of cellular systems. All of the early 1G systems were based on analogue radio transmission techniques.

Different 1G systems were developed and deployed in different countries around the world and few of them were compatible with each other. In most cases, this lead to limited economies of scale, high rollout and deployment costs and limited opportunities for roaming by users. The only notable exception to this was the NMT system developed in Finland and Sweden, which supported a form of international roaming.

Examples of 1G systems are illustrated in Figure 3.2 and include the TACS (Total Access Communications System) developed in the United Kingdom, the TZ-80x and JTACS (Japan TACS) developed in Japan, the AMPS (Advanced Mobile Phone System) developed in the United States, the C-NETZ/C-450 system developed in Germany, the Radiocom2000 system developed in France and the NMT (Nordic Mobile Telephone) system developed to serve the Nordic countries.

Americas	Europe	Middle East	Africa	Asia	Pacific
AMPS		AMPS			AMPS
	NMT				TZ-80x
	TACS		TACS		JTACS
	Radiocom				
	C-NETZ		C-NETZ		

Figure 3.2 First generation mobile networks

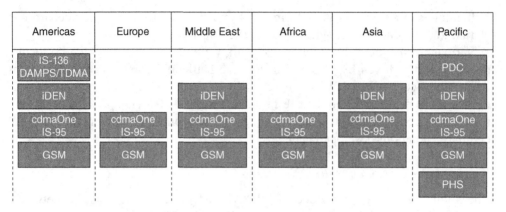

Americas	Europe	Middle East	Africa	Asia	Pacific
IS-136 DAMPS/TDMA					PDC
iDEN		iDEN		iDEN	iDEN
cdmaOne IS-95	cdmaOne IS-95	cdmaOne IS-95	cdmaOne IS-95	cdmaOne IS-95	cdmaOne IS-95
GSM	GSM	GSM	GSM	GSM	GSM
					PHS

Figure 3.3   Second generation mobile networks

3.2.3   2G – Second Generation Networks

It became apparent during the 1980s, as these 1G network types began to be deployed, that the quality, capacity and security issues related to analogue transmission coupled with the lack of compatibility and roaming were destined to become limiting factors to the widespread adoption of cellular services.

Several organisations began the process of developing Second Generation (2G) systems, all of which were designed to use digital radio transmission techniques.

When compared to their analogue 1G predecessors, digital 2G networks offered: better security, as digitised user traffic could be encrypted before transmission; higher capacity, as digital multiplexing techniques allow multiple users to share each radio channel concurrently; and more consistent call quality, as the process of digitising traffic effectively prevents reasonable levels of radio interference from influencing the content of a received signal.

Some of the more popular 2G networks types are shown in Figure 3.3 and include: GSM developed in Europe by ETSI; PHS (Personal Handyphone System) and PDC (Personal Digital Cellular) developed in Japan; D-AMPS (Digital AMPS; also known as IS54/IS136 and TDMA) developed in the United States; cdmaOne (also known as IS95) developed in the United States; and iDEN developed in the United States.

The first 2G networks started to be deployed in the late 1980s and began to go into use in the early 1990s, with GSM eventually becoming by far the most widely deployed system. The original 2G systems typically offered voice and text messaging services, with some also offering dial-up data and fax services.

Later enhancements to these networks, collectively known as 2.5G and 2.75G evolutions, added PS (Packet Switched), Internet-based data connectivity to the set of services provided by 2G systems.

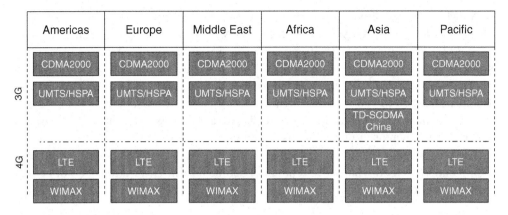

Figure 3.4   Third and fourth generation mobile networks

3.2.4   3G – Third Generation Networks

Development of Third Generation (3G) systems began in the mid1990s, just as access to the Internet was becoming widespread. 3G systems were therefore designed to offer much faster and more efficient data handling capabilities than had been the case with 2G networks.

One of the reasons that GSM came to dominate the 2G world was that it offered a common standard that could be deployed in many countries and which allowed users from one country to 'roam' to other countries that also had GSM networks. The widespread adoption of GSM also lead to large economies of scale developing in the manufacture of network equipment and mobile devices, making it a less expensive system to deploy and operate than some of its competitors.

These factors lead to a degree of consolidation in the industry when the development phase of 3G systems began.

As indicated in Figure 3.4, instead of the 10 or more types of 1G system or the five or more types of 2G system, there were really only three main types of 3G system.

These were: UMTS an evolution of GSM jointly developed in Europe and Japan; CDMA2000, an evolved version of cdmaOne/IS95 developed in the United States; and TD-SCDMA (Time Division–Synchronous Code Division Multiple Access) a variant of UMTS developed in China to meet the requirements of Chinese operators.

3G systems began to be deployed from around 2002 and, later, 3.5G enhancements progressively increased the capacity and maximum data rates that are achievable.

3.2.5   4G – Fourth Generation Networks and Beyond

Further consolidation took place in the eventual development of Fourth Generation (4G) systems, with only two network types being generally deployed: LTE, which was a further evolution of the GSM/UMTS family; and WIMAX, which developed out of a desire to create a 'wide area' version of WiFi.

WIMAX-based networks had been available since the mid2000s, but the development of 'Mobile WIMAX' saw the technology eventually adopted as an official 4G standard. LTE networks began to be deployed 5 years or more after WIMAX started to become popular. Although Mobile WIMAX was a popular and widely deployed technology, many of the operators initially adopting that technology have subsequently swapped to using LTE, which is now by far the dominant 4G technology.

An enhancement to LTE, known as LTE-Advanced, has been made available that offers a large increase in potential user data rates.

Cellular system evolution continues and early standardisation work is already being undertaken on what will eventually become a Fifth Generation (5G) of cellular mobile technologies.

3.3   3GPP Network Types

The 3GPP was formed to coordinate the development of 3G UMTS, but also inherited responsibility for 2G GSM. The timeline of 3GPP technology development is shown in Figure 3.5.

The 2G GSM networks began to be launched in the early 1990s. They offered voice, SMS text, fax and dial up data services.

GPRS (General Packet Radio Service) enhancements were added in the late 1990s, which added the ability to carry 'packet data' services like IP (Internet Protocol) that allowed Internet and e-mail traffic to be carried more efficiently. A further upgrade, known as EDGE (Enhanced Data rates for Global Evolution) was also added in the late 1990s and improved the data rate available for GPRS data services. GPRS and EDGE are regarded as 2.5G and 2.75G technologies respectively.

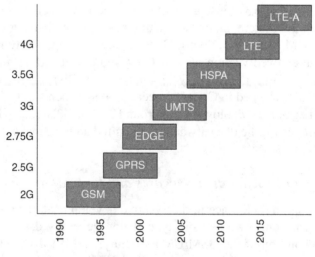

Figure 3.5   3GPP networks

The 3G UMTS was developed by 3GPP and was an evolution of 2G GSM/GPRS/ EDGE services.

The original version of UMTS (known as Release 99) supported voice and text services and also offered quite fast packet data services. A 3.5G enhancement developed in the mid-2000s was known as HSPA (High Speed Packet Access) that greatly increased the data rates available for packet data (e.g. Internet connection) services. A further evolution known as HSPA + is currently in use in 3G networks and offers very fast data services (40 Mbit/s or more).

3GPP was also responsible for coordinating the development of 4G LTE, which is a further evolution of GSM and UMTS technologies.

LTE offers very fast packet data services but did not originally offer a voice service, meaning that mobile phones had to use a technique known as Circuit Switched Fallback, which forced them to 'fall back' to a 2G or 3G cell to make a call. A 'native' voice service for 4G networks, known as VoLTE (Voice over LTE) began to be rolled out by some operators in late 2013.

An enhancement of LTE, known as LTE-Advanced (or LTE-A) has been developed, which provides much higher maximum user data rates and greater network capacity.

3.4   3GPP2 Network Types

The development of cellular networks in the United States was mainly undertaken by individual companies, for example cdmaOne was developed by Qualcomm and iDEN was developed by Motorola.

These activities were coordinated to some degree by the United States telecoms standards body, ANSI (American National Standards Institute), which was responsible for defining what were termed IS (Interim Standards), such as the IS54/IS136 standards that referred to the 2G D-AMPS/TDMA system and the IS95, IS95A and IS95B standards that applied to cdmaOne.

Following the successful collaboration between regional standards bodies to create 3GPP, which saw them collectively coordinate the development of the GSM family of technologies, it was decided to pass the responsibility for the development of the 3G IS2000/CDMA2000 standard, the successor to the IS95/cdmaOne system, to a new body known as 3GPP2.

The 3G CDMA2000 standard is now widely deployed in the United States and in many other countries around the world. It has undergone a number of evolutions and revisions, including 1x RTT (Radio Transmission Technology), which was the first iteration of the standard and various revisions of EV-DO (Evolution–Data Optimised), which offered a data-only mobile broadband service.

3GPP2 undertook initial development work on a 4G successor to CDMA2000 known as UWB (Ultra Wide Band), but development efforts were abandoned in favour of using LTE as the preferred 4G technology.

3.5   Other Types of Network

Other types of cellular network have been deployed and are (or have been) in use around the world. Although the vast majority of deployed commercial networks use the 3GPP/3GPP2 technologies described above, there is a possibility that a forensic radio surveyor may be asked to survey one of these other network types.

3.5.1   TD-SCDMA

TD-SCDMA (Time Division–Synchronous Code Division Multiple Access) is a 3G network type developed in China for use within its own domestic communications market. The impetus for developing this network type was for China to be able to adopt its own standard allowing it to be less reliant on technologies developed and controlled by other regions.

TD-SCDMA refers to the radio technology used within the system and the wider network is essentially the same as a UMTS network. TD-SDMA therefore offers services that are broadly similar to UMTS and CDMA2000 using broadly similar technologies and techniques, but it has been adapted to meet the density requirements of Chinese urban environments.

3.5.2   iDEN

iDEN (Integrated Enhanced Digital Network) is a 2G network type originally developed by Motorola that has been deployed in all world regions but most notably by Nextel/Sprint in the United States.

iDEN was designed to offer services that are a cross between the 'one to one' communications of standard telephony and the group communications techniques usually found in 'all informed' messaging systems such as those used by public safety organisations. iDEN handsets therefore offered both 'dial up' and 'press to talk' services to users.

3.5.3   WiFi

WiFi (Wireless Fidelity) is an umbrella term used to describe the family of WLAN (Wireless Local Area Network) technologies that have been developed since the mid1990s.

The development of these technologies is coordinated by the IEEE (Institute of Electrical and Electronics Engineers), which publishes WLAN specifications in a family of standards known as 802.11 [9].

Most types of survey device are capable of capturing WiFi/802.11 measurements and, thanks to the growing popularity of WiFi hotspots, the demand for this type of survey is also increasing.

3.5.4   Wireless Broadband

There are several broadband providers who employ radio to connect over the 'last mile' to their subscribers in a service known generically as FWA (Fixed Wireless Access) or BWA (Broadband Wireless Access). Most of these FWA/BWA providers use 'point to point' microwave transmission techniques, but some use cellular technologies and all of them are competitors to 'wired' broadband alternatives such as DSL (Digital Subscriber Line), which is carried over telephone lines, and DOCSIS (Data Over Cable Service Interface Specification), which is carried via cable TV connections.

There have been multiple FWA/BWA standards developed over the years, including LMDS (Local Multipoint Distribution Service) and MMDS (Multichannel Multipoint Distribution Service). Providers have also made use of UMTS, WIMAX and LTE as bearers for wireless broadband services.

3.5.5   WIMAX

WIMAX (Worldwide Interoperability for Microwave Access) was originally designed as an evolved FWA technology, but gradually developed to support mobile services too.

WIMAX, like WiFi, was developed under the coordination of the IEEE (although much of the development was actually undertaken by Intel and by the industry-sponsored WIMAX Forum) and was given the IEEE standards designation 802.16.

Mobile WIMAX (also known as 802.16e) offers high-speed mobile broadband data services that have been deployed in many countries around the world [10].

The original version of Mobile WIMAX was adopted as a 3G standard and a more recent evolved version (802.16 m) was adopted as a 4G standard.

3.5.6   Wireless Local Loop

WLL (Wireless Local Loop) services are a voice-only companion to FWA and aim to provide standard voice telephony services to fixed users via radio as an alternative to standard 'wired' landline services.

A number of WLL technologies were developed during the 1980s and 1990s, but many of the systems that were eventually deployed use adapted versions of standard mobile cellular technologies, particularly the GSM and cdmaOne/CDMA2000 1x standards, although the cordless telephony DECT (Digitally Enhanced Coreless Telephony) standard has also been used.

WLL has proved popular in the developing world, where basic voice services can be rolled out to new users quickly and without the necessity to dig up streets or install networks of telephone poles to distribute wired connections.

3.5.7   GSM-R

Based, as the name suggests, on standard 2G GSM technologies, GSM-R (GSM for Railways) networks are designed to offer radio communications services that have been tailored for the needs of railway networks. It allows, for example, railway control rooms to connect to train drivers and maintenance crews.

Although GSM-R networks use an adapted version of 2G GSM, the technology and the frequency bands it uses (such as the 876–880 MHz uplink and 921–925 MHz downlink resources assigned in the United Kingdom) are not accessible by 'normal' forensic radio survey devices, even though these frequencies fit around those used by GSM900 [11].

3.5.8   TETRA

TETRA (Terrestrial Trunked Access) is a 2G cellular technology designed for use by 'blue light' emergency services and other public services.

TETRA uses a heavily adapted version of GSM that offers 'group communication' functions in addition to standard 'one to one' connections [12].

Although TETRA employs the same basic cellular techniques as GSM, TETRA cells are not accessible to standard forensic radio survey devices.

The frequency bands typically assigned to TETRA networks are in a range between 380 and 395 MHz, with other allocations commonly made in the 450 MHz band. The relatively low frequencies assigned to TETRA deployments are designed to ensure that each base station can serve a comparatively large area (30 km or more in radius) allowing public safety networks to be deployed as cost-effectively as possible.

There have also been a number of commercial TETRA providers in some countries (such as the United Kingdom and Qatar), who offer the 'group calling' capabilities of TETRA to private sector users, such as taxi companies and airports.

3.6   Deployed Technologies by Region

The ITU coordinates spectrum use and spectrum policy through the WRC, which are held every 3–4 years and bring together policy makers, regulators, technology vendors and spectrum users in an attempt to ensure harmonised use of the global spectrum.

The WRC divides the planet into three administrative regions, each of which has its own spectrum use policies.

Region 1 covers Europe, the Middle East, Africa and Russia; Region 2 covers the Americas and Greenland; and Region 3 covers Asia Pacific.

Table 3.1 provides a basic overview of the extent to which the wireless technologies that have so far been discussed in this section are deployed across the world. The descriptions 'high', 'medium' and 'low' are intended to provide an indication as to how densely deployed or how popular each technology is.

Table 3.1 Regional deployment of popular wireless technologies.

	Region 1	Region 2	Region 3
2G GSM	High	High	High
3G UMTS/HSPA/HSPA+	High	High	High
4G LTE	Low	Medium	Low
2G cdmaOne/IS95	Low	High	Low
3G CDMA2000/IS2000	Low	High	Low
3G TD-SCDMA	None	None	China
2G iDEN	Low	Low	Low
WIMAX	Low	Low	Low
WiFi	High	High	High
GSM-R	Low	None	Low
TETRA	Medium	Low	Low
WLL	Low	Low	Low
FWA/BWA	Low	Low	Low

Sources: GSM Association [13], CDMA Development Group [14] and 4G Americas [15].

Table 3.2 Commonly-used cellular frequency bands by region.

Band (MHz)	Africa	Eastern Europe	Western Europe	Middle East	North America	South America	Asia Pacific
450	—	✓	—	—	—	✓	✓
700	✓	—	—	—	✓	✓	✓
800	✓	✓	✓	✓	—	—	✓
850	✓	✓	—	—	✓	✓	✓
900	✓	✓	✓	✓	—	✓	✓
1500	—	—	—	—	—	—	✓
1700	—	—	—	—	✓	✓	✓
1800	✓	✓	✓	✓	—	✓	✓
1900	✓	—	—	—	✓	✓	—
2100	✓	✓	✓	✓	✓	✓	✓
2300	✓	✓	✓	—	✓	—	✓
2500	—	✓	—	—	✓	—	✓
2600	✓	✓	✓	✓	✓	✓	✓
3500	—	✓	✓	—	—	✓	—

Sources: GSM Association [13], CDMA Development Group [14] and 4G Americas [15].

In general, 3GPP network types (GSM, UMTS and LTE) are typically quoted as having more than 80% of cellular user share globally, with CDMA2000 and TD-SCDMA making up most of the remaining numbers.

WiFi, a non-cellular technology, is the most widely deployed wireless standard of all.

3.7   Commonly-used Frequency Bands by Region

Each of the three WRC regions has its own subsets of radio bands dedicated to carrying cellular services and there are also some bands that are available in most parts of the world. Table 3.2 provides an overview of the usage patterns of the most commonly deployed cellular frequency bands around the world.

References

[1] International Telecommunications Union (2014) *Home Page*, http://www.itu.int/en/Pages/default.aspx (accessed 30 May 2014).

[2] International Telecommunications Union (2014) *World Radiocommunication Conferences*, https://www.itu.int/ITU-R/index.asp?category=conferences&rlink=wrc&lang=en (accessed 30 May 2014).

[3] European Telecoms Standards Institute (2014) *Home Page*, http://www.etsi.org (accessed 30 May 2014).

[4] Association of Radio Industries and Businesses (2014) *Home Page*, http://www.arib.or.jp/english/ (accessed 30 May 2014).

[5] Alliance for Telecommunications Industry Solutions (2014) *Home Page*, http://www.atis.org (accessed 30 May 2014).

[6] China Communications Standards Association (2014) *Home Page*, http://www.ccsa.org.cn/english/ (accessed 30 May 2014).

[7] Third Generation Partnership Project (2014) *Home Page*, http://www.3gpp.org (accessed 30 May 2014).

[8] Third Generation Partnership Project 2 (2014) *Home Page*, http://www.3gpp2.com (accessed 30 May 2014).

[9] Institute of Electrical and Electronic Engineers (2014) *IEEE Get 802 Program*, http://standards.ieee.org/about/get/802/802.html (accessed 30 May 2014).

[10] Institute of Electrical and Electronic Engineers (2014) *IEEE 802.16: Broadband Wireless Metropolitan Area Networks (MANs)*, http://standards.ieee.org/about/get/802/802.16.html (accessed 30 May 2014).

[11] UIC – International Union of railways (2014) *GSM-R Specifications*, http://www.uic.org/spip.php?rubrique874 (accessed 30 May 2014).

[12] European Telecoms Standards Institute (2014) *TETRA*, http://www.etsi.org/technologies-clusters/technologies/tetra (accessed 30 May 2014).

[13] GSM Association: GSMA Intelligence (2014) *Home Page*, https://gsmaintelligence.com (accessed 30 May 2014).

[14] CDMA Development Group (2014) *Worldwide Deployments*, https://www.cdg.org/worldwide/index.asp (accessed 30 May 2014).

[15] 4G Americas (2014) *3G/4G Deployment Status*, http://www.4gamericas.org/index.cfm?fuseaction=page&pageid=939 (accessed 30 May 2014).

4

Cellular Theory

4.1 Pre-cellular Radiotelephone Networks

The original radiotelephone networks, which went into service from the 1920s onwards, employed a single radio transmitter to provide service over a wide geographical area. The main limiting factor of these networks was the lack of capacity caused as a consequence of the large radio transmission areas used.

If a network operator employed just one very powerful transmitter to provide coverage for a city or a region, they would only ever be able to serve a tiny fraction of the potential user base in that area. This is illustrated in Figure 4.1.

Cellular mobile communications networks were developed to address this capacity problem.

As is shown in Figure 4.2, cellular network architecture does not provide just one transmitter for each region, but instead uses hundreds or even thousands of much smaller and less powerful radio transmitters to cover a region that would previously have been served by a single, large transmitter.

These smaller transmitters are known as base stations and the small geographical areas covered by their radio signals are known as radio cells.

In the same area previously covered by just one large transmitter, a cellular operator might site hundreds of base stations, each supporting several radio channels, which would increase the number of radio connections available to users by an order of magnitude.

Forensic Radio Survey Techniques for Cell Site Analysis, First Edition. Joseph Hoy.
© 2015 John Wiley & Sons, Ltd. Published 2015 by John Wiley & Sons, Ltd.

Figure 4.1   Single transmitter coverage

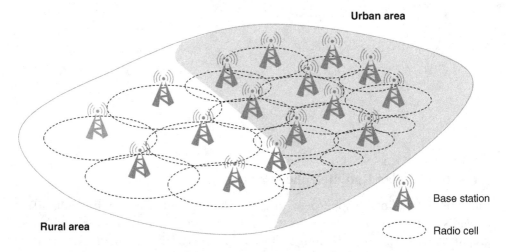

Figure 4.2   Cellular network coverage

The size of the cells used in a network can vary dependent upon such factors as geography and demand. Base stations serving rural locations with low demand for user services might be configured with cells that cover a large area and offer low capacity. Base stations covering high-demand areas such as city centres, business areas and airports might be configured to use very small cells, which each cover a limited area but collectively offer high capacity.

4.2   Radio Cells

The radio service in each cell in a network is supplied by a base station which, depending upon the type of network, can also be known as a BTS (Base Transceiver Station), Node B or eNode B (Evolved Node B or eNB). Each base station will be allocated one or more radio channels to use for customer connections in its cells.

The base stations in a network are all connected to a 'core network' and may be connected to a local 'access network' controller, which is employed to handle users' calls and control the main functions of the access network.

Most cellular network designs use a Frequency Division Duplex (FDD) air interface service, in which each cell supplies separate uplink (transmit path from mobile to base

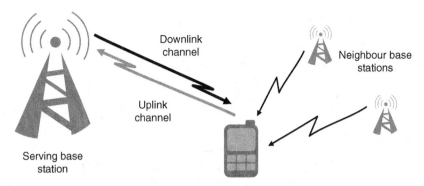

Figure 4.3   Cellular network operation

station) and downlink (receive path from base station to mobile) radio channels to serve
users. To simplify network radio planning, uplink and downlink channels are usually
implemented as a 'matched pair' of radio channels, so whichever uplink channel a
phone is allocated in a cell, it will always use the specific corresponding downlink
channel. The basic concepts of cellular network operation are summarised in Figure 4.3.

User mobility presents a number of problems to a telecoms network. The network
needs to be able to track a user's location as they move around, to ensure that incoming calls can be quickly routed to their phones. Mobile networks therefore employ
special databases to store and process user location information. A 'location update'
is sent to these databases when a mobile phone moves into a different area of the
network, even when there is no call in progress.

Mobility also causes a problem for the user's phone and the network's base stations
when a call is in progress. When a cellular customer makes or receives a call on their
phone, the network will allocate them a radio channel to use for their connection. If
the user is mobile they may eventually move out of range of the base station they are
currently using. To ensure that the call can continue, the phone needs a radio channel
to be allocated to it in the cell they are moving into. This process is known as 'handover' (or 'handoff') and is controlled by the network.

Whilst a call is in progress, a mobile phone will be taking a series of 'received
signal strength' measurements of the channel currently being used and also of channels in neighbouring cells. Each phone reports these measurements to the network at
regular intervals.

When the network decides that one of the neighbour channels reported by a phone
could provide a better quality connection, the phone is sent a handover instruction
informing it of the details of the new channel. The process of call handover should be
transparent to the user (unless something goes wrong) and takes place automatically.

In addition to the 'traffic channels' that carry user calls, a variety of 'control channels' are defined in each cell which are used to carry administrative information such
as handover instructions, measurement reports and call setup information between
the phone and the network.

4.3   Frequency Reuse

The major potential disadvantage of a cellular system, in which large numbers of cells are deployed in close proximity to each other, is that of interference.

If two neighbouring base stations used the same radio channel, the users in those cells would experience interference. This would be classed as 'co-channel' interference. An example of this could be that Cell A and Cell B are neighbours, both use Channel 1 and both experience co-channel interference from each other.

It is also possible to experience 'adjacent channel' interference, which is caused by the use of neighbouring channels (in spectrum terms) in close proximity. An example of this could be that Cell A and Cell B are neighbours, Cell A uses Channel 1 and Cell B uses Channel 2 and both experience 'adjacent channel' interference from each other.

These scenarios for co- and adjacent channel interference are summarised in Figure 4.4. They are issues for all generations of mobile technology but are especially problematic for some 2G (second generation) network types such as GSM (the Global System for Mobile Communications). 2G networks types such as cdmaOne and most 3G (third generation) and 4G (fourth generation) systems have been provided with the means to operate effectively in the presence of such interference, but some types of system are unable to function properly if there is too high a level of interference.

To minimise this problem for susceptible network types, sophisticated network frequency planning tools are employed to ensure that neighbouring cells are not allocated the same (or adjacent) radio channels. However, as the radio signals from low power base stations only travel a limited distance, the channels used in one cell can

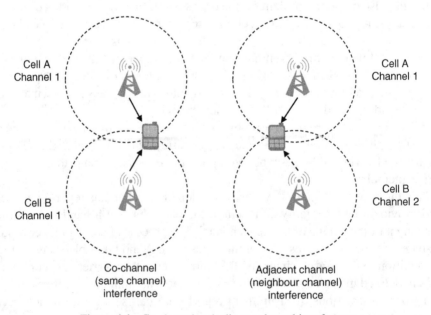

Figure 4.4   Co-channel and adjacent channel interference

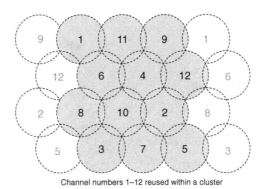

Channel numbers 1–12 reused within a cluster

Figure 4.5 Example of a frequency reuse pattern

be used again but in cells that are further away in the network. The 'frequency reuse' provided by this concept allows cellular systems to operate effectively and efficiently even in circumstances in which network operators have only been allocated a limited amount of radio spectrum.

Generally, there are a limited number of frequencies available to each network operator and they must be distributed between all cells to ensure a balanced coverage is achieved throughout the network.

If the channels in a cellular system are not properly distributed, the result can be a high level of interference caused by overlapping deployments of the same frequencies. To avoid this, many network types, including 2G GSM networks, include specifications that define frequency reuse patterns, an example of which is presented in Figure 4.5.

Cells in this type of access network are gathered into 'clusters' and the same frequency will not be repeated within a cluster. Network planners will attempt to use and reuse the cluster plan in a 'cookie cutter' fashion across an area of the access network, safe in the knowledge that sufficiently large cluster should ensure that a minimum frequency reuse distance is always maintained between repetitions of the same channel.

4.4 Cell Size and Coverage

A low frequency radio signal can typically be received from further away than a signal with a higher frequency transmitted at the same power level. Therefore, a frequency of 900 MHz can provide cellular coverage to a larger area than a cell transmitted on a frequency of 1800 MHz, which means that the lower 900 MHz band is ideal for creating large cells to serve rural areas. Conversely, a cell using a radio frequency of 1800 MHz would be more suited to providing small cells in a densely populated city, as smaller cells enable operators to reuse frequencies much more often.

Figure 4.6 illustrates the difference in general cell coverage areas for a range of common cellular frequencies. This is an illustration only and is not intended to provide an exact guide to the differences in actual coverage areas for the different frequency bands indicated.

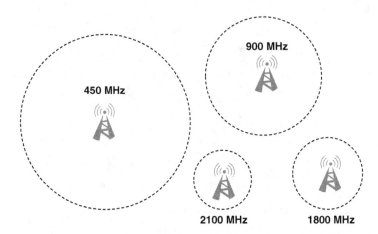

Figure 4.6   Cell size and frequency

Lower frequency radio signals can form large cells; however such cells potentially provide coverage across a large area and their capacity may have to be shared by large numbers of users. That means each cell might allow only a small proportion of served users to make simultaneous calls. A network (or a region of a network) based on large cells may therefore offer a relatively low user capacity.

Higher frequency radio signals travel shorter distances and therefore produce smaller cells. A small cell might have the same capacity as a large cell, in terms of the number of simultaneous calls that can be carried, but this capacity will be focused in a much smaller area. To provide contiguous coverage across a region using smaller cells, the network will be required to deploy a large number of cell sites, meaning that, overall, a network (or region of a network) that is based on small cells may offer very high user capacity.

There are generally more radio channels available in the higher frequency bands (especially for GSM, where the 1800 band has several times more channels available than the 900 band has) so networks based in higher frequency bands often have greater amounts of radio capacity available, allowing them to deploy more channels to each cell site and cover densely populated areas more effectively.

This further means that, as a general rule, networks based on higher radio frequency bands – those using 1800 or 2600 MHz for instance – will typically provide more capacity for subscribers than networks based on frequencies of 900 MHz or lower due to the greater number if available channels and the increased potential for frequency reuse associated with the use of small cells. There are, however, techniques available to improve the capacity and reuse potential even of low frequency networks, so this should only be taken as a guideline rather than a rule.

Many networks employ a mix of frequency bands: low frequency 900 MHz cells (or 700 or 850 MHz cells, depending upon the region in which they operate) are used for rural coverage and also provide wide-area 'umbrella' cell coverage across urban areas, while 1800 MHz cells (or 1500, 1700 or 1900 MHz cells, again depending upon region) are used to provide high capacity coverage in towns and other high demand areas.

4.5 Duplex Techniques

Mobile phone networks were primarily (or at least, were originally) designed to carry voice services, which require a duplex connection. Duplex is the ability to speak and listen (or transmit and receive) simultaneously, which is a key feature of conversational communication.

The two separate channels needed to provide a duplex radio connection (one for transmit and one for receive) can be configured in two main ways – frequency division or time division, both of which are depicted in Figure 4.7.

4.5.1 Frequency Division Duplex

In FDD systems, two separate radio channels are assigned to serve users – an uplink channel that carries the transmitted signal from the phone 'up' to the base station and the downlink channel which carries the received connection back 'down' to the phone. The division between duplex paths is therefore defined by using different frequencies for each, hence 'frequency' division.

2G GSM and 3G CDMA2000 networks are a based on FDD techniques, with each cell consisting of two paired 'carrier' channels, an uplink carrier and a downlink carrier. 3G UMTS (Universal Mobile Telecommunications System) and 4G LTE (Long Term Evolution) also have FDD variants that operate in much the same way as GSM; most 3G UMTS networks use only the FDD variant and many 4G LTE networks are based on FDD techniques, although both of these network types also support TDD (Time Division Duplex) operation.

The terminology employed in most FDD networks types labels the network to mobile device connection as the 'downlink' and the device to network connection as the 'uplink'. 3GPP2 (Third Generation Partnership Project 2) network types (cdmaOne and CDMA2000) use the terms 'forward link' and 'reverse link' to describe their equivalents of these connections.

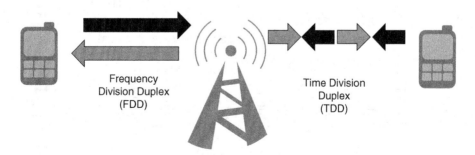

Figure 4.7 Duplex techniques

4.5.2 Time Division Duplex

In a TDD system, the transmit and receive directions are configured to use the same frequency but to occur at different points in time – hence 'time' division duplex. These systems divide a single radio channel between uplink and downlink and assign different 'timeslots' or periods of time to each.

The theory is that once a mobile device has synchronised with a cell it can discover the pattern of uplink and downlink traffic flows that share the single TDD channel.

3G UMTS has two TDD variants: UMTS-TDD has not been widely deployed, except as a form of FWA (Fixed Wireless Access), but TD-SCDMA (which is also known as UMTS$_{LCR}$) is extensively deployed in China. 4G LTE also has a TDD (TD-LTE) variant, which has been much more widely deployed. The 4G WIMAX (Worldwide Interoperability for Microwave Access) system is TDD only.

4.6   Multiple Access Techniques

In general terms, the most obvious difference between the various generations of mobile network has been the set of radio techniques employed on their air interface connections – 'air interface' is the term commonly used to describe the radio connection between a mobile device and its serving base station. Figure 4.8 provides an overview of the most common air interface techniques.

These techniques are classified by the ways in which they allow 'multiple access' to cellular resources. This means that they all offer some method by which more than one user can access the resources of a cell at any one point in time and they work by dividing the available radio resource on a channel by frequency, time or some other property.

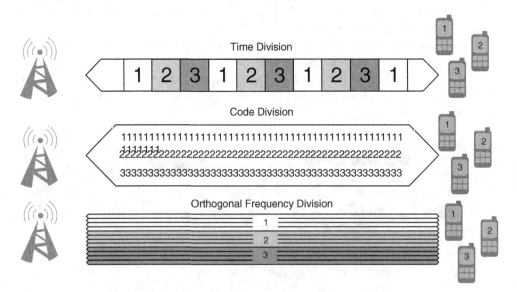

Figure 4.8   Multiple access technologies

4.6.1 Time Division Multiple Access

Most 2G network types, for example, employed an air interface based on a combination of FDMA (Frequency Division Multiple Access) and TDMA (Time Division Multiple Access) techniques in which each base station supported a number of separate radio channels (which provided the FDMA 'frequency division' component) and each radio channel was divided into a set of repeating 'time slots' (which is where TDMA 'time division' was used).

Network types that employ TDMA include GSM, IS136/TDMA, PDC and iDEN.

4.6.2 Code Division Multiple Access

Most 3G network types (and 2G cdmaOne) employ an air interface based on CDMA (Code Division Multiple Access) techniques, in which each cell (on the downlink) and each mobile device (on the uplink) transmit radio signals that have been 'scrambled' using unique digital codes, each producing a radio signal that has distinct and distinguishable characteristics that allow them to be separately received and understood even among the set of other signals that are sharing the same channels.

3G UMTS networks, which were developed in Europe and Japan but which are used around the world, employ an air interface based on WCDMA (Wideband Code Division Multiple Access) techniques that use relatively wide 5 MHz radio channels for each cell. The 3G CDMA2000 based networks used in the United States and other countries also employ CDMA, but in a configuration based on 1.25 MHz radio channels that are narrower than those employed by WCDMA. The Chinese 3G TD-SCDMA is also based on CDMA techniques and uses radio channels that are 1.6 MHz wide.

4.6.3 Orthogonal Frequency Division Multiple Access

4G systems such as LTE and WIMAX employ a technology known as OFDMA (Orthogonal Frequency Division Multiple Access) on the air interface. OFDMA is an evolved version of the FDMA concept employed by earlier network types. It works by transmitting multiple parallel, low bandwidth 'subcarriers', each of which is associated with a particular carrier frequency within the frequency band covered by the overall channel. Traffic to be transmitted across the radio connection is broken up into smaller streams, each of which is mapped to a different subcarrier ready for parallel transmission.

Traditional radio techniques, as employed by TDMA and CDMA interfaces, can be thought of as like a serial cable, with the bits of transmitted data following each other in sequence across a single radio carrier. An OFDMA air interface, conversely, can be thought of as more like a ribbon cable, with separate streams of traffic sharing the same radio channel simultaneously and in parallel.

4.6.4   Multiple Radio Access Technologies

The variety of air interface technologies employed in different generations and versions of cellular systems are commonly grouped by the term RAT (Radio Access Technology) or, less commonly, by the term RTT (Radio Transmission Technology).

The term 'multi-RAT' (Multiple Radio Access Technologies) device is often used to describe a mobile device that is able to operate using two or more radio technologies: for example, most modern smartphones are able to access 2G GSM, 3G UMTS and 4G LTE services and would therefore be described as multi-RAT devices. If a device is capable of accessing multiple RATs it should be possible for networks to perform handovers from one access type to another, a GSM to UMTS handover for example. This type of network activity is commonly known as ISHO ('inter-system' handover) or IRAT ('inter-RAT') handover.

4.7   Generic Network Architecture

Cellular networks are generally divided into three main areas, as illustrated in Figure 4.9 [1]:

- Mobile devices – the phones, tablets, computers and other devices that use cellular services;
- Radio access network – which is home to the cells, base stations and other radio and access control elements;
- Core network – which is home to the network's central administrative and interconnection services.

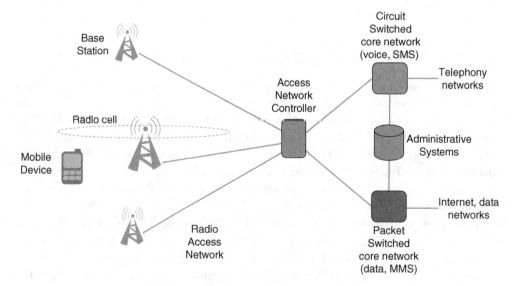

Figure 4.9   Generic network architecture

4.8   Mobile Devices

Mobile devices are the 'user facing' aspect of cellular networks that users are generally most familiar with.

Different types and generations of network have their own specific terminology that they use to describe mobile devices – for example, 2G GSM networks refer to the MS (Mobile Station), while 3G UMTS and 4G LTE networks refer to the UE (User Equipment).

4.8.1   The Third Generation Partnership Project Mobile Devices

3GPP (Third Generation Partnership Project) networks (GSM, UMTS and LTE) separate the mobile device into two elements:

- SIM (Subscriber Identity Module);
- ME (Mobile Equipment).

The SIM contains details of the subscriber's account – it stores the IMSI (International Mobile Subscriber Identity) that uniquely identifies each subscription and also stores details of the MSISDN (Mobile Subscriber Integrated Services Digital Network phone number(s) that are linked to the account.

The SIM also stores details of the security settings for the account, which allows the operator to check the user's authenticity when attempting to register with a network or use its services.

The ME can be any type of device, as long as it is capable of accessing cellular services. The categories of mobile device expand every year and, whereas there were originally just mobile phones, there are now network-connected tablets, computers, dongles and a host of other device types.

Each ME is uniquely identified by its IMEI (International Mobile Equipment Identity), which can be checked by a network during registration and which can be used to prevent stolen or faulty devices from accessing network resources.

The separation of 'subscription' from 'device' in this way was intended to make it easier for users to change, replace or upgrade their handset by simply moving (or 'plastic roaming' as it is sometimes known) the SIM from one device to another.

An additional benefit of this approach is that it gives investigators two separate identities that can be used to track an individual's usage of cellular services – details of the use of a SIM (including records of call made) can be gained by examining records related to an IMSI, while details of the usage of a handset (including tracking which SIMs had been inserted into it and used) can be gained by examining records related to an IMEI.

4.8.2   Other Network Types

3GPP2 networks, starting with IS-95/cdmaOne, did not originally employ the separated mobile device architecture of 'SIM + ME' and instead associated the ESN (Electronic Serial Number) of each device with the subscription to which it was linked. Later iterations of the 3GPP2 network, including CDMA2000, began to optionally support a version of the SIM concept.

Most other network types are based on the monolithic architecture of 'handset + serial number' and do not support a concept like the SIM.

4.9   Radio Access Networks

Each type and generation of cellular technology has a different RAN (Radio Access Network) configuration, mainly because each generation employs a different type of radio access technology.

A generic RAN consists of:

- Radio cells – which provide radio connectivity to mobile devices;
- Base stations – which control a set of cells;
- Access network controllers – which control a set of base stations and allocate resources (although not used in 4G networks);
- Backhaul – which connects RAN elements to each other and to the core network to carry user traffic and network control signalling.

4.9.1   Cells and Base Stations

A cellular base station is designed to generate radio cells that allow it to transmit and receive user and control traffic over the radio path or 'air interface' radio channels that connect to users' mobile devices.

A base station contains sets of radio transmitter/receiver units (also known as transceivers or TRXs) which each cover a certain geographical area of the operator's network. A base station may generate one cell or several cells and communicates with the operator's access network controllers and core network via a transmission link known as the 'backhaul' connection.

Base station configurations fall into two basic categories that are depicted in Figure 4.10:

- Omnidirectional sites or cells – which transmit their radio signals in all directions from one antenna.
- Sectorised sites or cells – which transmit their radio signals in sectors, each sector being generated by a different directional radio antenna. The traditional sectorised cell configuration uses three antennas to create three different cells that between them provide 360° coverage around the site.

Figure 4.10 Omnidirectional site and sectorised site base stations

Some cell sites are configured to employ a mix of omnidirectional and sectorised cells, whereas most sites will use one technique or the other. Sectorised sites are much more common than omnidirectional ones as they offer much more potential to configure the amount of capacity applied to each area of coverage.

The capacity of a base station can be increased by adding extra radio channels to cells or sectors and there are a variety of configuration options.

If, for example, a sectorised site had three sectors and two transceivers on each sector, it would be known as a $2 + 2 + 2$ site and would be able to support six radio channels in total. A higher capacity base station configuration could be deployed as a $4 + 4 + 4$ site, giving a total of 12 radio channels to be used by customers.

The transmit power of a base station varies depending upon the technology in use, but can range from as little as few milliwatts per sector to several hundred watts per sector. The size of a cell as it extends out from a site (known as the 'cell radius') varies too, with small 'femto' and 'micro' cells covering just a few tens of metres up to large 'macro' cells which can extend for several kilometres.

4.9.2 Location Areas

The cells in a mobile network are grouped into administrative areas, known variously as Location Areas (LA – for 2G/3G voice services), Routing Areas (RA – for 2G/3G data services) or Tracking Areas (TA – for 4G services), each of which is identified using its own specific index or code number: LAs are identified by a LAC (Location Area Code), RAs by a RAC (Routing Area Code) and TAs by a TAC (Tracking Area Code). These different area types are shown in Figure 4.11 and can be generically referred to as 'LA' for the purposes of this explanation.

The LA is used to track the approximate current location of mobile devices when they are in Idle Mode, which is the state they are deemed to be in when they are switched on but are not actively engaged in exchanging user traffic with the network.

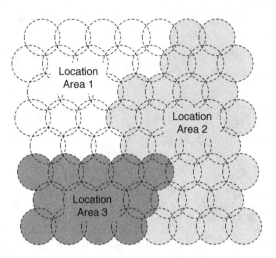

Figure 4.11   Location areas

The network uses the stored LA information to allow it to 'page' for an idle mobile device if it needs to re-establish contact – for example if an inbound call needs to be connected to a phone.

A mobile device will perform a location area update to inform the core network if it moves into a new area, this ensures that the network will page for it in the correct area if contact needs to be re-established.

Mobile devices are also provided with a 'periodic update timer' to use when idle. The timer is started when the device drops into Idle Mode. This timer is stopped if the device communicates with the network but, if it has not contacted the network for any other reason by the time the timer expires, it will connect and perform a 'periodic' location area update to let the network know that it is still available.

The specific forms of area update performed in each generation of 3GPP technology, and the core network database with which mobile devices communicate, are as follows:

- 2G/3G voice services – Location Area Updates (LAU) are made to the mobile's current VLR (Visitor Location Register).
- 2G/3G data services – Routing Area Updates (RAU) are made to the mobile's current SGSN (Serving GPRS Support Node).
- 4G services – Tracking Area Updates (TAU) are made to the mobile's current MME (Mobility Management Entity).

Non-3GPP technologies employ similar processes and use similar terminology.

The core network elements that handle these updates each maintain local mobility management databases that track the locations of a subset of network users served by a particular area of the access network. The subscriber data held in these databases is downloaded from the HLR (Home Location Register) when each local database assumes responsibility for a user.

The process of performing location updates can be quite expensive in terms of using mobile battery power and of consuming network signalling resources and so the network is usually configured to discourage unnecessary area changes.

Using a 'broadcast' control channel (BCC; meaning a channel that is broadcast within a cell and is therefore accessible by any mobile device), each cell advertises a set of selection and reselection parameters. One of the reselection parameters is a negative offset that is only applicable to phones that see that cell as a neighbour and that are currently in a different location area from that cell. The phone will apply this negative offset to any measurements it takes of the target neighbour cell, which has the effect of making the neighbour less attractive and therefore less likely to be reselected. This ensures that a mobile device will only initiate a move to a cell in a different area when it has little alternative and therefore consumes the additional battery power and network signalling capacity to send a location update only when it absolutely must.

When undertaking forensic radio surveys, this location area reselection offset sometimes has the effect of showing a very strong neighbour cell that is never selected as a serving cell, even if it is sometimes stronger than the current serving cell. These situations are simple to spot, as the 'strong but never serving' neighbour cell will have a different LAC/RAC/TAC to the cell(s) that have been selected as serving. Cells affected by this process are often classed in forensic radio survey reports as 'not detected as serving but having the potential to serve'.

4.10   Core Networks

Core networks provide a common set of functions for users served by the various types and regions of access network maintained by an operator.

The core network provides interconnection services, allowing voice, data and messaging transactions to be connected to the required destination user or service. Operators also host the 'inter network' connections (or 'interconnects') that allow them to exchange calls and other user traffic with other operators' networks.

4.10.1   2G/3G Core Networks

2G and 3G core networks are divided into three main areas, as described in Figure 4.12:

- The CS (Circuit Switched) core, which deals with 'real time' services such as voice and video telephony and also typically deals with 'value added' services like voice-mail and SMS text messaging.
- The PS (Packet Switched) core network, which deals with 'non real time' data services such as Internet connections, e-mail, instant messaging and MMS.
- A shared administrative area that hosts subscriber databases, the billing system, network management systems and so forth.

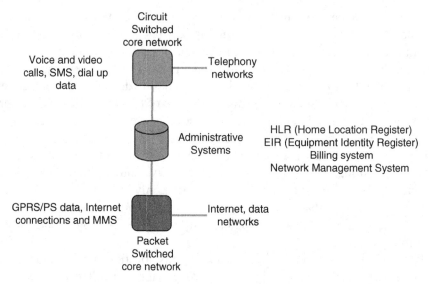

Figure 4.12   2G/3G core networks

4.10.2   4G Core Networks

4G LTE networks only support data services, or more accurately, only support services that can be carried by IP (the Internet Protocol). This means that an LTE network does not support services like voice calling that are carried using traditional CS techniques – instead, LTE networks can only carry VoIP (Voice over IP) services that use Internet technologies to carry voice traffic.

Ultimately, this means that an LTE network has no requirement for separate CS and PS core networks and only supports a PS core.

Although 2G/3G and 4G networks require different 'user traffic handling' elements, it is perfectly possible for them to share administrative systems such as the HLR/HSS (Home Location Register/Home Subscriber Server), billing system, customer care system and so on. It is also possible to interconnect networks of different generations via their core networks to permit mobile devices to perform handovers between different generations of access network – from a 4G cell to a 3G cell, or from 2G to 3G, for example. This kind of 'inter-system handover' activity is a crucial means of ensuring that mobile devices connect to the type of access network best suited to carry the services each user requires.

4.11   Subscriber and Device Identifiers

In order for a cellular network to operate effectively there are several subscriber and device identities that must be employed, including the following:

- MSISDN
- IMSI
- TMSI
- IMEI.

The identifiers listed above are specific to 3GPP type networks [2]. Non-3GPP network types may employ similar types of identifier.

4.11.1 Mobile Subscriber Integrated Services Digital Network

The MSISDN (Mobile Subscriber Integrated Services Digital Network) is a telephone number that is uniquely linked to a subscription in a mobile network. It is the telephone number allocated to (or 'paired' with) an IMSI (explained below) and it is the MSISDN which is the number normally dialled to connect a call to the mobile phone. The 'ISDN' part of the abbreviation refers to the Integrated Services Digital Network, which was the correct technical description for a modern digital telephone network at the time that GSM was developed.

The format of an MSISDN is shown in Figure 4.13.

The MSISDN is based on the standard E.164 numbering format, which allocates a maximum of 15 digits for each phone number. The E.164 standard, which was published by the ITU (International Telecommunications Union), specifies the accepted format for international telephone numbers and mandates that numbers start with a country code (e.g. 44 for the United Kingdom), followed by a network or area code, followed by a subscriber number [3].

4.11.2 International Mobile Subscriber Identity

The purpose of the IMSI is to uniquely identify a subscriber within the mobile network environment. The IMSI number is used for registering and identifying a subscriber in the Public Land Mobile Network (PLMN) with which they are registered [4].

When a new subscription is set up (or when a replacement SIM is created), the network's HLR (Home Location Register) has to associate (or 'pair') an MSISDN with the IMSI that uniquely identifies a mobile subscriber. The subscribers themselves are generally unaware of their IMSI number, which is held on the SIM card and is communicated to the network during 'attach' and authentication processes.

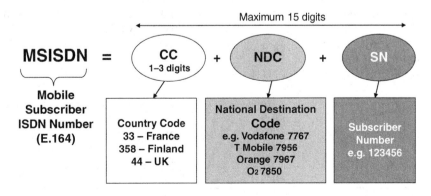

Figure 4.13   The MSISDN

An IMSI is always 15 digits long (when rendered in decimal, it is 64-bits long in binary) and it consists of the format shown in Figure 4.14:

The MCC (Mobile Country Code) number uniquely identifies the country in which the issuing PLMN operates. MCC numbers are issued and controlled by the ITU, which is an agency of the UN that coordinates global telecoms activities. MNCs (Mobile Network Codes) are generally administered by each individual country's telecoms regulator and uniquely identify a network within an MCC area. Table 4.1

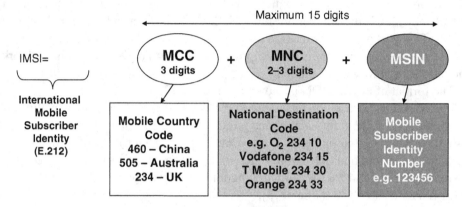

Figure 4.14   The International Mobile Subscriber Identity (IMSI)

Table 4.1   Examples of United Kingdom MCC and MNC.

Country	MCC	Operator	MNC
		O2	10
United Kingdom		Vodafone	15
Main operators	234	Three	20
		T Mobile/4GEE	30
		Orange	33
		BT	00
United Kingdom		Jersey/Guernsey Airtel	03
		COLT Jersey	05
		Network Rail	12/13/95
Selected other operators	234	Jersey Telecom	50
		C&W Jersey/Guernsey	55
		Airwave	78
Selected other MCCs	310	United States	
	311		
	313		
	316		
	419	Kuwait	
	505	Australia	
	470	Bangladesh	
	732	Colombia	
	648	Zimbabwe	

provides examples of MCC allocations and also provides details of some of the MNCs assigned in the United Kingdom.

The ITU is the main body responsible for assigning MCCs – the structure and usage of MCCs is defined in ITU recommendation E.212. The ITU website provides access to a master list of MCCs in report number 1005 and a list of MNCs per country is contained in report number 1019 – neither of the lists currently on the ITU website has been updated since early 2012, however, and other resources may be found to be more up to date.

A full list of current MCC/MNC allocations is provided in Chapter 9 and can also be obtained from the GSMA (GSM Association) via their website – www.gsmsintelligence.com – for those that have a login or via a range of other Internet resources, such as Wikipedia.

4.11.3   Temporary Mobile Subscriber Identity

A subscriber's IMSI is a sensitive piece of information that could, in theory, be captured by a hacker as it travelled across a radio control channel and could be used to 'spoof' a network into providing unauthorised access to services. Networks therefore attempt to ensure that the IMSI is transmitted only when absolutely necessary [5].

Following a successful 'attach' process, which allows a mobile device to register with a network and receive service, the local mobility database node serving the device (which in 3GPP networks will be the VLR, SGSN or MME) will assign a TMSI (Temporary Mobile Subscriber Identity) to be used to temporarily and anonymously identify the user. The mapping of the TMSI to the subscriber's IMSI is known only to the mobile device and to the database that assigned it and the TMSI will be 'refreshed' at regular intervals to increase the level of security provided.

Each 3GPP local mobility management database assigns a different type of temporary identity: VLRs assign TMSIs to devices that are attached for 2G/3G CS services; SGSNs (Serving GPRS Support Nodes) assign P-TMSIs (Packet TMSIs) to devices that are attached for 2G/3G PS services; and MMEs (Mobility Management Entities) assign an M-TMSI (MME TMSI) to devices that are attached for 4G services. Each of these identifiers is unique only within the node that assigned it or within the 'pool' of core network nodes that the assigning node belongs to.

Once a TMSI of whichever kind has been assigned to a device it is able to identify itself to its serving core network node without needing to transmit its sensitive IMSI over a radio connection.

4.11.4   International Mobile Equipment Identity

The IMEI is a number unique to every 3GPP-specified mobile device (other types of identifier are used by some other network types). It is usually found printed on the phone on the back case or underneath the battery and can also be displayed onscreen by dialing the sequence *#06# into the phone [6].

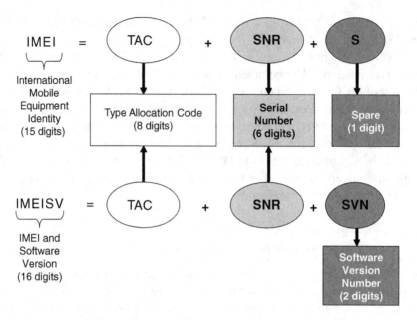

Figure 4.15   Structure of the IMEI and IMEISV

The IMEI number is used by a network in conjunction with a core network element known as the EIR (Equipment Identity Register) to identify valid devices before allowing them to be used and can therefore be employed to prevent a stolen phone from gaining access to network resources. For example, if a mobile phone is stolen, the owner can call his or her network provider and instruct them to 'block' the phone using its IMEI number. This renders the phone useless, regardless of whether the phone's SIM is changed or not.

The IMEI structure is outlined in Figure 4.15 and is composed of the following elements (the length of each element below is expressed in decimal digits):

- Type Allocation Code (TAC; eight digits).
- Serial Number (SNR) is an individual serial number uniquely identifying each equipment within each TAC – (six digits).
- Sparc digit – this digit is set to zero, when transmitted by the mobile device.

The IMEI (14 digits) is complemented by a check digit.

The first two digits of the TAC identify the Reporting Body through which the device type was registered. The remaining six digits of the TAC code are assigned to device manufacturers and each TAC should uniquely identify both the manufacturer and the specific handset model.

The check digit is used as an error checking mechanism and allows for a simple check of the integrity of a transmitted IMEI. The IMEI captured in billing records often has the check digit set to 0 and may therefore not exactly match the IMEI printed on the mobile device. As long as the leading 14 digits match then that is sufficient.

A variation of the IMEI is provided by the IMEISV (IMEI and Software Version), where the last two digits of the IMEI indicate the software version currently running on the device.

4.12   Network Databases

3GPP-based cellular networks rely on three types of database to manage subscribers, authorise their services and maintain security; other types of network will use alternative (but similar) resources.

4.12.1   Home Location Register/Home Subscriber Server

The most important core network database is the HLR, which is known as the HSS in evolved networks. The HLR/HSS is the main repository of subscriber data within a network and there will be one logical HLR/HSS database for each network (although the database might be distributed over several physical nodes for capacity, redundancy and resilience reasons).

Each subscriber's details are held in the HLR/HSS, with the database key being provided by their IMSI. The database record also holds details of the MS-ISDN (mobile phone number) that has been paired with the account and lists the set of services (international roaming, call diversions, call barring, etc.) that the user has set or is permitted to use. The database also shows each subscription's current availability state (attached or detached) and indicates an attached device's approximate current location by identifying the local core network node (VLR, SGSN or MME) that is currently serving it. This allows inbound calls and messages to be directed quickly to the correct part of the network.

The HLR/HSS also manages user security. When a mobile device 'attaches' to the network, the HLR retrieves a copy of the secret subscription security key (which is known as 'K'). The user's SIM card also has a copy of 'K' and so it forms a 'shared secret' between the HLR and the individual SIM. The shared secret allows the network to 'challenge' the mobile by generating a random number and asking the phone to process that number using an authentication algorithm and the SIM's stored copy of 'K'. Only a SIM that has the value of 'K' that is valid for the subscription/IMSI will be able to respond to the challenge with the correct answer, allowing the network to quickly and simply check the authenticity of all users before they are allowed to access network services.

4.12.2   Local Mobility Management Database

The HLR/HSS is a central store for subscriber data and it downloads a copy of that data into the 'mobility management' database of whichever core network node is currently serving each subscriber. The local mobility management nodes employed differ depending upon the type of service or network the user is connected to. 3GPP networks support several different types of mobility management database:

- For 2G/3G voice/CS services the user will be managed by a VLR.
- For 2G/3G PS data services the user will be managed by an SGSN.
- For 4G services the user will be managed by an MME.

The VLR/SGSN/MME will also have copies of the current set of security keys that were created as part of the authentication process, allowing ciphering (encryption) and integrity protection to be invoked on traffic and control connections for each phone.

4.12.3   Equipment Identity Register

The EIR is an operator's database of known mobile devices and their IMEIs.

The operator registers in the EIR the IMEI of each device it supplies to users, which allows the IMEI to be checked when a device attempts to connect to the network.

The EIR holds IMEIs in one of three areas of its database:

- The 'white list' contains IMEIs of devices that are permitted to use the network.
- The 'grey list' contains details of IMEIs that are permitted to use the network but that should be monitored, possibly due to a 'fraud flag' or because they are suspected of having a fault.
- The 'black list' contains details of IMEIs that are not permitted to use the network, normally because they have been registered as stolen or because they are known to be faulty.

In some countries, the United Kingdom for example, all of the cellular operators have been obliged to interlink their EIRs in an attempt to ensure that no stolen phones or other cellular devices will be permitted to connect to any of those operators' networks. The theory behind this being that if a stolen device cannot be used there is little point in stealing it.

4.12.4   Network Types – MNOs and MVNOs

There are two types of network operator:

1. An MNO (Mobile Network Operator) owns their own base stations, radio access network and core network and supports a full range of mobile services.
2. An MVNO (Mobile Virtual Network Operator) offers mobile services to their customers but does not own its own physical network; instead they 'piggyback' on the facilities of an MNO.

MVNOs can be configured in a range of ways, with the simplest being just an alternative 'brand' operated by an MNO, whilst the most sophisticated examples operate their own customer care and billing systems and some even operate their own HLR.

There is an intermediate class of operator, which mediates between MNOs and MVNOs and which builds or manages the infrastructure on behalf of MVNOs. These entities are known as MVNEs (Mobile Virtual Network Enablers) and MVNAs (Mobile Virtual Network Aggregators).

From a forensic point of view, call records for MVNOs and their customers are almost always generated or disclosed by the host MNO, which means that there will not necessarily be additional call record formats for cell site analysts to deal with, even if a country hosts a large number of MVNOs.

4.13   Cell Sites

The term 'cell site' is used to refer to the location of a base station – it is the site from where cells are transmitted. The elements deployed at a cell site include the base station itself and its radio antennas, plus the infrastructure required to support the site such as power and transmission equipment and potentially also a tower upon which to mount the antennas.

4.13.1   Channels and Carriers

The term 'radio carrier' (or sometimes just 'carrier') is used to describe the combination of radio channels that provide service in a cell.

Traditionally, an FDD radio carrier consists of two radio channels, as shown in Figure 4.16:

1. An uplink channel, used by mobile devices to transmit to the base station (3GPP2 networks call this the 'reverse link').
2. A downlink channel, used by the base station to transmit to the mobile devices (3GPP2 networks call this the 'forward link').

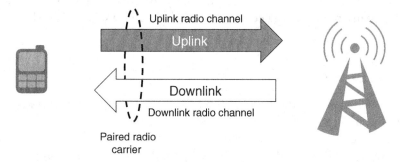

Figure 4.16   Channels and carriers

TDD cells use just one channel that is split, using a time division scheme, between uplink periods and downlink periods.

The whole set of FDD radio channels available for a particular network type is administratively divided into separate blocks of uplinks and downlinks. A particular uplink channel is usually permanently 'paired' with a particular downlink channel to create each carrier. A simple ID number is then assigned to each carrier, which provides a 'shorthand' method of assigning and describing radio resources. These numbers are known as ARFCNs (Absolute Radio Frequency Channel Numbers).

2G GSM ARFCNs are assigned from a range from Channel 0 to Channel 1023, although not all possible ARFCNs are employed.

3G UMTS radio carrier IDs, known as UARFCNs (UMTS Absolute Radio Frequency Channel Numbers), take the form of a four- or five-digit number such as 2963 or 10761. A simple calculation can be used to equate a UARFCN with the centre frequency of each carrier (see Ref. [7]).

2G cdmaOne and 3G CDMA2000 radio resources are defined in a series of band classes, each of which supports a set of numbered channels. The channel number ranges are reused within each band class and typically range from zero to a maximum value, which varies with the bandwidth of each band. This means that the centre frequency of a channel cannot be determined from knowledge of the channel number alone, the band class must also be known. The highest CDMA2000 channel number, used in Band Class 2, is 2047.

4G LTE EARFCNs (EUTRAN ARFCNs) also take the form of a 4 or 5 digit channel number such as 1617 or 38249. A less simple calculation can be used to derive carrier frequency from EARFCN (see Ref. [8]).

Each 2G GSM cell will typically consist of several carriers, one of which will carry the BCCH (Broadcast Control Channel) while the others are used as 'traffic carriers' or 'expansion carriers' and provide additional capacity. All of the carriers in such a cell (the BCCH carrier plus the expansion carriers) will share the same Cell ID and BSIC (Base Station ID Code; as these are carried on the BCCH).

It is not possible to deploy multiple carriers to one cell in 3G UMTS, so each carrier will have its own BCCH and a separate cell ID.

It is also currently not possible to deploy multiple carriers to a cell in 4G LTE, so each carrier has its own BCCH and Cell ID, but a proposed evolution of LTE will allow 'expansion carriers' to be created that will share a Cell ID with a BCCH carrier.

4.13.2   Cells and Sectors

There are two main configurations for cell sites: omnidirectional and sectorised. There is also a third option which is a hybrid of the two.

An omnidirectional site uses radio antennas that 'broadcast' their signal through 360°, allowing all of the radio channels supported by that base station to potentially be available anywhere within the site coverage area.

A sectorised site, in contrast, employs directional antennas, which direct transmission of each radio channel to a specific part of the base station coverage area. Sectorised sites can be configured in a variety of ways, depending on the type of antenna employed:

- Antennas with a coverage arc of 120–130° can be used to provide up to a three-sectored site.
- Antennas with a coverage arc of 60–65° can be used to provide up to a six-sectored site.

These concepts are illustrated in Figure 4.17.

The third option is a hybrid of omnidirectional and sectorised and is commonly known as OTSR (Omnidirectional Transmit Sectorised Receive). The standard sectorised configured can therefore also be known as STSR (Sectorised Transmit, Sectorised Receive). These concepts are illustrated in Figure 4.18.

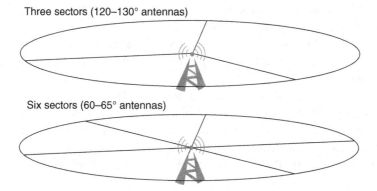

Figure 4.17 Cellular configurations

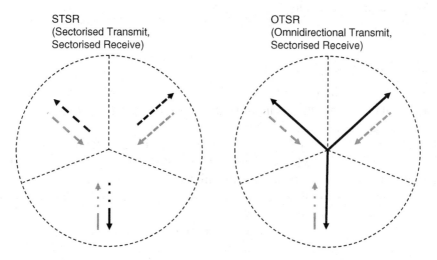

Figure 4.18 Omnidirectional transmit, sectorised receive

In an OTSR configuration, a sectorised base station will be provided with one radio unit per sector but only one of them will be set to transmit downlink/forward link traffic.

In a three-sector OTSR site, the output of the transmitter is passed to a 'splitter' device, which creates equal copies of the signal to pass to three transmit antennas. Even though such a site might have three antennas arranged in a typical sectorised configuration, each antenna (and therefore each site) carries the same downlink signal. OTSR therefore provides an omnidirectional service using a sectorised antenna configuration.

The 'receive' side of the three antennas will each be connected to a different radio receiver unit, and will therefore work independently of each other in receiving traffic that originated in only one of the site's sectors, so the site has sectorised receive.

The benefit of the OTSR model is that networks are able to operate a low capacity base station, with a single transmitted cell, but are able to improve the performance and sensitivity of the uplink/reverse link, as each sector antenna only needs to deal with user traffic against the background of the noise and interference generated in a portion of the surrounding coverage area.

The STSR model retains the benefits of sectorised receive but has the extra capacity of sectorised transmit. STSR is by far the most commonly employed cellular configuration.

Given the variety of cell configurations that exist, the terminology used in cellular networks that describes coverage in terms of 'cells' and 'sectors' has the potential to be misleading as the meaning is partly dependent on the configuration that has been used.

The capacity of an omnidirectional cell can be described as being 'one cell'.

The capacity of an OTSR site can be described as 'one cell with X number of sectors'. A three-sector site would be classed as one cell.

The capacity of an STSR site can be described as 'one base station supporting X number of *cells*'. A three-sector site would therefore be classed as three separate cells hosted by the same base station.

4.13.3   Cell Capacity

The capacity of a cell or sector is partly determined by the number of radio carriers that are available within it. It is intuitive that a cell or sector with two carriers would have greater capacity than a cell or sector with only one.

Standard cellular terminology describes the capacity and configuration of a site using 'N-notation'. This allows the number of sectors/cells to be counted and then defines the number of carriers (uplink/downlink pairs) available in each. For example, a three-sectored site that uses one carrier per sector/cell would be described as a 1 + 1 + 1 site. This is illustrated in Figure 4.19.

Higher and lower capacity cellular configurations are possible:

- 1 + 0 + 0 – is a way of describing an omnidirectional site: one carrier, one sector.

- 1 + 1 + 0 – describes a two-sectored site. This would typically be used to provide coverage along a stretch of railway line or motorway with each sector aligned 180° from the other.
- 2 + 2 + 2 – three sectors/cells and two carriers per sector.

Cells covering very busy or high-demand areas can be deployed with six sectors, rather than the more traditional three-sector design. In these circumstances the '$n + n + n$' terminology can become unwieldy and some networks instead employ an '$n \times y$' description: for example, a base station with six sectors, with three carriers per sector would be described as a '6×3' site.

4.13.4 Cell Identifiers

In order for a mobile device to be able to identify, measure and select a cell it must have a way of discriminating between the various cells it can detect and more widely, operators need a way of individually identifying each cell in their network. There are two levels of identifier that are used, as shown in Figure 4.20:

- Physical-layer Cell Identifiers (PCI) provide a simple way of discriminating between local cells based on information that is transmitted at the physical layer – for

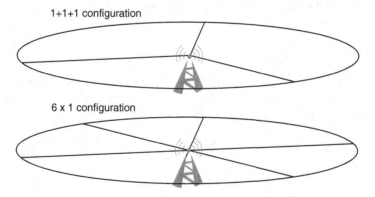

1+1+1 configuration

6 x 1 configuration

Figure 4.19 Cellular capacity

Radio cell

Physical layer ID
Locally unique, identifies a cell from its immediate neighbours, different formats for 2G, 3G and 4G

CGI - Cell Global Identifier
Globally unique, identifies network and cell, different formats for 2G, 3G and 4G

Figure 4.20 Cell identifiers

example information that is embedded in the cell's radio signal. These IDs will be unique only within a small part of an access network and may be reused by other cells elsewhere in the same network. There are different PCI types for each different type or generation of network.

- Cell Global Identifiers (CGI) provide a globally unique ID for each cell, which identifies the country (MCC), network (MNC), area or base station identity and Cell ID. There are different CGI formats for each different type or generation of network.

4.13.5   Physical-layer Cell Identifiers

Each generation of cellular network employs a different type of PCI:

- 2G GSM uses the BSIC.
- 3G UMTS uses the PSC (Primary Scrambling Code).
- 2G cdmaOne/3G CDMA2000 use the PN (Pseudo Noise) offset.
- 4G LTE uses the PCI.

All of these methods are different but share the concept that they are carried within the structure of the physical radio signal being transmitted by a base station, allowing phones to detect them without needing to fully synchronise with the cell and decode its BCCH. The PCI offers a quick and simple method of local cell discrimination.

4.13.6   Cell Global Identifiers

The CGI provides each cell with a fully unique global ID, which allows an individual cell to be identified among all of the cells in all 2G, 3G and 4G networks globally. The structure of the CGI differs for each generation of technology:

- 2G GSM CGIs consist of MCC-MNC-LAC-CI, where LAC is Location Area Code and CI is Cell ID.
- 3G UMTS CGIs consist of MCC-MNC-RNC ID-CI, where RNC ID is the ID of the Radio Network Controller responsible for managing the cell and CI is Cell ID.
- 2G cdmaOne/3G CDMA2000 CGIs consist of MCC-MNC-SID-NID-BSID, where SID/NID identify an area of the network and the BSID (Base Station ID) component identifies the base station and the sector.
- 4G LTE CGIs consist of MCC-MNC-eNB ID-CI, where eNB ID is the identity of the eNB base station responsible for managing the cell and CI is Cell ID.

Each of these formats is discussed in more detail in the relevant sections in Chapters 5 and 6.

CGIs are carried on a cell's BCCH and are transmitted in binary. Survey equipment will present the CGI in either decimal or hexadecimal and some network operators also present cell IDs in hexadecimal in CDRs (Call Detail Records).

4.13.7 Decimal, Binary and Hexadecimal

Numbering systems can be devised that can contain any number of unique characters. The decimal system has 10 characters (0, 1, 2, 3, 4, 5, 6, 7, 8, 9) was adopted originally because human beings have ten fingers. Table 4.2 provides an overview of three numbering systems that are commonly used in digital telecoms.

As digital equipment such as computers use pulses of electricity or light or use a radio modulation scheme to represent numbers, they operate using the binary system, which has only two conditions, ON and OFF. These two conditions are represented by the characters 1 and 0.

Large numbers in binary have many 1s and 0s, which can be difficult to represent in a 'human readable' format. For example, 27 in decimal is represented in binary as 11011, while a larger number such as 1 000 000 would be represented as 1111010001000000.

Hexadecimal (or 'hex') is a numbering system of 16 characters (10 digits and 6 letters). It is used to condense the long strings of zeroes and ones in large binary numbers into a more manageable form. This base-16 numeric notation is frequently used to specify addresses in computer memory as it makes life simpler for programmers. The decimal numbers 0–9 are represented by the decimal digits 0–9 and the decimal numbers 10–15 are represented by the letters A–F.

Therefore, if a byte (eight bits) in binary is 01001101 then this is represented as 4D (4 = 0100 and D = 1101) in hex, which equates to 77 in decimal.

A single hexadecimal digit is known as a 'nibble' and represents just four bits of binary information. For example, the hex digit/nibble '4' is equal to '0100' in binary and the nibble F is equal to 1111 in binary (15 in decimal).

A long string of binary digits can therefore be broken up into groups of four bits, each of which can then be represented using a hex digit.

Table 4.2 Comparison of decimal, binary and hexadecimal notation.

Decimal (base 10)	Binary (base 2)	Hexadecimal (base 16)
0	0000	0
1	0001	1
2	0010	2
3	0011	3
4	0100	4
5	0101	5
6	0110	6
7	0111	7
8	1000	8
9	1001	9
10	1010	A
11	1011	B
12	1100	C
13	1101	D
14	1110	E
15	1111	F

Table 4.3 Highest and lowest hexadecimal CI values.

Use	Binary	Decimal	Hexadecimal
Lowest possible Cell ID	0000000000000000	0	0
Highest possible Cell ID	1111111111111111	65 535	FFFF

Cell ID information is usually presented in decimal, but some network operators have elected to present cell IDs in hex in call records and other network documentation, which can cause confusion.

The CI (Cell ID) component of 2G and 3G CGIs is transmitted on the cell's BCCH in binary and consists of a 16-bit binary number. As shown in Table 4.3, when converted to decimal, a 16-bit binary field has a range of values from 0 to 65 535 and, in hexadecimal, has a range of 0 to FFFF.

Networks often use Cell ID 65 353/FFFF to mean 'no cell used' or 'cell not known'.

The cell ID component of a 4G eCGI is transmitted as a 28-bit binary number, which equates to a decimal range of 0–268435455 and a hex range of 0–FFFFFFF.

4.14   Antennas and Azimuths

Most sectorised cellular antennas are of a type known as a 'stacked' or 'bayed' dipole and produce a radio beam in a fan shape. The idealised coverage of such a beam can be thought of as being in a 'cloverleaf' pattern when seen from above, as depicted in Figure 4.21.

The typical horizontal coverage arc for three-sector site antennas is 130° with a vertical beamwidth of about 4–6°.

Directional antennas like those used at sectorised cell sites have a radiation intensity peak in the direction of the boresight (the centre of the beam). Beamwidth is the angular distance between the points on two opposite sides of the boresight where the radiated signal intensity drops to half of the intensity in the boresight direction (known as the 3 dB or 'half power' points, as a loss of 3 dB is equivalent to a halving of signal strength).

In a typical three-sectored site, each sector is 130° wide and the antenna will have a beamwidth of 65°. This means that by 32.5° either side of the boresight the power has dropped by 3 dB (half power). The power continues to get progressively weaker towards the edge of the sector but can still provide a signal strong enough to carry communications.

The term 'azimuth' is used to describe a mathematical concept defined as the angle, usually measured in degrees, between a reference angle and another point. In navigation, the reference angle is typically true north which is considered 0° azimuth. Moving clockwise, a point due east would have an azimuth of 90°, south 180° and west 270°.

With a three-sectored cell site, the azimuth angle of each antenna indicates the central pointing angle of the beam – the boresight. For example, if one of the antennas has an azimuth of 0° then the other sectors would normally be pointed at 120 and 240°. Examples of cell site azimuths are shown in Figure 4.22.

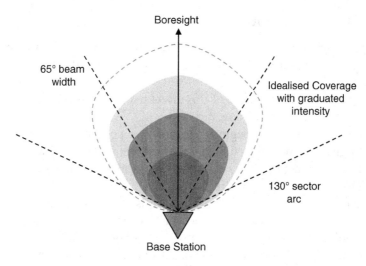

Figure 4.21   Radio beam coverage from an antenna

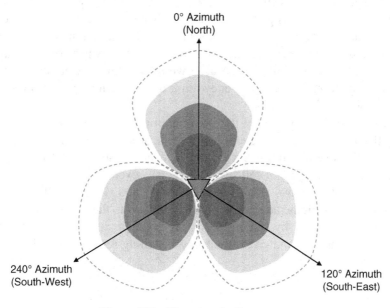

Figure 4.22   The azimuth of an antenna

From a forensic point of view, if the location of a cell site is known and if the width of a sector (60° or 120°) and its azimuth are known, then an approximate area of use can be inferred for phones that used particular cells. This technique can be termed 'high-level cell site analysis'. It cannot be used to 'pinpoint' a phone's potential location (cell site analysis can rarely claim to be that accurate) but it can help to provide a degree of localisation. For example, high-level analysis could help an analyst to conclude that a target phone could potentially have been 'in an area to the north-east of the cell site' at the time of a call.

4.15   Uptilt and Downtilt

The radio signal generated by an antenna is radiated perpendicular to the orientation of that antenna – so if a panel antenna is mounted vertically the radio signal will radiate horizontally.

Network planners can make use of this effect to control the propagation of a cell's signal and thereby influence its coverage area.

If the antenna is tilted a few degrees forwards (i.e. pointing down from the vertical) its radiated signal will be directed below the horizontal plane and will hit the ground before it reaches the natural horizon created by local geography or the curvature of the Earth. This is known as 'downtilt'.

If the antenna is tilted a few degrees backwards (i.e. up from the vertical) its radiated signal will be directed above the horizontal plane, away from the ground. This is known as 'uptilt'. These concepts are illustrated in Figure 4.23.

Downtilt offers an additional method of limiting the propagation of a radio signal (the other method being to limit the signal's transmit power) and therefore of controlling the size of a cell. By 'firing the signal into the ground' using downtilt, a planner can ensure that a cell that is planned to cover only a small area does not extend much beyond that area. Downtilt is commonly employed in urban areas, where cells are generally required to offer high capacity coverage that is focused on small areas. The main benefit of using techniques such as downtilt is to limit the effects of co-channel interference and to reduce the frequency reuse distance within which the same channel can be used for further cells.

Uptilt, conversely, is typically used to maximise the coverage area of a cell by ensuring that much of its signal avoids 'hitting the ground'. Uptilt is usually employed in rural locations (that require large cells) and may also be used in hilly areas to ensure that areas upslope of a cell site still receive coverage.

Tilt can be applied in two ways – mechanical tilt and electrical tilt.

With mechanical tilt the antenna is physically tilted to the required angle. This can either be done manually by sending a 'rigger' to each site to move the antenna or, increasingly, it can be handled remotely by installing remote control 'pan and tilt' motors to each antenna.

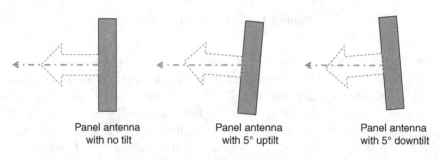

Panel antenna   Panel antenna   Panel antenna
with no tilt   with 5° uptilt   with 5° downtilt

Figure 4.23   Uptilt and downtilt

Electrical tilt works by altering the phase alignment of the signals applied to each of the elements within the antenna, which has the effect of bending the radiated signal away from the horizontal plane. Electrical tilt may also be adjusted on site or remotely.

4.16   Cell Types and Sizes

A typical radio cell, of any generation, has a finite capacity.

With large cells this finite capacity is spread across a wide area and might need to be shared by a large number of users, which has implications for the density of coverage (in terms of the numbers of concurrent connections that can be supported vs. the population of the area served by the cell) and may also limit the data rates that are available to individual users. With smaller cells this finite capacity is focused on a smaller area, which implies high density of coverage and potentially high data rates per user.

On the other hand, a few large cells can cover a large area, lowering the cost of providing service to that area, while a large number of small cells would be required to cover the same sized area, which would increase the cost of deploying that service.

Cellular planning is focused on managing the balance between cell size/capacity of coverage versus cost of deployment. Cellular operators are therefore very careful about planning the size and number of cells they deploy to match the expected customer demand (and therefore the income potential) in each area.

The range of cell types that operators can choose from is outlined in Figure 4.24 and is generally categorised as follows:

- Macrocells – outdoor sites that provide wide area coverage with typical cell radius measurements of 1–20 km or more.
- Microcells – outdoor sites that provide more focused hotspot coverage with typical cell radius measurements of 0.5–1 km.
- Picocells – can be deployed as outdoor sites, in which case the cell radius can be up to 500 m, or as indoor sites in offices, shopping centres or airports with a typical cell radius of 20–30 m.
- Femtocells – can be deployed as outdoor sites or indoor sites with a typical cell radius of 10–20 m.

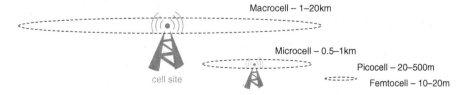

Figure 4.24   Cell types and sizes

There are no rigidly defined standards for cell descriptions. The descriptions provided above are representative of general industry opinion, but should be viewed as suggestions rather than rules.

4.17   Cell Site Types and Uses

Cell sites are usually deployed in the optimum location to serve the required area and operators will design each site to make best use of the local resources.

In general, macrocell sites that are designed to provide wide area 'umbrella' coverage will be deployed at some form of high point – on top of a hill, on the roof of a tall building or on a tall mast or tower.

Microcell sites are typically designed to provide limited areas with focused coverage and are usually deployed at a lower level – on the side or roof of a low building, strapped to a lamppost or mounted on a streetside mast.

Small cells, such as picocells and femtocells, are designed to cover very limited areas, those deployed to provide outdoor coverage are often mounted on the wall of a building and some designs are even incorporated into lampposts. Small cells can also be deployed to provide indoor coverage, in which case they are usually generated by devices that are the size of a typical Internet router that are plugged into the user's broadband connection.

The majority of sites are deployed using sectorised cells, with three sector and six sector configurations being common. Sites designed to provide specialist coverage – a site that serves a stretch of motorway or a railway line, for example – might be deployed with just two sectors and might not deliver 360° coverage. There are even examples of single sector sites which are designed to provide coverage to a specific location, for example on a site that is designed to provide coverage from land to an offshore location such as an oil rig. Examples of typical cell sizes and uses are shown in Figure 4.25.

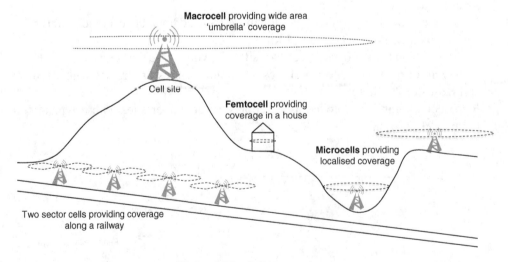

Figure 4.25   Cell types and uses

4.18   Single and Multi Frequency Networks

Some technology types – 2G GSM, for example – must employ complex frequency reuse patterns as they are not permitted to reuse the same or an adjacent channel in neighbouring cells. The resulting access networks therefore employ many different radio channels and can be termed 'multi frequency' networks.

Other technologies – 2G cdmaOne, 3G CDMA2000, 3G UMTS and 4G LTE, for example – are able to employ exactly the same radio carriers in neighbouring cells. In fact it is perfectly possible to build an entire nationwide network using just one radio carrier, which is repeated in every cell. These access networks are known as 'single frequency' networks.

4.18.1   Single Frequency Networks

In an SFN (Single Frequency Network), as used in CDMA-based networks and in LTE, neighbouring cells can all use the same radio channel(s) and would therefore all be co-channel neighbours of, and interferers to, each other.

With reference to Figure 4.26, which illustrates the SFN concept, Cells A, B and C are co-channel neighbours and all use the same radio channel (Channel 1). Each cell is separately distinguishable however as each uses a different PCI for its downlink transmissions.

A mobile device served by a single frequency network will only be able to reliably distinguish between signals transmitted by different cells if the 'signal to noise ratio' (how strong a 'wanted' signal of a serving call is compared to all of the 'unwanted' radio noise of neighbour cells) if each cell's received signal strength is sufficiently good.

Still with reference to Figure 4.26, when a mobile device served by an SFN is very close to a base station, the signal from that cell is often (but not always) so strong that it overwhelms or 'drowns out' the signals from any co-channel neighbours, meaning that the device will only be able to measure signals from that 'dominant' cell. When a mobile device is closer to a cell boundary it is more likely that the signals from neighbour cells will be closer in strength to that of that serving cell, meaning that the mobile device should be able to distinguish between and take measurements of the signals transmitted by several different cells.

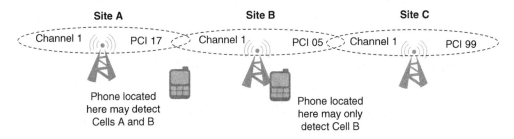

Figure 4.26   Single frequency networks

A radio survey device is subject to the same operating principles, meaning that a survey device that is located close to a base station will sometimes only capture details of that site's cells and may not be able to capture details of any neighbour cells. Surveys conducted in areas of less dominant coverage will usually be able to detect several neighbouring cells and will be able to list their details.

A particular issue related to radio surveys in SFNs is that a survey device may be able to detect signals from neighbouring cells and take measurements of each signal's strength, but it may not be able to capture any coherent information, such as a Cell ID, from those neighbour cell signals. Being able to detect a cell but not knowing its Cell ID is less than useful from a forensic point of view.

Surveys undertaken in areas where no cell is dominant and where all detected cells have very low signal strengths may be able to detect the details of a large number of cells.

4.18.2   Multi Frequency Networks

Most 2G, TDMA-based technologies must deploy MFN (multi frequency net-works) that employ strict frequency reuse plans. Although the SFN configuration is the standard deployment option for CDMA-based networks and for 4G LTE, operators that have sufficient amounts of bandwidth are also able to opt for MFN designs.

A MFN deployment, of the type depicted in Figure 4.27, works using similar frequency reuse concepts as are used in most 2G networks (such as GSM, PDC and IS136/TDMA); cells will be assigned frequencies that are not used by their neigh-bours but which may be reused by cells deployed further away in the network.

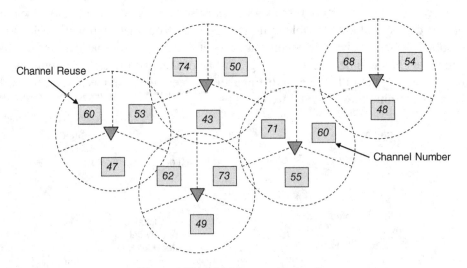

Figure 4.27   Multi frequency networks

A typical multi frequency deployment based on traditional three sector sites will need a minimum of three carriers to achieve a reasonable level of frequency reuse.

To use 4G LTE as an example, the implications of multi frequency deployment are that an operator would need to have been assigned at least 2 × 60 MHz of spectrum (to support three 20 MHz FDD carrier pairs) if they wanted to offer the highest capacity LTE service. Most operators do not possess this amount of spectrum, meaning that 4G MFN deployments would generally need to be based on much smaller channel sizes (5 or 10 MHz for example), which in turn would limit their overall data throughput potential.

4.18.3 Multi Carrier SFNs

A much more common approach for CDMA and 4G operators is to deploy a 'multi carrierSFN', as outlined in Figure 4.28. In this concept, the network has a number of radio channels available and deploys all of them to all base stations, which creates several parallel SFNs made up from multiple frequency 'layers'. This produces what are commonly referred to as 'stacked cells' or 'stacked sectors'.

The multi carrier SFN configurations employed by operators of CDMA-based networks allow them to, for example, define an SFN on one carrier to be used for voice/SMS services and a separate SFN on a different carrier to be used for data services.

3G UMTS operators might use this concept to deploy so-called 'Release 99' voice and SMS services on one layer and HSPA/HSPA + on another. 4G LTE operators deploy multi layer services in much the same way and might plan to use one carrier for VoLTE (Voice over LTE) and other low bit rate services and other carriers for higher bandwidth data services.

A real-world example of this is provided by Orange UK, which in its 3G UMTS deployment traditionally used carrier 10811 as the SFN layer for voice/SMS services and carrier 10836 as the SFN layer for HSPA.

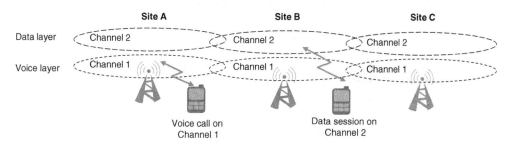

Figure 4.28   Multi carrier single frequency networks

4.19   Cell Coverage Concepts

At any point in time a mobile device in Idle Mode will only select one serving cell upon which to camp for paging and connectivity purposes. Forensic radio survey devices will typically operate in the same way and the serving cell details they capture will be influenced by the number of detected cells at a location and the strength of their signals.

4.19.1   Dominance

If the measurements taken at a location indicate that one cell consistently provides the strongest signal, a mobile device being used at that location can be expected to remain camped on that cell all of the time. In such a scenario, the consistently serving cell can be said to be 'dominant' at that location, or to put it another way, the surveyed location can be said to be within the cell's 'area of dominance'. This is illustrated in Figure 4.29.

In multi frequency TDMA-based networks (such as GSM or PDC), a mobile device may be able to detect multiple neighbour cells when camped on a serving cell, due to those neighbours each using different radio channels. In urban areas those neighbour cell sites are likely to have been deployed in close proximity to each other and may provide signals of roughly equivalent signal strength to a surveyed location, meaning that the location may have several serving cells, each providing the strongest signal (and being selected as the serving cell) for a short period of time before a different cell takes over and is in turn selected as serving. This further means that areas of single cell dominance in urban areas can be comparatively rare.

In suburban or rural areas there is likely to be a far greater distance between cell sites and less overlap between their cells, allowing a greater opportunity for cells to offer areas of dominance.

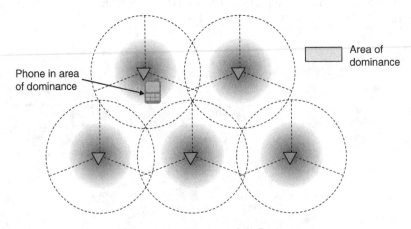

Figure 4.29   Dominant Cell Coverage

In the single frequency environments employed by CDMA-based networks and 4G LTE systems, mobile devices are often unable to detect any neighbours when camped on a serving cell. This is due to the fact that when a mobile device in a single frequency environment is located close to a base station and is receiving a strong or very strong signal from it, the 'noise' of the signal being generated by the serving site may well overwhelm or dominate any signals being transmitted on the same frequency by neighbouring but more distant sites. In short, if a mobile device is receiving a strong signal from a nearby site it often will not be able to detect signals from any other sites.

Except in areas where cells are very closely packed together and have a large degree of overlap, SFN cells can be expected to provide a dominant service across the area immediately surrounding the site and for several tens or hundreds of metres beyond. Areas of dominance are therefore more likely to be encountered during surveys of SFNs.

Forensic surveys undertaken of SFN types in an area of dominance quite often provide details of just one dominant serving cell and fail to detect any neighbour cells. In other cases there will be one reported serving cell with a number of weaker neighbours reported.

4.19.2   Non-dominance

If there are two or more cells covering an area with roughly equivalent received signal strengths, a mobile device can be expected to reselect between those cells on a regular basis. A location with such a coverage profile can be said to be in an area of non-dominance, as depicted in Figure 4.30.

Forensic radio surveys undertaken of SFNs and MFNs in areas of non-dominance can be expected to indicate that more than one cell serves at a surveyed location. A mobile device can only select one cell as an Idle Mode serving cell at any one time,

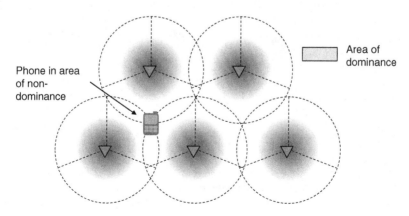

Figure 4.30   Non-dominant cell coverage

but in areas of non-dominance the normal reselection processes will ensure that the device camps on whichever cell currently provides the strongest signal.

Results of surveys taken in an area of non-dominance, where there is no single serving cell, will typically show details of two or three serving cells and it is not uncommon to see survey results that list five, six or more serving cells.

Some cell site experts employ the term 'best serving cell' in these situations and use it to identify the cell, out of the set of detected serving cells, that serves most often, offers the most consistently strong signal or is selected most often during test calls. Not all experts agree on the use of this term, however, and even those that do employ it have differing definitions of its meaning.

4.19.3   Poor Coverage Areas

No cellular operator offers completely contiguous network coverage and there are numerous places, especially on the edges of areas that are otherwise well served, where connectivity is still possible but where coverage is poor.

It is often found that such areas receive poor coverage from numerous cells, all providing comparably moderate or poor signals. The operation of the reselection algorithms in such scenarios means that a mobile device could reselect much more often than would usually be expected, as very small differences in received signal strength cause neighbours from amongst a large group of cells to become momentarily more attractive than any of the others.

Forensic radio surveys of poor coverage areas, such as the one depicted in Figure 4.31, are generally characterised by listing details of multiple serving cells – with sometimes 10 or more cells being selected over the survey period. Results of this kind can be misinterpreted as being representative of an area that is well served by numerous strong cells, whereas in reality such results are usually indicative of the

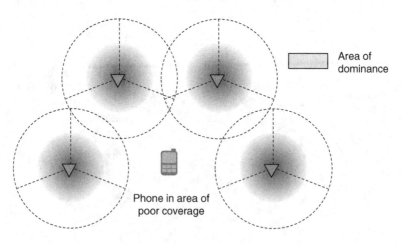

Figure 4.31   Poor cell coverage

opposite. It is important therefore in these situations to look at the recorded average signal strengths for the set of cells that have been reported as serving. If the signal strengths are all in the moderate, low or poor categories it is indicative of a survey being undertaken in an area of poor coverage.

In reality, mobile devices used in areas that receive generally poor coverage from one network type (e.g. 3G) can be expected to fall back to another network type (e.g. 2G) if alternative coverage is available.

A distorting factor of the way in which network surveys are often conducted is that the survey device is usually locked to a particular radio access type. A device used for forensic surveying of a UMTS network, for example, would be 'locked' (e.g. forced to select only 3G cells) to 3G before the survey commenced. If the survey was of an area that received consistently poor 3G coverage the NEMO would continue to survey 3G channels long after the point where a 'normal' mobile device would have abandoned 3G and reselected to 2G or 4G coverage instead.

Cell site conclusions based on forensic radio survey results obtained in areas of poor coverage generally need to consider the range of coverage provided by the set of serving network technologies, especially if a target phone made calls using a variety of network types.

4.20 Small Cells and Closed Subscriber Groups

3G and 4G networks are able to deploy femtocell or 'home' base station nodes, which each provide a small 'bubble' of cellular coverage at a user's home or office. Although 2G femtocells are technically possible, there is no standard specification for them and only limited numbers of examples have been manufactured or deployed.

Femtocells belong to a class of cell types collectively known as 'small cells', which also includes cellular picocells, microcells and some configurations of WiFi hotspot. An overview of small cell concepts can be gained from Ref. [9].

The 'pico' and 'micro' grades of small cell are usually deployed by a network operator and are configured to offer 'open' access that is available to carry connections for any network user. Femtocells are usually deployed in a private 'closed' access mode.

To control access to a private femtocell it is possible for the network to define a CSG (Closed Subscriber Group), which is essentially a list of the mobile devices that are permitted to use that cell. Each CSG is assigned a unique CSG ID. Details of CSG membership are written to each permitted user's mobile device and each 'closed' femtocell will advertise its CSG ID on its BCCH.

When performing cell selection or reselection, a 3G or 4G connected mobile device must check to see if there are any CSG cells in the area. If there are, and if it is a member of one of those CSGs, it will either inform its user that a registered CSG cell is available and ask if they want to connect or will simply connect automatically. Mobile devices must therefore check for CSG cells before attempting to select a 'normal' cell to use.

An earlier mechanism that was developed to control access to femtocells, before the CSG concept was defined, saw operators create specific LACs for femtocells. If a phone attempted to select to a femtocell that it was not permitted to use, the network would inform it that the femtocell's LAC was 'forbidden'. In which case the phone would stop attempting to access the cell for a period of time and would stop logging measurements of cells within that LAC.

Most femtocells are deployed in the 'closed' state, meaning that they will only accept connection requests from devices that are registered as belonging to the cell's CSG. Devices that are not registered with the CSG will not be able to make calls via the femtocell, apart from emergency calls, and this includes forensic radio survey test calls. Femtocell surveys will almost always have to be conducted in Idle Mode only, unless the survey device can be added to the list of permitted CSG devices.

One further issue associated with femtocells is related the Cell IDs captured in CDRs during calls.

Cells belonging to the most commonly deployed technologies, even femtocells, broadcast a CGI on their BCCH and all CGIs are theoretically unique. Even if the CGI broadcast by a femtocell is unique, the Cell ID captured in any CDRs that result from calls made via a femtocell may not be.

Some network operators assign real, unique CGIs to their femtocells and capture these unique IDs in CDRs, allowing there to be a correlation between the 'used' cell ID and the 'recorded' cell ID. Other operators, however, either use a generic cell ID in CDRs to identify all femtocells in a region of their network or the CDR captures the cell ID of the macrocell that serves in the area in which the femtocell is located when a call is made.

All of this means that forensic radio surveys of femtocells will be problematical: first, because 'closed' configurations make normal surveys difficult and test calls impossible; second, in networks that still employ the old method of managing femtocell access (using 'forbidden LAC' messages) a survey device might not even take measurements of a cell even in Idle Mode; third, because it may often be difficult to determine that particular calls were even made via a femtocell due to the 'used' and 'captured' cell IDs not matching.

Femtocells, in short, are a problem for forensic radio surveys and will continue to be so.

4.21   Network Activities

The activities detailed below are specifically related to 3GPP-type networks, but similar activities are undertaken in other network types, often using the same procedures or terminology.

4.21.1   Mobile Device Activities at Power On

When the mobile device is switched on or returns to network coverage, it searches for a suitable signal from a permitted mobile network (a PLMN). The general concepts

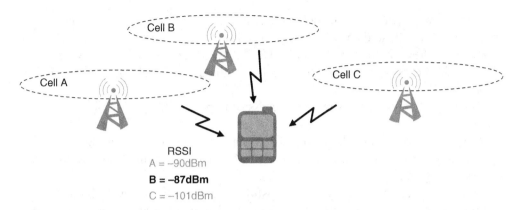

Figure 4.32   Cell selection

related to cell selection and reselection are outlined in Figure 4.32. An example of the procedures employed in GSM is provided in Ref. [10].

Each cell transmits its own BCCH. In multi-frequency network types, like GSM, each sector may employ several carriers but only one of these per sector will typically carry a BCCH. This is known as the 'BCCH carrier' (and is sometimes known as the 'beacon channel') and it is this frequency that a mobile device first searches for. In single frequency networks there are typically only a few channels per network to choose from and all of them will be BCCH carriers.

During normal operation, a mobile device will store the channel number of the currently serving cell on the SIM. When the mobile device detaches or powers down, the 'current' serving cell information is retained on the SIM as details of the 'last used' cell.

After powering on again, the mobile device will check for stored details of the 'last used' channel and will attempt to 'camp on' to any cells that it finds on the channel. Operators are optionally also able to write a list of 'preferred' BCCH carrier channel numbers to SIM cards, so if the 'last used' cell is not available the mobile device will start to work down the list of preferred carriers looking for a viable cell instead.

If none of the 'stored details' BCCH carriers is available, the mobile device is obliged to begin a search of all the radio channels within its bands of operation and will compile a list of candidate cells found there. Each network type specifies a minimum number of cells that need to be detected by this process before a choice of serving cell can be made, but the typical value is around 20–30 cells.

Once a list of local candidate cells has been compiled, the mobile device will invoke the network selection process. If the device is operating within its home country then network selection is based on removing cells from the candidate list that do not belong to the home PLMN. If the device is roaming then it will check to see if the user has configured 'automatic' or 'manual' network selection; if manual selection is configured the device will present the user with a list of detected networks and ask them to choose one; if automatic selection is enabled the device usually chooses the strongest local cell of whichever network and attempts to access it.

Once network (PLMN) selection has taken place the mobile device will process the remaining list of candidate cells through a cell selection algorithm – for 2G GSM this is known as the C1 algorithm and for 3G UMTS/4G LTE it is known as the S algorithm. The C1/S processes can be summarised as 'list the detected cells and choose the one that offers the strongest signal'.

The cell selection process will be successful if the following conditions are met:

- The cell selection will only be successful if the cell belongs to an 'allowed' PLMN – for example the home PLMN or a visited PLMN with which the home network has a roaming agreement.
- The cell selection will only be successful if the user is not in a 'forbidden' LA – these are often used to control the parts of a visited network that are accessible to roamers.
- The cell selection will only be successful if the cell has no 'access control' barring restrictions set.

Once a cell has been successfully 'camped' on, the mobile device recovers local operating parameters from the cell's BCCH, which allows it to initiate 'random access' to request an active radio connection from the serving cell.

4.21.2   Attach and Detach

A mobile device must identify itself to the selected network before it can make use of network resources. The identification information can be given in two ways:

1. By sending an IMSI;
2. By supplying temporary identification data that has previously been assigned by the serving PLMN.

The latter case is the preferred means of identification since it does not compromise the confidentiality of the subscriber. However, if the registration is new or if the device has been disconnected from the network for some time, identification must be made via the IMSI.

As depicted in Figure 4.33, the mobile device attempts to 'attach' to the network by sending either a Location Update message (in GSM and UMTS) or an Attach Request (in GPRS and LTE), which is passed to the local mobility management node (VLR/SGSN/MME, depending on network type) for analysis. The Attach message will contain either the device's IMSI or a temporary identifier (TMSI, P-TMSI, S-TMSI) previously issued by the serving network.

For an initial attach, on receipt of the IMSI, the PLMN can work out which HLR (or HSS in evolved networks) stores the required subscription data for the user based on the fields in the IMSI. For 'home' users this data will be in the network's own

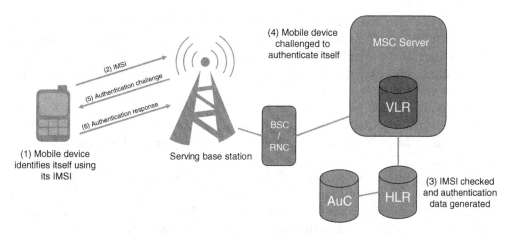

Figure 4.33   Attach (in GSM networks)

HLR, but for roaming visitors the data will be in their home network's HLR. It is possible for 'visited' network nodes to interrogate another network's HLR if a roaming agreement exists between the operators.

4.21.3   Authentication and Ciphering

Networks are able to employ an additional level of access security by requiring mobile devices to 'authenticate' themselves before allowing them to access network resources. Authentication is an optional process, but if employed is usually invoked during an Attach, when a device requests access to the network, and is also often triggered during events such as location update and call setup.

To initiate an authentication 'run' the local mobility management node that is serving a device will transmit an Authentication Info Request, containing the subscriber's IMSI, to the relevant HLR, which responds with a set of authentication 'vectors'.

The HLR hosts a secure database, known as the AuC (Authentication Centre), which stores details of the secret authentication key (known as 'K') that is associated with each subscription.

The value of K associated with the supplied IMSI is passed through an authentication algorithm along with a random number (known as RAND) generated by the AuC. The result of this is a numerical value known as the Expected Response (XRES).

K, RAND and a different group of algorithms are also used to generate a set of encryption keys for use by the network and the mobile device and the whole set of 'vectors' – RAND, XRES and the encryption keys, along with a final vector known as AUTN (Network Authentication Token) – are passed back to the serving network node.

The serving VLR/SGSN/MME stores the authentication vectors and sends RAND to the mobile device as an 'authentication challenge'.

The device also has a copy of the 'K' value associated with the subscription, securely stored on the SIM. RAND is passed to the SIM, which uses it, K and the same authentication algorithm used by the AuC to generate a RES (Response). The mobile device also uses RAND and K plus other algorithms to generate its own copies of the encryption keys to be used.

RES is transmitted back to the VLR/SGSN/MME, which compares it with the expected response (XRES) supplied by the AuC. If RES matches XRES then the user is deemed to be authentic and the mobile device is permitted to access network resources.

K, RAND, RES and XRES are all 128-bit binary numbers. Such numbers have approximately 3.4 undecillion (3.4 trillion trillion trillion) possible values, which makes the chance of guessing the correct value of RES for a given value of RAND sufficiently unlikely to ensure that the authentication system can adequately protect access to network resources.

4.21.4   TMSI Allocation

The final process required for the Attach is for the PLMN to allocate a local temporary identity.

A subscriber's IMSI is a sensitive piece of information that could theoretically be used to clone a user's network service, so the IMSI is only sent over the air interface if absolutely necessary. In all other situations, a temporary identification is used, which the mobile device is assigned during the attach process, in the form of a TMSI.

TMSIs are assigned by the current serving local mobility management node – TMSIs are assigned by VLRs for 2G/3G CS services; P-TMSIs are assigned by SGSNs for 2G/3G PS services; M-TMSIs or S-TMSIs (MME-TMSI or Serving TMSI) are assigned by MMEs for 4G services – and will change if the device is passed on to another core network node. They are also changed at regular intervals even if the device stays registered with the same core network node.

4.21.5   Detach

Detach is the process of deregistering a device from a mobile network to which it was previously attached. The detach procedure informs the network that the device is about to become unreachable.

A mobile device will send a detach message when it is being powered off (unless the power off is achieved by pulling the phone's battery out, in which case it does not have an opportunity to send a detach message) and also if the device is just about to run out of battery power. These are examples of 'explicit detach', where the mobile device informs the network that it intends to detach.

If roaming, a mobile device will detach from one visited network before reattaching to another, if it decides that cells of a different network will offer a better service.

It is also possible for the network to initiate a detach if, for example, a mobile device in Idle Mode has not contacted the network for a period of time and has failed to perform an expected periodic location update. The network can decide that the device is no longer active and can 'implicitly' detach it.

Once detached, a mobile device cannot use network resources until it re-attaches. The network does not attempt to page for a device that is logged as detached and will not attempt to establish connect to it. Inbound calls for detached devices will typically be diverted directly to voicemail and inbound text messages will be stored for later delivery.

4.22   Idle Mode and Connected Mode

Following successful cell selection and attach processes, a mobile device will be registered with a network and will be capable of requesting traffic connections.

Mobile devices that are powered on and attached to a valid network will generally be in one of two modes:

1. Idle Mode, where devices are not currently connected but are available to make calls.
2. Connected (or Dedicated) Mode, where devices are connected to a base station, have radio resources assigned and are able to actively exchange user traffic with the network.

There are also a number of 'limited service' modes that devices could be in, typically when normal connections are not available (such as 'emergency calls only' mode). These modes are summarised in Figure 4.34.

4.22.1   Idle Mode Procedures (Cell Selection and Reselection)

Mobile devices are typically battery-powered, which requires system designers to incorporate ways of saving power during periods when the device is not actively connected to the network. The Idle Mode concept employed in most types of cellular network has been designed to serve this purpose.

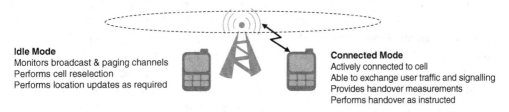

Idle Mode
Monitors broadcast & paging channels
Performs cell reselection
Performs location updates as required

Connected Mode
Actively connected to cell
Able to exchange user traffic and signalling
Provides handover measurements
Performs handover as instructed

Figure 4.34   Idle mode and connected mode

When a device is powered on but is not engaged in a call, exchanging text messages or actively exchanging data with a data service, it will generally drop into Idle Mode to preserve power.

The generic activities undertaken by a mobile device in Idle Mode are set out in Ref. [10]; they are:

- Network selection
- CSG selection
- Cell selection and reselection
- Location registration.

A mobile device's main responsibility when in Idle Mode is to search for and select a local cell that it is permitted to access and to 'camp on' to that cell. Camping on means that the device chooses that cell as its 'serving' cell and will monitor the signalling channels broadcast by the selected cell, checking for broadcast control information and paging messages.

A device in Idle Mode will take regular measurements of the serving cell and also of other local cells to see if any of them offer a better (e.g. stronger) service than the serving cell. If a more suitable neighbour is detected the device will reselect and will camp on the neighbour cell instead.

In 3GPP-defined networks, 2G devices use the C1 algorithm to manage initial cell selection and the C2 algorithm to manage cell reselection, while 3G and 4G devices use the S and R algorithms for the same purposes. The C1/S algorithms can be paraphrased as 'select the strongest available permitted cell' and the C2/R algorithms can be paraphrased into a set of instructions that say 'if a neighbour cell is strong enough (e.g. stronger than the current serving cell), for long enough (usually 1–5s) then the device must reselect, unless any temporary or permanent offsets need to be applied to the serving or neighbour cells which alter their attractiveness'.

Reselection Offsets are illustrated in Figure 4.35 and are used to make serving cells artificially more attractive (e.g. a positive offset is applied to make the measured signal seem stronger) or to make neighbour cells artificially less attractive (e.g. negative offsets are applied to make measured signals seem less strong).

Positive offsets are usually applied to a new serving cell for a short period of time following a reselection. This is designed to apply a process known as 'hysteresis', which seeks to prevent the 'flip-flopping' that can occur if a device jumps from one reselection to another in quick succession. The hysteresis offset that is added to the serving cell's measurements after a reselection makes the new cell artificially more attractive for a short period and should ensure that the device does not reselect away from that cell again too quickly (unless the signals received from the serving cell become very weak, very quickly).

Negative offsets are usually applied by devices to measurements taken of neighbour cells on a permanent basis. For example, if a device is camped on a cell belonging to one location area and it takes measurements of a cell belonging to a different location area, then it will apply a negative offset to make the neighbour cell artificially

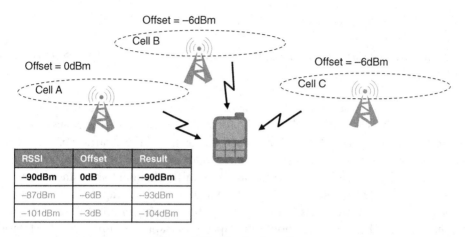

Figure 4.35 Reselection offsets

less attractive, therefore making it less likely that the device will initiate a reselection that will trigger a location update process.

Cell reselection measurements can also be affected by so-called 'Idle Mode behaviour' instructions provided to a device by the network, which allow the network to influence the priority the device places on cells that use particular frequencies or that belong to particular network generations.

The instructions provided to a mobile device as it enters Idle Mode might, for example, order it to prioritise 4G cells over 2G or 3G cells. When taking cell reselection measurements a device so instructed will attempt wherever possible to reselect to a 4G cell, even if 2G or 3G cells are available.

A further example of this function can be seen in 3G networks; especially where an operator has one carrier layer that is predominately used for voice and another that is mainly used for HSPA data. Mobile devices would routinely be provided with Idle Mode instructions that prioritised the 'voice' layer higher than the 'data' layer, meaning that the device would be more likely to camp on a voice layer cell when in Idle Mode rather than camping on the data layer.

4.22.2 Connected Mode Procedures (Connection Setup and Handover)

When a mobile device is required to establish a radio connection to its serving cell it moves from Idle Mode into Dedicated or Connected Mode – Dedicated Mode was the term originally used in 2G GSM networks, Connected Mode is the term used in 3G and 4G networks – both terms mean essentially the same thing. The most recent versions of GSM now also use the term 'Connected Mode'.

The generic functions performed by a device when in Connected Mode include:

- Transfer of user traffic and control information to/from the network;
- Capturing serving and neighbour cell measurements;

- Reporting radio measurements and traffic quality information to the serving base station;
- Performing handovers (or handoff) to neighbour cells if instructed to by the network.

In Connected Mode, a mobile device will be assigned radio capacity in the current serving cell. This capacity might take the form of a 'dedicated' traffic channel, which would typically be used to carry a voice call, or a 'shared' traffic channel, which would be used to carry a data connection. SMS text message traffic is usually carried on a control channel and does not require a full traffic channel to be established for it.

Devices that are in Connected Mode take a continual and rapid series of received signal strength measurements of the serving cell and some or all of the current set of neighbour cells. Each device should be capable of capturing up to several hundred different measurements every second.

Depending on the type and generation of network technology, a mobile device may be required to report its captured measurements back to the current serving base station somewhere between every 0.5s and every 10s.

Unlike Idle Mode mobility, where cell reselection is managed individually by each mobile device using the C2/R algorithms, Connected Mode mobility is managed by the access network, but decisions are still based on measurements provided by the mobile device.

Each cell will have a set of handover parameters defined for each of its 'adjacent' neighbours within its controlling access network node. These parameters typically specify the threshold power levels that need to be reported by mobile devices in order to trigger a handover. They will also include a 'time to trigger' period, which indicates the length of time the measurements have to be reported as being above the threshold before a handover can be initiated. This essentially means that handovers will only be initiated towards neighbour cells that have been strong enough for long enough to offer a better service.

During a handover (which is the term used in 3GPP networks) or handoff (the term employed by 3GPP2), the 'old' base station or access controller communicates with the nodes responsible for the selected 'target' cell and arranges for the mobile device and its active traffic connections to be passed on.

4.22.3   Transition from Idle Mode to Connected Mode

One of the key functions of cell site analysis is to attempt to determine the choices that a mobile device would make when instructed to transition from Idle Mode to Connected Mode (i.e. to make a phone call) in a certain location.

When a mobile device in Idle Mode is required to establish a connection it will typically attempt to use the cell it is currently camped on Depending on the technology being used, this could be the cell with the strongest signal, or it could be any cell with a signal that lies above some threshold value.

One of the key tasks of forensic radio surveying, therefore, is to capture details of the set of serving and non-serving cells that provide coverage at a location and attempt to determine which of those cells a mobile device could choose to use if it were required to connect to the network.

4.23   Cell Access Control

GSM (and subsequent 3G UMTS and 4G LTE systems) was designed with an access control mechanism known as SIM Access Priority, which is designed to limit access to certain cells during periods of high demand or during emergencies.

Every SIM card is assigned an ACC (Access Control Class) in the range 0–15, which is written to the card when it is configured. Ordinary subscribers are randomly assigned to one of the levels between 0 and 9 when their SIM is created. Details of SIM ACC are shown in Table 4.4 [11].

During exceptionally busy periods, a network operator is able to limit access to cells to only a subset of SIM access priority groups, by advertising the appropriate access restrictions on a cell's BCCH, therefore reducing the traffic load in those cells. For example, if an operator decided to invoke the overload control mechanism and mandated that only users with ACC 4 could access Cell ID 12345, phones assigned to other classes would not attempt to access that cell, reducing the burden on it and mitigating the congestion.

SIM Access Priority levels 10–15 have a wider scope.

Access Control Class 10 is set on a cell's BCCH to indicate that it is not able to accept emergency calls (in addition to any other restrictions that have been set) and ACCs 11–15 are used to identify 'special' types of user. Special users must have the appropriate ACC written to their phone's SIM card by the operator when the SIM is created.

If any of the ACCs between 11 and 15 are set to 'false' on a cell's BCCH it indicates that the corresponding class of special user is not barred and is able to access the corresponding cell. ACCs 11 and 15 are usually reserved for SIMs assigned to a

Table 4.4   3GPP SIM ACC.

ACC	Use	Assignment method
0–9	'Normal' subscribers	Randomly assigned and written to SIM during configuration
10	Indicates cells that will not accept emergency calls	Not written to SIM cards, only indicated on a cell's BCCH
11	For PLMN user	Must be specifically written to SIM card during configuration
12	Security services	
13	Public utilities	
14	Emergency services	
15	PLMN staff	

From 3GPP TS 22.011:4.2 – used with permission from 3GPP.

network's engineering staff. ACC 12 indicates 'security services' users. ACC 13 is used to identify phones belonging to defined 'public utilities' personnel and ACC 14 is used to identify members of the emergency services.

ACCs 11–15 are intended to allow ordinary users to be 'barred' from some or all cells in a network to leave them clear for the authorities and first responders during an incident or emergency. When this happens, ACCs 0–9 will be set to 'true' (indicating that barring is enabled for those classes) on the BCCH, but ACCs 11–15 will remain set to 'false' (indicating that barring is not enabled for those classes and access is possible). If any of the classes within the group ACC 11–15 is also to be barred then the corresponding bit in the BCCH access control message is set to 'true'.

Cell barring of this sort is comparatively rare. SIM cards must be specially coded to allow Class 11–15 operation and the rules surrounding the use of these restrictions are very tight. Several schemes have been developed, including the United Kingdom's MTPAS (Mobile Telephony Privileged Access Scheme), which provide a framework for how and when operators should invoke special barring.

In general, the access control processes are most commonly used when a cell is first being commissioned, as they allow the cell to go on air to be tested by the network without getting clogged up with user connections or emergency calls. In this scenario, ACCs 0–9 will be set to 'true' to bar ordinary users, ACCs 12–14 will be set to 'true' to bar special users apart from network engineers and ACC 10 will be set to 'true' to indicate that emergency calls are not permitted for any of the barred classes. The resources of the cell will then be available only to network engineers (ACCs 11 and 15).

4.23.1   Cell Barring

More general forms of cell barring also exist, in addition to the specific or special barring options provided by the ACC scheme.

In 3GPP network types, for example, the BCCH in a cell, which carries the cell's current ACC class permissions, also carries two cell reservation flags – operators can set the *cell Barred* flag, which denies access to all phones, and also the *cell Reserved for Operator Use,* which reserves access for ACC 11–15 phones only [12].

If a phone detects that a cell has a reservation flag set it will remove that cell from its reselection candidate list for 300s before checking to see if the flag status has changed.

Some cells, especially femtocells, are set up with permanent access restrictions which limit the set of users that can access the cell to a defined CSG.

4.23.2   Forbidden LAC/TAC

As well as being able to bar access to individual cells, operators are also able to bar access to entire regions of the network for some or all subscribers. They do this by defining certain LACs (2G/3G LAC) or TACs (4G TAC) as 'forbidden' to some users.

If a user attempts to perform a Location/Tracking Area Update in an area that the network does not wish them to use, the network will send a rejection message with a cause code of 'forbidden LAC/TAC'. The phone will add this to a list of forbidden LACs/TACs held on the SIM and will remove cells belonging to the same area from its list of reselection candidates. It will recheck after 10 min if it is still within range of cells belonging to the LAC/TAC.

Forbidden LACs are a common method of controlling Roaming and National Roaming (NR), by limiting those areas in which roaming users are able to connect to network resources. Access to some 3G femtocells is also managed in this way.

NR is employed as a way of allowing users of one network to gain access to the resources of another. In normal circumstances, the users of one PLMN are barred from using the resources of other PLMNs in the same country, except when a NR agreement exists. This barring is usually managed by placing all other national PLMNs on the 'Forbidden PLMN' list on SIM cards.

NR can be managed at the SIM level or the network level. If SIM-managed NR is used, then the SIMs belonging to the 'NR user' network are edited to remove the 'NR donor' network from the 'Forbidden PLMN' list. If managed at the network level, mobile devices from the 'NR user' network are informed in Attach Accept or Location Update Accept messages that the 'NR donor' network should be considered an 'Equivalent PLMN'. In both cases mobile devices are then able to consider cells belonging to the 'donor' network as reselection candidates.

4.24 Location Updating (Idle Mode Mobility)

Cellular access networks are administratively divided into areas, each of which will consist of several contiguous cells. Each specific network technology has its own particular type of access network area, which is variously known as LA, RA, TA, and so on. The generic term 'location area' will be used for the purposes of this explanation and is illustrated in Figure 4.36.

As a mobile device moves around the network, it recognises when it has entered a new LA by observing the change in the LAC component of a new cell's CGI. If such a change is detected, the mobile device informs its current serving local mobility management database node (such as VLR) accordingly. This is known as a 'location area update'.

Each mobility management database will typically serve multiple location areas, and each of these will be defined as the area in which the search for an Idle Mode subscriber is carried out if the network needs to establish a connection to their phone. This search process is called 'paging'.

During paging, the network sends a paging message to each base station in the device's currently registered LA. Mobile devices monitor the PCCH paging channel in their current serving cell, if they detect their paging ID being broadcast they initiate a connection via the current serving cell. This allows the network to determine their location and proceed with the connection setup.

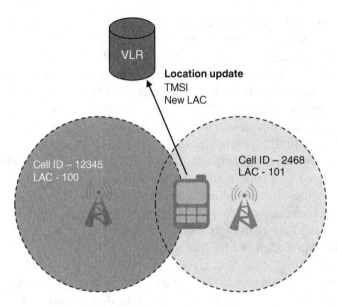

Figure 4.36   Location updating

During Idle Mode cell reselection, a mobile device monitors the BCCH of each new cell that it camps on to check the cell identity and access information that is broadcast. The cell-specific broadcast data includes the identity of the LA the cell belongs to. In order to keep track of its approximate location, the mobile stores in the SIM the location area identity of the area in which it is currently registered. Every time the network broadcasts the area identity in the current cell, the mobile compares this information to the identity stored in its memory.

When the two identities are no longer the same (e.g. following a cell reselection into a new LAC), the mobile sends the network a location update request. If the change of LA also entails a change of mobility management node (as each node typically looks after only a subset of cells) then the network initiates the process that transfers responsibility for the mobile device to a new mobility management node. The HLR is informed about the mobile device's new location area and of the new mobility management node that is serving it and the data concerning the subscriber is cleared from the previous mobility management node.

Location updates are also performed on a 'periodic' basis if an Idle Mode phone has not contacted the network for any other reason during a configurable period of time. A typical value for a periodic location update timer is around 2 h.

2G/3G LAs are subdivided into one or more RAs which perform the same location tracking function as the LA but are used for GPRS/PS services. Each LA will support at least one RA but could be subdivided into up to 256 RAs. RA numbers range from 1 to 256 in every LAC, but because many operators do not bother to define additional RAs in each LAC it results in all RACs being equal to 1.

4G access networks employ a similar concept to the LA, although in LTE these aggregations of cells are known as TAs.

4.25   Handover (Connected Mode Mobility)

Handover is the term used to describe the act of Connected Mode mobility that allows a mobile device and its active traffic connections to be passed from one cell to another. 'Handover' is the term used by 3GPP-specified network types (GSM, UMTS, LTE) and most other systems. 3GPP2 networks (e.g. cdmaOne, CDMA2000) prefer to use the term 'handoff'.

The type of handover performed when a mobile device is in Connected Mode depends upon the generation of network the device is connected to and the frequency reuse configuration being employed. An overview of generic handover types is provided in Figure 4.37.

4.25.1   Inter-carrier Handover

Devices that are connected to multi-frequency network types (such as GSM, PDC and some LTE networks) will always perform an 'inter-carrier handover'. This entails changing the frequencies on which the device is transmitting/receiving to match the carrier employed by the new cell, as neighbouring cells in multi-frequency networks are not permitted to use the same carriers.

Devices connected to multi-carrier SFN that are being handed over between frequency layers (e.g. from a 'voice layer' carrier to a 'data layer' carrier in 3G), will also be required to retune to a new radio carrier to allow the connection to continue.

Inter-carrier handover is also employed for handovers between technology types – for 2G to 3G handover or a 3G to 4G handover, for example – as different technologies typically employ different frequency bands and definitely employ different carrier types and radio techniques.

4.25.2   Intra-carrier Handover

In network types that employ a SFN configuration (2G cdmaOne, 3G and 4G networks), it is possible to hand over from one cell to another whilst remaining on the same frequency, as long as both cells are on the same frequency layer.

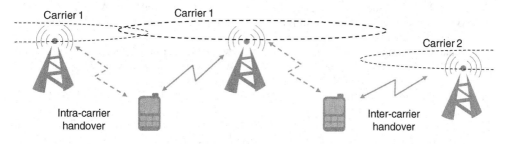

Figure 4.37   Types of handover

For power control and connection management reasons, CDMA-based (but not TD-SCDMA-based) networks must use a handover technique known as 'soft handover' in which the connection to the 'new' cell is established *before* the connection to the 'old' cell is released. This means that, for a brief period of time during a handover, a mobile device will be connected to two (and sometimes, three) cells simultaneously (in what is known as the device's Active Set'), as long as all cells are using the same frequency carrier.

For a variety of technical reasons, the 'soft handover candidate' cells employed during CDMA handover are sometimes drawn from a set of neighbours that do not show up in Idle Mode surveys, meaning that details of those cells would not be captured if Idle Mode surveying alone was employed.

4G network types (LTE and WIMAX) are also able to support intra-frequency handover between cells on the same carrier, but the OFDMA technologies used by these network types do not use soft handover and so are able to release the 'old' connection before establishing the 'new' one.

In theory there is also a class of handover known as 'intra-cell' in which the phone is instructed to go through the motions of a handover but ends up connected to the same cell it started out on. This process is mainly used as a way of resetting some counters and security parameters for cell connections that have lasted for a long time.

References

[1] 3GPP Technical Specification (2013) *Network Architecture*, TS 23.002 v11.6.0, www.3gpp.org (accessed 25 July 2014).

[2] 3GPP Technical Specification (2013) *Numbering, Addressing and Identification*, TS 23.003 v11.6.0, www.3gpp.org (accessed 25 July 2014).

[3] 3GPP Technical Specification (2013) *Numbering, Addressing and Identification*, TS 23.003 v11.6.0 Section 3.3, www.3gpp.org (accessed 25 July 2014).

[4] 3GPP Technical Specification (2013) *Numbering, Addressing and Identification*, TS 23.003 v11.6.0 Section 2.2, www.3gpp.org (accessed 25 July 2014).

[5] 3GPP Technical Specification (2013) *Numbering, Addressing and Identification*, TS 23.003 v11.6.0 Section 2.4, www.3gpp.org (accessed 25 July 2014).

[6] 3GPP Technical Specification (2013) *Numbering, Addressing and Identification*, TS 23.003 v11.6.0 Section 6, www.3gpp.org (accessed 25 July 2014).

[7] 3GPP Technical Specification (2013) *Base Station (BS) radio transmission and reception (FDD)*, TS 25.104 v11.7.0 Section 5, www.3gpp.org (accessed 23 July 2014).

[8] 3GPP Technical Specification (2013) Evolved Universal Terrestrial Radio Access (EUTRAN); Base Station (BS) radio transmission and reception, TS 36.104 v12.2.0 Section 5.5, www.3gpp.org (accessed 24 July 2014).

[9] Small Cell Forum (2014) *Home Page*, http://www.smallcellforum.org (accessed 1 June 2014).

[10] 3GPP Technical Specification (2013) *Non-Access Stratum (NAS) Functions Related to Mobile Station (MS) in Idle Mode*, TS 23.122 v11.4.0 Section 2, www.3gpp.org (accessed 25 July 2014).

[11] 3GPP Technical Specification (2013) *Service Accessibility*, TS 22.011 v11.3.0 Section 4.2, www.3gpp.org (accessed 23 July 2014).

[12] 3GPP Technical Specification (2013) *User Equipment (UE) Procedures in Idle Mode and Procedures for Cell Reselection in Connected Mode*, TS 25.304 v11.4.0 Section 5.3.1.1, www.3gpp.org (accessed 23 July 2014).

5

3GPP Network Types

The Third Generation Partnership Project (3GPP) was formed to coordinate the development of 3G (Third Generation) UMTS (Universal Mobile Telecommunications System), but also inherited responsibility for 2G (Second Generation) GSM (Global System for Mobile Communications).

The set of network types that are coordinated by 3GPP includes:

- 2G GSM
- 2.5G GPRS
- 2.75G EDGE
- 3G UMTS
- 3.5G HSPA/HSPA+
- 4G LTE
- 4G LTE-Advanced.

Technical details of these network types, with an indication of the most important aspects of each of them from a forensic radio survey perspective, are detailed below.

Technical specifications related to 3GPP network types are freely available from their website – www.3gpp.org.

Commercial and interoperability aspects of GSM family network operations are coordinated by the GSMA (GSM Association). Interoperability standards can be downloaded from their website – www.gsma.com.

Forensic Radio Survey Techniques for Cell Site Analysis, First Edition. Joseph Hoy.
© 2015 John Wiley & Sons, Ltd. Published 2015 by John Wiley & Sons, Ltd.

5.1   2G GSM Networks

2G GSM networks began to be launched in the early 1990s. They offered voice, SMS text, fax and dial-up data services [1].

GPRS (General Packet Radio Service) enhancements were added in the late 1990s, which added the ability to carry 'packet data' services like IP (Internet Protocol) that allowed Internet and e-mail traffic to be exchanged more efficiently. A further upgrade, known as EDGE (Enhanced Data rates for Global Evolution) was also added in the late 1990s and improved the data rate available for GPRS data services over the GSM radio interface. GPRS and EDGE are regarded as 2.5G and 2.75G technologies respectively [2].

Development work continues to take place within GSM and recent years have seen the release of further upgrades, such as EDGE, which further enhance the data rates available over 2G radio connections.

GSM networks have been deployed in over 200 countries or territories and 3GPP and GSMA assert that it is the most successful cellular radio standard in terms of active numbers of users.

5.1.1   2G GSM Access Networks

The GSM access network was originally known as the BSS (Base Station Subsystem) but is now known as the GERAN (GSM/EDGE Radio Access Network) or even just the 2G RAN (Radio Access Network).

The GERAN contains all of the radio access devices needed to allow subscribers to make and receive mobile calls plus the backhaul connectivity that connects each site to the core network.

The elements that comprise a BSS/GERAN are illustrated in Figure 5.1 and include [3]:

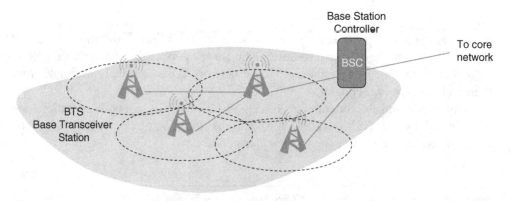

Figure 5.1   Base Station Subsystem

- BTS (Base Transceiver Stations)
- BSC (Base Station Controllers)
- Backhaul connectivity.

BTSs manage radio transmission and reception duties and host the cells that provide network coverage. BSCs manage the activities of a group of BTSs and are responsible for RRM (Radio Resource Management) duties. Backhaul links provide a transmission path to carry user and network traffic from a BTS to its controlling BSC and from a BSC to the operator's core network.

5.1.2   2G GSM Radio Interface

A GSM phone does not send a continuous stream of data across the radio 'air interface'. Instead it stores up information over a brief period of time (20 min) and then transmits it in a compressed burst. Each GSM mobile device will transmit around 217 bursts of information every second.

GSM can allow more than one mobile device to share the same radio channel by interleaving the burst transmissions from different devices, as long as each device transmits its burst of information at a different and defined point in time. This technique is known as Time Division Multiple Access (TDMA) and is depicted in Figure 5.2 [4].

To allow TDMA to operate, each GSM radio channel is divided into frames. One frame contains eight timeslots numbered from Timeslot 0 to Timeslot 7 and around 217 frames are generated every second. When users make or receive a mobile phone call, they are allocated timeslots on both an uplink and a downlink radio channel over which they can transmit and receive data.

GSM was designed primarily to support voice services, which require a duplex connection that has the ability to both transmit and receive simultaneously. In GSM, this is achieved by using FDD (Frequency Division Duplex).

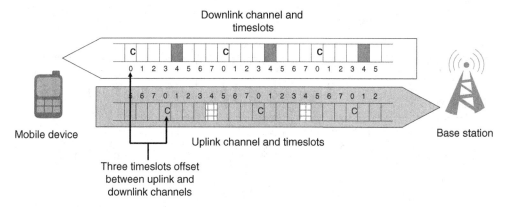

Figure 5.2   Time Division Multiple Access

In FDD-based systems, uplink (transmit from mobile device to base station) and downlink (base station back to mobile device) connections are transmitted on different radio channels and the combination of an uplink and a downlink make up one duplex GSM radio carrier.

To ensure that GSM devices can be constructed with only one radio transceiver (to reduce costs), downlink and uplink frames on a carrier are offset from each other by three timeslots. This allows a device to receive traffic on the assigned downlink timeslot and return to the uplink channel ready to transmit, all via the same transceiver.

Timeslot 0 in each GSM frame (on BCCH Carrier channel) is typically used as a CCH (Control Channel) to carry signalling and control information. This leaves up to seven timeslots to be used as TCH (Traffic Channels), which carry customers' voice calls. GPRS data connections are carried in a set of PTCH (Packet Traffic Channels). The timeslots on a GSM carrier are often shared between voice and data services by assigning some to carry TCHs and others to carry PTCHs. SMS information is carried in a CCH.

5.1.3   GSM Channel Configuration

GSM uses multi-frequency network techniques and each cell or sector generated by a BTS can consist of one more separate radio carriers. The radio carriers are known as 'physical channels' as they provide the physical carrying capacity for each cell. The type of information mapped into each timeslot represents a 'logical channel' – there are logical channels that carry user traffic and others that carry network control signalling.

One carrier in each cell will be assigned to carry the cell's BCCH (Broadcast Control Channel), in addition to other control channels and a set of TCH/PTCHs. A GSM radio channel that is assigned to carry a BCCH logical channel is known as a BCCH Carrier or 'beacon channel'. The other carriers in a cell will usually only carry TCH/PTCHs, leaving control channel duties to the BCCH Carrier. Figure 5.3 provides an overview of the logical channels employed on a GSM radio carrier.

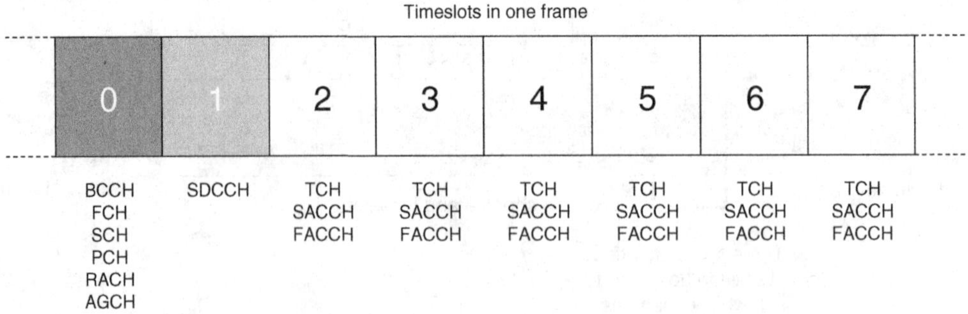

Timeslots in one frame

0	1	2	3	4	5	6	7
BCCH FCH SCH PCH RACH AGCH	SDCCH	TCH SACCH FACCH	TCH SACCH FACCH	TCH SACCH FACCH	TCH SACCH FACCH	TCH SACCH FACCH	TCH SACCH FACCH

Figure 5.3   Logical channel distribution

The BCCH and CCCH (Common Control Channel) generally share iterations of Timeslot 0 on a BCCH Carrier channel.

The SDCCH (Standalone Dedicated Control Channel) is generally carried in Timeslot 1 of a BCCH Carrier channel, although additional SDCCH instances can be carried in other timeslots and carriers in cells that have a particularly high level of administrative or SMS traffic. The SDCCH is used for connection setup negotiations and for carrying SMS text message transactions.

SACCH (Slow Associated Control Channel) and FACCH (Fast Associated Control Channel) are multiplexed across TCH timeslots along with user data, following a timeslot-sharing pattern known as a 'multiframe'. The SACCH/FACCH carry dedicated control signalling for a specific phone and are used to transfer things like signal strength measurement reports on the uplink and handover instructions on the downlink.

Mobile devices in Idle Mode monitor Timeslot 0 on the BCCH Carrier in the selected serving cell.

The downlink BCCH functions carried on the BCCH Carrier provide the mobile device with all of the information required to operate in that cell, including: information that allows the device to synchronise with the channel and calibrate its receiver to the channel's frequency; the country and network identity [MCC (Mobile Country Code) and MNC (Mobile Network Code)], LAC (Location Area Code) and CI (Cell ID); random access arrangements; the range of services supported in the cell; paging messages; and responses to Random Access requests.

The uplink half of Timeslot 0 on a BCCH Carrier provides the opportunity for mobile devices to transmit 'random access' requests, via the RACH (Random Access Channel), to the BTS when attempting to establish a connection, when responding to a page or when performing administrative functions such as Attach or Location Update. The BCCH Carrier also carries the FCH (Frequency Correction Channel), SCH (Synchronisation Channel), PCH (Paging Channel) and AGCH (Access Grant Channel), which all share Timeslot 0 in a repeating pattern during a multiframe.

All GSM operators define a subset of their assigned channels as BCCH Carriers. Information about the BCCH Carriers being used by neighbouring cell is advertised in the BCCH Allocation (BA) list carried on the BCCH. This aids in the search for neighbour cells during Idle Mode reselection and Connected Mode handover operations.

5.1.4   2G GSM Cell Selection

Initial Cell Selection – C1

Following power on or a return to network coverage, once a suitable PLMN has been selected, a mobile device must find a useable cell upon which to 'camp' and through which to perform a Location Update and Attach [5, 6].

$$C1 = A{-}B \text{ [or 0, whichever is higher]}$$

Figure 5.4   C1 equation

In GSM, initial cell selection is controlled by the C1 algorithm, which allows the mobile device to compare locally available cells to determine which of them is likely to offer the best quality service.

The C1 equation is summarised in Figure 5.4:

A represents values related to the measured cell and consists of:

RLA_C – the rolling 5 s average signal strength of the target cell as measured and calculated by the mobile device.

RXLEV_ACCESS_MIN – a BCCH broadcast parameter for the target cell that specifies the minimum signal strength the cell must present to the measuring mobile device for that cell to be considered a viable selection candidate. Essentially, this parameter says 'If you measure my signal as lower than X, do not consider me'.

B represents the transmission capabilities of the mobile device and consists of:

MS_TXPWR_MAX_CCH – another BCCH broadcast parameter for the target cell which sets the maximum transmit power the mobile device is permitted to use on a CCH in the cell.

P – the maximum transmit power the mobile device is physically capable of producing.

All of the above values are measured in dBm.

A more complete rendering of the C1 equation would therefore be:

C1 = [(RLA_C – RXLEV_ACCESS_MIN) – (MS_TXPWR_MAX_CCH – P)] or [0] whichever is higher

To be considered as a selection candidate, a cell must have a C1 value that is greater than 0 – any negative values of C1 are replaced by 0. If more than one cell is available the one with the highest positive C1 value will be selected.

The process by which C1 values are compared can be paraphrased as 'choose the cell that consistently offers the strongest signal after eliminating cells that are too weak and after taking the mobile device's transmit power limits into consideration'.

A GSM mobile device is only required to collate measurements for the six strongest cells that it can detect, but some devices collect information on more cells than this.

When a mobile device selects a cell with the highest C1 value it 'camps' on that cell. This means that it synchronises with the cell's BCCH Carrier, reads its BCCH information, performs Location Update and Attach processes if necessary, and monitors the cell's PCH for inbound call indications.

The common terminology used to describe a cell that has been selected in this way is that it is a SERVING CELL.

A cell is therefore classed as a serving cell if the mobile device has elected to camp on it, making it the mobile device's choice as to which cell it is served by when in Idle Mode.

C1 – Worked Example

A worked example of the C1 calculation is illustrated in Figure 5.5 and is detailed below and may provide an insight into its operation:

Assume a mobile device has detected Cell A. Cell A's current averaged received signal strength (RLA_C), as measured as a rolling average over a 5 s period by the mobile device is –85 dBm.

Cell A's BCCH specifies an RXLEV_ACCESS_MIN of –100 dBm and an MS_ TXPWR_MAX_CCH (maximum permitted transmit power) of 29 dBm.

The mobile device has a maximum output power of 29 dBm (GSM900 power class 5).

In relation to the C1 calculation, the mobile device will use the following broadcast parameters:

- RXLEV_ACCESS_MIN = –100 dBm
- MS_TXPWR_MAX_CCH = 29 dBm
- P = 29 dBm.

The mobile device measures the following signal strengths from the three detected local cells:

1. Cell A RLA_C = –85 dBm
2. Cell B RLA_C = –89 dBm
3. Cell C RLA_C = –98 dBm.

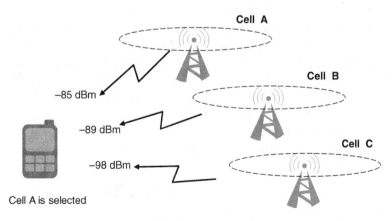

Figure 5.5 C1 calculation

The full C1 equation is as follows:

**C1 = [(RLA_C – RXLEV_ACCESS_MIN) – (MS_TXPWR_MAX_CCH – P)]
or [0] whichever is higher**

So,

Cell A

C1 = (–85 – –100) – (29 – 29)

C1 = (–85 + 100) – (0)

C1 = (15) – (0) = 15

Cell B C1 = (–89 – –100) – (29 – 29) = 11

Cell C C1 = (–98 – –100) – (29 – 29) = 2

A C1 value of 15 is comparatively high and is higher than the C1 values of the other local cells, so Cell A will be selected.

Cell Reselection – C2

Once a mobile device has initially selected a serving cell on which to camp, it must continue to ensure that it is camped on the best available local cell. The C2 algorithm is used for GSM serving cell reselection.

C2 is depicted in Figure 5.6 [6]:

Offsets are additional parameters that are used to influence the 'attractiveness' of a cell. There are offsets that can be permanently applied to a cell and which are advertised on the cell's BCCH. Other offsets are applied on a temporary basis following particular events or activities.

The CELL_RESELECT_OFFSET parameter is used to apply a permanent offset of a cell's C2 value. This may be used, for example, to allow a base station's P-GSM900 (Primary GSM900 band) and GSM1800 cells to be differentiated or to prevent mobile devices selecting cells set aside for capacity expansion. This parameter may apply positive or negative offsets.

A TEMPORARY_OFFSET, as the name suggests, is applied for a limited period, which is defined by the PENALTY_TIME parameter.

Temporary offsets are most usually applied after reselections have taken place. For example, if a mobile device had been camped on Cell A but then reselected to camp on Cell B it would apply a temporary positive offset to the readings taken of Cell B to prevent the C2 algorithm immediately switching back to the previous cell.

Irrespective of any temporary offsets or penalty timers that may be in effect, a C2 reselection will generally only take place if a 'new' cell is seen to have a better C2 value than the current cell for a period of around 5s. This is due to the workings of the RLA_C parameter, which is based on a rolling 5s average.

$$C2 = C1 + Offsets$$

C2 = C1 + CELL_RESELECT_OFFSET - TEMPORARY OFFSET * H(PENALTY_TIME - T)

Figure 5.6   C2 equation

Reselections in Idle Mode are only of concern to the mobile device itself (the network does not get involved in Idle Mode decisions) and the act of reselecting simply means that the device begins to monitor the BCCH and PCH resources of a new cell.

However, if the 'new' cell is part of a different Location Area to the 'old' cell, a reselection will also trigger a Location Update, which will involve not just the mobile device but also the base station, the BSC and core network nodes such as the VLR (Visitor Location Register).

The CRH (CELL_RESELECT_HYSTERESIS) offset attempts to ensure that inter-LAC reselection and the associated consumption of network resources is discouraged.

Reselection to a cell in a new LAC will only be initiated if the new cell's C2 exceeds that of any cell in the current LAC by the standard reselection margin (typically 5 dB) plus the CRH offset for a period of time usually set at 5s.

After an inter-LAC handover has occurred the mobile device is usually forbidden from selecting back to a cell in the 'old' LAC for at least 15s.

C2 – Worked Example

A further worked example, shown in Figure 5.7, might illustrate these concepts more readily.

A mobile device is currently camped on Cell A, which has a C1 value of 15.

The device can detect two neighbour cells: Cell B has a C1 value of 19 and Cell C has a C1 value of 20.

Although both neighbour cells have C1 values that exceed that of Cell A, the mobile device elects to stay camped on Cell A for the time being.

The reasons that reselection to Cells B or C has not taken place are as follows:

Cell B has a higher C1 (and therefore C2) value that Cell A, but this has so far not been measured for more than 5s.

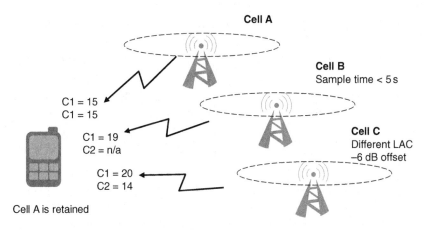

Figure 5.7 C2 calculation

Cell C has a far higher C1 value than Cell A, but it belongs to a different LAC and is specifying a CRH (Cell Reselection Hysteresis) offset of –6 dB on its BCCH. Although the C1 value for Cell C is 20, after applying the CRH offset this falls to 14, below the current value for Cell A.

If Cell B continues to offer a higher signal than Cell A, a reselection may be triggered within a few seconds. If Cell C's signal increases to above a C1 value of 21 then it too might be reselected.

5.1.5 2G GSM Reselection Candidates

In normal circumstances, GSM mobile stations are not permitted to choose reselection candidates freely, they are instead provided with a list of 'approved' channels from which to choose.

The list specifies the BCCH Carrier details of neighbouring cells and is known as the BA List. Carriers are identified by a combination of their ARFCN (Absolute Radio Frequency Channel Number) and BSIC (Base Station ID Code), as illustrated in Figure 5.8.

In theory a network operator can produce one generic BA List that identifies all of the defined BCCH Carrier frequencies employed across the network, allowing mobile devices to select from whichever of these is actually available in each area. Alternatively, the operator may decide to compile bespoke BA Lists that show only details of the specific BCCH Carriers deployed in particular areas or regions.

Limiting the set of channels that need to be considered as candidates has the effect of simplifying and speeding up the selection and reselection process.

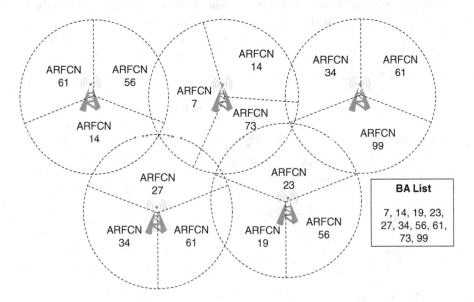

Figure 5.8 BA list and distribution of BCCH carriers

The BA List only contains details of the channels employed as BCCH Carriers; it does not provide details of any non-BCCH Carrier channels (i.e. TCH-only carriers) deployed to cells; these are irrelevant to the Idle Mode selection/reselection process as selection/reselection is only made to channels that carry a BCCH.

Some radio survey devices allow users to 'lock' the device to a specified cell by using a function known as a cell lock. This means that the device is forced to camp on a cell that it might not otherwise have selected as the serving cell. When this happens, the device is also forced to accept the 'lock' cell's BA List.

When the cell lock is removed and the device is able to reselect freely again, it might take a few seconds for it to reselect to a new cell and capture that cell's BA List, meaning that any reselection decisions made in the 5 or 10 s following the release of a cell lock must be regarded with suspicion. The survey device might, during this period, select serving cells that would not ordinarily be selected as serving at that location due to being forced to use the 'wrong' BA List for that area.

A similar concept to the BA List is employed in 3G and 4G networks, which are optionally able to transmit a variety of NCL (Neighbour Cell Lists) to idle mobile devices.

5.1.6   2G GSM Cell Configurations

The capacity of the BCCH Carrier is shared between administrative functions and traffic carrying. Each cell on a BTS can have only one BCCH carrier channel but, in busy areas, more than one cell could be configured to the same sector, each with its own BCCH Carrier. For example, an operator might elect to deploy a GSM900 cell (with a BCCH Carrier and zero or more expansion carriers per sector) and a GSM1800 cell (with its own BCCH and set of expansion carriers) on the same sector azimuth. This would be regarded as one physical sector supporting two different cells and is some times described as being a 'stacked sector'.

Additional expansion carriers may be deployed in a cell that carry only TCH. All carriers in a cell share the same CI and BSIC but will be transmitted on different ARFCNs, making the various carrier resources in a cell easily distinguishable.

Mobile devices monitor the BCCH Carrier of the serving cell when in Idle Mode and use the RACH facilities of that channel to establish connections. Mobile devices might be assigned a TCH on one of the expansion carriers in a cell once a connection had been established. This means that, when examining survey data that includes test calls, the call setup phase may be linked to one carrier (ARFCN 67 + BSIC 17) but the connected phase of the call may be linked to a different carrier (ARFCN 45 + BSIC 17) and both carriers will be linked to the same CI.

The TCHs configured on a carrier are often a mix of GSM TCH that carry voice and GPRS/EDGE PTCHs that carry packet data. Base stations are typically able to dynamically switch a timeslot from TCH use to PTCH use as required by current traffic demand levels.

Figure 5.9 provides an example of a typical GSM cell configuration.

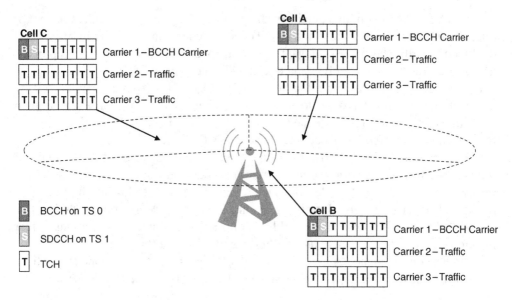

Figure 5.9   Typical 3 + 3 + 3 carrier deployment

Some operators, like Vodafone and O2 in the United Kingdom for example, who have been assigned channels from a mix of radio bands, employ more complex carrier deployment strategies.

In these scenarios, a physical sector of a BTS may consist of some 900 MHz P-GSM900 carriers along with some E-GSM900 (Extended GSM900 band) or GSM1800 carriers. The BCCH Carrier would be on one of the P-GSM900 frequencies, with the others configured as 'traffic only' TCH carriers. Different resources within the same cell could therefore be transmitted in different frequency bands.

In even more advanced configurations, a BTS may support separate P-GSM900 and E-GSM900/GSM1800 cells on the same sector. Both the P-GSM900 and E-GSM900/GSM1800 cells would be allocated individual CIs and each would transmit its own BCCH on one of its carriers. Only the P-GSM900 cell would offer an SDCCH though and only the P-GSM900 cell would be available for call set up; the other cells would be available purely for cell capacity extension purposes.

One further configuration example is where the operator elects to deploy a P-GSM900 cell using a standard Sectorised Transmit Sectorised Receive configuration, but overlays this with a single E-GSM900 or GSM1800 cell that is transmitted using Omni-directional Transmit, Sectorised Receive techniques. The main coverage of the site would therefore be provided using sectorised resources, with an omnidirectional umbrella cell providing site-wide capacity expansion.

To ensure that mobile devices do not attempt to establish calls via the E-GSM900/GSM1800 cells, in the above scenarios, the operator can configure them to appear 'unattractive'; this can be achieved by ensuring that the capacity extension cells have artificially low C1 or C2 values.

Cell access can be discouraged using C1 parameters by setting the cell's RXLEV_
ACCESS_LEVEL_MIN parameter to an unreasonably high value, of say –48 dBm.
Mobile devices would rarely receive a signal from a macrocellular site at such a high
signal strength, meaning that the RXLEV_ACCESS_LEVEL_MIN value for that site
would be unlikely to be met and that, therefore, no mobile device would choose it
during idle mode cell selection or reselection.

Cell access can be discouraged using C2 parameters by specifying a large negative
offset in the cell's BCCH-broadcast C2 parameters. This could, for example, reduce
a cell's 'real' C2 value as measured by a mobile device from a healthy and attractive
25 to a less attractive 5.

These techniques are only applicable to Idle Mode and the capacity extension carriers and cells controlled in this way would be available for handover once a call had
been established using the cell's P-GSM resources.

From a forensic point of view, cells that have had such parameters applied to them
would rarely appear as Idle Mode serving cells in survey results, but would appear as
strong or very strong non-serving neighbours. These cells would also rarely appear as
'start cells' in call records, as calls usually start on cells that are selected as Idle Mode servers, but they would appear as 'end cells' having been handed over to after the start of a call.

5.1.7   2G GSM Channel Numbering

Absolute Radio Frequency Channel Numbers

Each GSM radio channel nominally occupies 200 kHz of spectrum (the true figure is
more like 270 kHz) and each channel is ultimately defined by its central 'carrier'
frequency. These are spaced at 200 kHz intervals.

A mobile device in Connected Mode is instructed to use a particular radio channel
by the serving base station using signalling messages. These messages could become
unnecessarily long if they were required to carry details of each radio channel using
its exact carrier frequency, so GSM instead employs a set of 'channel index numbers'
as a shorthand way to identify each carrier [7].

The ARFCN specifies a certain radio FDD frequency pair used for transmission
and reception within the GSM network. The ARFCN is associated with both an uplink
and a downlink radio channel to form a carrier, as depicted in Figure 5.10.

The channel numbers allocated to the various forms of GSM are shown in Table 5.1.

P-GSM900 channels were first made available in Europe, part of WRC Region 1, in
the early 1990s, when the 900 MHz band was still being shared in many countries with
the earlier 1G (First Generation) mobile networks. When 1G networks were finally
switched off, their spectrum was taken over by GSM and was reclassified as the
E-GSM900 band. Spectrum in the 1800 MHz band was originally set aside for a
scheme known as DCS (Digital Communications System) in Europe and was therefore
known as DCS1800. In reality, the only technology that was put forward to be used in
this band was GSM, and the technology/use of the band become known as GSM1800.

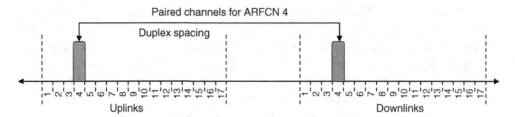

Figure 5.10 GSM Frequency Division Duplex and ARFCNs

Table 5.1 GSM ARFCN ranges.

Band name	Uplink range	Downlink range	Downlink UARFCN		Deployment
			Low	High	
GSM450	450.4–457.6	460.4–467.6	259	293	Not used
GSM480	478.8–486.0	488.8–496.0	306	340	Not used
GSM700	698.0–716.0	728.0–746.0	Dynamic allocation		Not used
GSM700	747.0–763.0	777.0–793.0	438	511	Not used
GSM850	824.0–849.0	869.0–894.0	128	251	Americas
E-GSM900	880.0–890.0	925.0–935.0	0, 955	1023	Global
P-GSM900	890.0–915.0	935.0–960.0	1	124	Global
GSM-R	873.0–890.0	918.0–935.0	940	1023	Europe
GSM-R	890.0–915.0	935.0–960.0	0	124	Europe
GSM1800	1710.0–1785.0	1805.0–1880.0	512	885	Global
GSM1900	1850.0–1910.0	1930.0–1990.0	512	810	Americas

Based on 3GPP TS 45.005:2 used with permission from 3GPP.

When deployments of GSM began to spread beyond Europe additional radio bands had to be defined to fit in with Region 2 and Region 3 allocations. GSM1900 was created to serve the United States PCS (Personal Communications Service) concept and operates in the 1900 MHz band (which it shares with cdmaOne deployments). GSM400, GSM700 and GSM850 were also developed to match the spectrum allocations in use in a variety of countries and regions.

GSM-R (GSM for Railways) was developed as an adapted version of GSM900 to provide 'group communication' services for railway networks.

There is an overlap between the channel numbering assigned to some bands – there is overlap between GSM1800 and GSM1900, for example, and between P-GSM900/ E-GSM900 and GSM-R – so it is necessary to also have knowledge of the radio band or GSM variant being used to fully determine ARFCN assignments in these cases.

Duplex Spacing

Voice services, which GSM was designed primarily to support, need a duplex connection – which gives them the ability to simultaneously transmit and receive. This is achieved by using FDD techniques. The uplink and downlink are transmitted on different radio channels and the combination of an uplink and a downlink make up one duplex GSM radio carrier.

The fact that the uplink and downlink channels in a GSM carrier are permanently 'paired' means that, for each version of GSM, there is a fixed spacing between the paired channels covered by the same ARFCN. This is known as the 'duplex spacing' of the system.

In the 900 MHz band the duplex spacing is 45 MHz and in the 1800 MHz band the duplex spacing is 95 MHz. Other versions of GSM have their own specific spacings, which are determined by the overall width of each radio band and the separation that is inserted between the highest uplink channel and the lowest downlink. It is interesting to note that in most FDD systems, the uplink always uses the lower frequency range because the mobile phone has limited power resources (being battery powered) and lower frequencies have greater range.

5.1.8   2G GSM Cell Identifiers

Physical Layer Identifier

2G GSM cells are identified at the physical layer using the BSIC, which is a six-bit identifier transmitted on a BCCH Carrier's SCH. The most significant three bits form the NCC (Network Colour Code) and the least significant bits are the BCC (Base Station Colour Code) [8].

The NCC is used to quickly distinguish between cells belonging to different operators, as each operator in a country will be assigned a different NCC. A mobile device can therefore quickly group detected cells by network simply by recovering their BSICs.

The BCC is used to distinguish between local cells belonging to the same network; neighbouring cells should be assigned different BCCs. In practice, with only eight BCCs (000 to 111 in binary) available for each network, these codes get reused with great regularity, so cell discrimination usually takes place using a combination of BSIC and channel number (ARFCN). Network planners seek to ensure that the same ARFCN/BSIC pair is not reused in the same region of their network.

The SCH, which carries the BSIC, is transmitted approximately 25 times per second on the BCCH, which provides mobile devices with multiple opportunities to quickly discover enough information about detected cells to enable a basic level of cell discrimination to take place.

Cell ID

2G CGIs (Cell Global Identifiers), as detailed in Figure 5.11, contain all the details necessary to uniquely identify an individual cell out of all the cells in all the networks in the world [8].

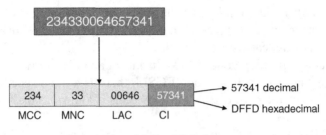

Figure 5.11 2G Cell ID

In most cases however it is enough to be able to identify a cell only amongst the other cells in its home network, in which case only the last ten digits (in decimal) of the CGI are important.

Take for example the full figure CGI 234330064657341.

234 is the MCC for the United Kingdom.
33 is the MNC for Orange UK.
00646 is the LAC within the network.
57341 is the CI.

The CI component of the CGI is 16 bits long in binary (five digits in decimal), which gives 65 535 different CIs. These must be uniquely assigned to a cell but are only unique within a LAC and may be reused in other LACs. In the early days of GSM network deployment, most networks had fewer than 65 000 cells and the CIs that were allocated were unique within their home network, irrespective of LAC. As networks became larger and denser it became necessary for some operators to begin to reuse CIs in different LACs. It is therefore important when analysing call records or surveying cells belonging to larger networks to take the LAC component of the CGI into account as well as the CI, to ensure that the correct cell is identified.

Cell ID Formats

Most networks present CIs in decimal format in call records and other documentation, but a minority of operators elect to present them in hexadecimal format.

The EE (Everything Everywhere) network in the United Kingdom, for example, elects to present the CIs of cells belonging to their T-Mobile and 4GEE brands as four digit hex codes.

For example, a 2G T-Mobile cell with a decimal CI of 57341 would be presented as DFFD in hexadecimal.

In reality, the CGI is stored in base stations in binary, with the CI component occupying 16 bits, so the difference between decimal and hexadecimal formats is simply a presentational choice. A mix of CI numbering formats can make it difficult to

Table 5.2 Comparison of decimal and hexadecimal CI.

Rank	Binary	Decimal	Hexadecimal
Lowest CI	0000000000000000	0	0
Highest CI	1111111111111111	65 535	ffff

produce coherent cell site call tables, however, and thought should be given to how call data containing a mix of formats should be processed and represented in cell site reports.

When comparing decimal and hexadecimal to binary, the lowest and highest possible values for a 16 bit CI or LAC field are as shown in Table 5.2:

5.1.9   2G GSM Cell Discrimination

Cells are uniquely identified at the 'logical' level using CIs, which are broadcast and advertised on the BCCH. All channels within the same sector of a GSM site will share the same CI (unless multiple cells have been deployed on the same sector) – which may lead to multiple channels being assigned the same CI in network surveys if test calls are made [9].

To determine a cell's identity using this method it is therefore necessary for a mobile device to synchronise with the cell and monitor its BCCH Carrier until the required information elements have cycled past; all of which could take several tens of milliseconds to occur. This process would be far too time and energy-consuming a process for mobile devices to constantly perform when in Idle Mode.

A simpler set of 'physical layer' identifiers is therefore employed to allow mobile devices to quickly and easily discriminate between local cells. This method is based on a combination of the ARFCN of the cell's BCCH Carrier and the cell's BSIC and is illustrated in Figure 5.12.

BSICs are reused with great regularity, as there are only eight available per network. The frequency reuse patterns employed by GSM network planners should ensure that each channel reuse occurs at a safe distance from other cells using the same channel. Network planners ensure that when a channel is reused it uses a different BSIC and BCC to any other cells sharing the same channel in that region of the network.

2G cell discrimination should, therefore, be a reasonably simple process, as no two cells in the same general area of a network should share both an ARFCN and a BSIC. There are exceptions to most rules, however, and surveyors will sometimes encounter cases where the cell planning process has failed and the same ARFCN/BSIC pair from different cells is detectable at the same location. In these situations it is often the case that measurements from neither cell is captured correctly.

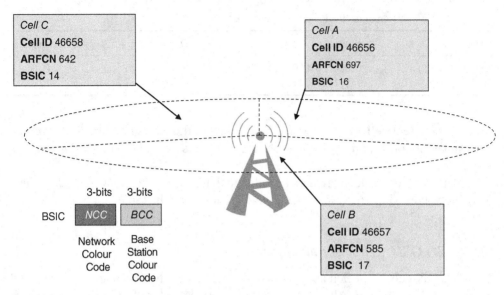

Figure 5.12   Cell discrimination using BSICs

5.1.10 2G GSM Radio Measurements

GSM base stations and mobile devices are required to collect and evaluate data related to radio signals and their relative strengths and quality levels to enable functions such as cell selection and handover to take place.

Received Signal Level

GSM signal strength measurements are taken in absolute terms using the dBm scale, but for reporting purposes they are usually equated to a scale known as RXLEV (Received Signal Level) which is detailed in Table 5.3 [10].

The RXLEV scale assigns a six-bit code to represent each integer value between –48 and –110 dBm.

The RXLEV scale is employed partly for efficiency reasons as a six-bit RXLEV code takes less bandwidth to transmit in a measurement report than the full dBm value would.

Received Signal Quality Level

Determinations of signal quality are primarily made using BER (Bit Error Ratio) calculations [11].

The BER of a signal can be computed by comparing the bit rate of a connection (e.g. the total amount of data bits transmitted) by the number of bits that were received in error.

Table 5.3 RXLEV reporting values.

RXLEV	Lower boundary (dBm)	Upper boundary (dBm)
0	<–110	–110
1	–110	–109
2	–109	–108
…	…	…
62	–49	–48
63	–48	>–48

From 3GPP TS 45.008:8.1.4 used with permission from 3GPP.

In fast-moving, real-time transmission systems such as the GSM air interface, BER levels are generally estimated and expressed as 'probabilities', which is faster and simpler than attempting to calculate rigorously accurate ratios.

RXQUAL is reported using values from RXQUAL_0 (highest quality, <0.2% errored) to RXQUAL_7 (lowest quality, >12.8% errored).

Typical Values

Measurements of a GSM cell will typically provide the following data:

- RXLev in dBm
- C1 (cell selection criterion) as a positive integer
- C2 (cell reselection criterion) as a positive integer.

RXLev is received signal strength on one ARFCN; it is measured in dBm and has a reporting range of –48 (very strong) to –110 dBm (very weak).

The two other values typically calculated are C1 and C2, which are used to inform cell selection and reselection activities. C1 and C2 will be positive integer values and the higher the value the more likely it is that the corresponding cell would be selected.

RXLev is the primary measurement for GSM as far as Idle Mode cells selection and Connected Mode handover are concerned and is also sometimes known as RSSI (Received Signal Strength Indicator).

5.2 3G UMTS/HSPA Networks

3G UMTS networks were developed by 3GPP as an evolution of the popular 2G GSM standard. Like other 3G systems, UMTS was designed to support 'legacy' services such as voice and text messaging, but was also provided with enhanced capabilities such as video calling and fast Internet data connectivity [12].

Like other cellular systems, a UMTS network is divided into the RAN and core network environments.

A 3.5G evolution of UMTS, known as HSPA (High Speed Packet Access), was developed by 3GPP and there have been several iterations of HSPA and then HSPA + released.

Although GSM was developed in Europe, the initial development of UMTS was shared between European and Japanese standards bodies before being assigned to 3GPP. UMTS is now widely deployed globally. It is also known as FOMA (Freedom of Mobile Access) in Japan.

3GPP publish its specifications grouped in to 'Releases'. The first full set of UMTS specifications were published in 1999 and were known as 'Release 99' (R99). Subsequent releases were decoupled from any specific year and were known as Release 4 (R4), Release 5 (R5) and so on. At the time of writing (Summer 2014) 3GPP were preparing to 'freeze' further developments of Release 12 specifications ready for final publication.

5.2.1 3G UMTS Access Networks

The UMTS access network is known as the UTRAN (UMTS Terrestrial Radio Access Network); it is detailed in Figure 5.13 and consists of the following elements [13]:

- The Node B base station, which manages one or more individual cells;
- The RNC (Radio Network Controller), which manages groups of Node Bs;
- The backhaul connectivity that connects RAN elements together and to the core networks.

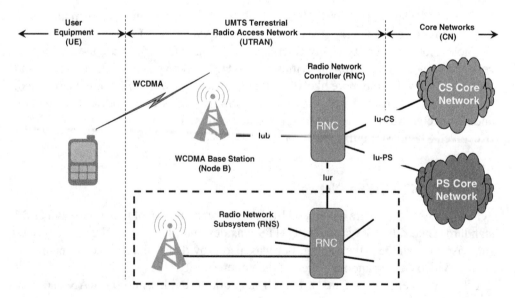

Figure 5.13 UTRAN interfaces and elements

The RNC performs RRM and traffic routing functions for the Node Bs, mobile devices and subscribers in its area. These functions include:

- Radio Resource Control – radio connection setup and management;
- Admission Control – which estimates whether a new call can be connected without affecting other calls in the cell;
- Code Allocation – allocates the WCDMA (Wideband Code Division Multiple Access) codes to be used for each call;
- Power Control – accurate and rapid power control of each mobile device is important as all devices in a cell transmit simultaneously using the same uplink carrier;
- Handover Control – several Handover methods are used in UMTS under the control of the RNC.

An RNC in control of a user connection is known as the Serving RNC. One RNC and its group of Node Bs combine to form one RNS (Radio Network Subsystem).

In the air interface, users' signals are separated using WCDMA, which supports SFN (Single Frequency Network) techniques. This means that, within a frequency layer, UMTS has a Frequency Reuse Factor of 1, so the same radio frequency can be reused in adjacent cells. This makes frequency planning for UMTS considerably less time consuming than for GSM; it does however make capacity and service planning far more complicated.

5.2.2   3G UMTS Radio Interface

The generic radio technique chosen for use in UMTS is CDMA, which is outlined in Figure 5.14.

Like the FDMA and TDMA systems used in earlier generations of mobile technology, CDMA is designed to allow multiple subscribers access to the same base station

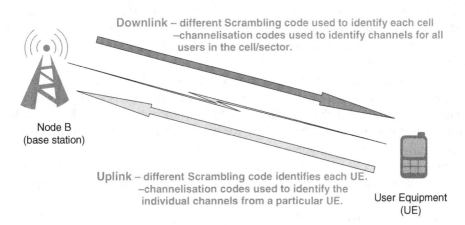

Figure 5.14   Code Division Multiple Access

at the same time. Unlike the earlier systems, CDMA allows all of these subscribers to concurrently use the same radio channel when accessing a cell.

There is generally no frequency or time-division taking place in a standard CDMA system [although time division is a factor in the Chinese TD-SCDMA (2G received signal level system)], so user channels are separated by using sophisticated digital coding techniques instead.

Each mobile device in a CDMA cell is allocated a different digital code, known in UMTS as a 'scrambling code'. This code, when combined with the users' data, produces an uplink radio signal with distinct characteristics that can be 'found' and recovered by a base station.

On the UMTS downlink, each cell in an area also uses a different digital scrambling code, which allows mobile devices to distinguish between the different cells that are sharing the same SFN carrier.

If each mobile device is allocated a different scrambling code (which happens dynamically at the start of each call or connection) and if all of the codes are distinct enough from each other, the base station will have no difficulty in receiving and correctly interpreting the data transmitted by a large number of devices in the same cell. The same is true of the downlink; as long as each cell in an area is assigned a different scrambling code, mobile devices have a chance of discriminating between the signals generated by each base station.

As with other cellular systems, UMTS users can each have several logical traffic or control channels active simultaneously. The first stage of WCDMA coding therefore uses 'channelisation' or 'spreading' codes to separately identify the different channels being used by the same device on the uplink or by different devices on the downlink. The second stage of coding involves the use of 'scrambling codes'.

Different uplink scrambling codes are dynamically allocated to each Connected Mode mobile device in an area of the network. There are over 16 million uplink scrambling codes available in UMTS. The Node B base stations can identify different users' signals on the shared uplink channel by looking for their allocated scrambling code. The Node B first recovers the signal using the scrambling code and then uses the original channelisation code for each channel to recover the individual streams of user traffic and control data.

In the downlink direction, PSC (Primary Scrambling Codes) are used to identify the transmissions in each cell and are required because all cells in an SFN layer will be using the same radio channel. There are 512 different PSCs in the downlink direction and this number should be enough, for cell planning purposes, to ensure that no two cells in the same geographical area are assigned the same code.

Each active user connection in a cell is then allocated a particular channelisation code on the downlink. A receiving mobile device looks first for the required cell scrambling code on the downlink radio channel and then recovers its own data using its assigned channelisation code.

UMTS systems typically operate with a radio channel bandwidth of 5 MHz (although variants that operate in channels with bandwidths of 1.6 and 10 MHz have also been defined). Five Megahertz was regarded as being a relatively 'wide' radio bandwidth when the system was developed, which is why the specific version of CDMA employed by UMTS is known as WCDMA.

5.2.3   3G UMTS Cell Selection

UMTS mobile devices employ the S (Selection) algorithm to perform initial cell selection [14].

The S algorithm is the equivalent of the GSM C1 algorithm and essentially instructs the mobile device to select the cell that offers the highest quality, although quality is actually determined by signal strength.

The S equations are detailed in Figure 5.15:

The requirements of the S algorithm are deemed to be met – a cell can be selected for example – if

$$Squal > 0 \text{ and } Srxlev > 0.$$

The procedure states that the mobile device will search for the strongest cell and will then select it, which implies that if signals from more than one cell are detected during the process the cell with the highest value of Squal/Srxlev will be selected.

The parameters employed by the S process include:

S_{qual} – cell selection quality value. Calculated by comparing a target cell's readings with a broadcast minimum value

S_{rxlev} – cell selection RX level value. Calculated by comparing a target cell's readings with a broadcast minimum value and a value related to the maximum permitted and possible transmit power capabilities.

Although expressed in fairly complex terms, the S criteria essentially work in the same way as the GSM C1 algorithm and allow the mobile device to select the local cell that provides the best received signal strength values.

The selected cell in UMTS is termed the ACTIVE cell, which is equivalent to the 'serving' cell concept used in 2G networks.

$$\mathbf{S}_{qual} = Q_{qualmeas} - Q_{qualmin}$$

$$\mathbf{S}_{rxlev} = Q_{rxlevmeas} - Q_{rxlevmin} - P_{compensation}$$

Figure 5.15   The UMTS S cell selection equations

Cell Reselection

Cell Reselection analysis is performed according to the R (Reselection) criteria [15].

The reselection process is only triggered if the measured signal provided by the current active cell drops below a broadcast minimum value, known as S_x (which is based on the current S_{qual} value).

In normal cells, the R criteria are outlined in Figure 5.16:

The Q_{map-s} and Q_{map-n} parameters relate to the measured quality of the current active cell and a candidate neighbour cell respectively; whilst Q_{hyst} and Q_{offset} are offset parameters that can be broadcast in a cell to permanently alter the perceived attractiveness of that cell as a reselection candidate.

TO is a Temporary Offset applied during a Penalty Time period and is used to temporarily alter the perceived attractiveness of a target cell. This is usually applied after a reselection has occurred to prevent the mobile device reselecting back to the 'old' cell too quickly.

There is no direct equivalent of the GSM CELL_RESELECT_HYSTERESIS parameter that is used to control inter-LAC handovers. Where required such functions can be achieved by applying Q_{hyst} and Q_{offset} values to cells along location area boundaries.

5.2.4 Active and Monitored Cells

A UMTS mobile device in Idle Mode will select the 'best' (or, at least, an acceptably strong) local cell to 'camp' on – this means that the device will monitor the cell's BCCH and PCH. This cell is known as the 'active' cell.

When the mobile device enters Connected Mode it will attempt to use the current Active cell to establish a connection.

If some forms of Connected Mode handover are triggered then the mobile device may enter 'soft handover' and find that it is concurrently connected to more than one active cell in what is known as its 'Active Set'.

In both Idle and Connected modes the mobile device will take regular neighbour cell measurements in an attempt to identify viable reselection or handover candidates. Neighbour cells (those that don't belong to the Active set but that were included in a Neighbour Cell List) are known as 'Monitored' cells and the list of handover candidates complied by the mobile is known as its 'Monitored Set'. Other detected cells will be reported as being part of the 'Detected Set'.

An illustration of the 'active' and 'monitored' concepts is provided in Figure 5.17.

$$R_s = Q_{map-s} + Q_{hyst-s}$$

$$R_n = Q_{map-n} - Q_{offset-s,n} - TO_n$$

Figure 5.16   UMTS R cell reselection equations

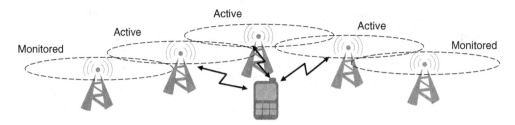

Figure 5.17   Active and monitored cells

A radio survey device will typically compile regular measurement 'events' that each consist of details of the currently active cell(s) plus details of the neighbour cells that make up the current monitored and detected sets.

5.2.5   Single and Multi Frequency 3G UMTS Networks

3G UMTS networks use WCDMA on the air interface. CDMA-based systems can distinguish between cells based on their PSC, meaning that there is no absolute necessity to deploy neighbouring cells on different radio channels in order to differentiate between them, instead the differentiation can be based on PSC.

This allows 3G operators to deploy SFNs, where the same 5 MHz radio carrier pair is used for all cells. This is especially useful considering the comparatively wide radio channels employed by UMTS and means that a reasonably high capacity network can be deployed within a limited allocation of spectrum.

Many 3G networks were initially rolled out using a single carrier, which was generally enough to serve the initial levels of demand. As user numbers increased, operators had the option of rolling out services on additional carriers, if they had licences for additional spectrum.

The deployment of additional 3G channels coincided with the release of HSPA. This allowed operators to plan to deploy their additional carriers as 'data' channels, reserved for fast mobile broadband services.

As they developed, most 3G UMTS networks evolved to support two or more carriers, with one carrier typically assigned to carry basic 'R99' traffic (voice, text messaging, low rate data), while any other carriers were configured to support HSPA and carry high data rate mobile broadband services. Each of the additional carriers employed by an operator was deployed as a separate SFN 'layer'. Multi-carrier SFN configured in this way can be thought of as having been deployed as a set of separate but associated SFN layers, as depicted in Figure 5.18.

Each 3G carrier on a sector is configured as a full cell in its own right, with a unique CI and its own BCCH and any carrier is, in theory, available as a selection or reselection candidate for phones in Idle Mode to camp on.

The need to support an 'active' BCCH, for example, one that is actively used by phones to perform random access and that must populate and transmit a PCH, imposes

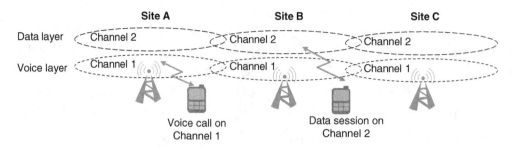

Figure 5.18   Multi-layer networks

a processing overhead on the host base station and it is desirable to limit the number of instances of active BCCH carriers on each Node B. Networks that define separate 'R99' and 'HSPA' frequency layers often attempt to ensure that the administrative activities related to active BCCH services are confined only to a single layer, which is usually the R99 layer. Many networks therefore use a combination of techniques to make their HSPA layer carriers less 'attractive' to idle phones.

One way of doing this is to configure the HSPA cells to advertise permanent negative offsets within the set of access control parameters broadcast on their BCCH. A negative offset is subtracted from any RF measurements made of the associated cell, making the received signal seem less powerful.

Another method employed by networks is to define a set of 'Idle Mode behaviour' parameters, which are passed to phones when they move into Idle Mode and offer a set of radio carrier or network technology priority levels. These parameters influence the phone's choice of cells when performing Idle Mode reselection. Using this method, a 3G phone might be instructed to prioritise the network's 'R99' carrier above any 'HSPA' carriers, which then means that phones will try to camp on to an R99 cell whenever possible. It is also possible to add priority levels for different generations of access network type, so a phone might be instructed to prioritise GSM over UMTS when in Idle Mode, which would therefore mean that it is more likely to camp on to a 2G cell than a 3G cell.

All of these features are designed to allow a network to control the paging and random access load that is associated with BCCH carriers, but it inadvertently presents challenges for forensic radio surveyors. If a surveyor has instructions to take measurements of all 3G cells in an area, and one of the 3G carriers is an HSPA channel which has been optimised not to allow Idle Mode camping, then the survey equipment will be unlikely to detect cells on that layer as serving (as the RF survey equipment would be discouraged from reselecting to it) and may not even be able to obtain reliable signal strength measurements of it when in 'free running' mode.

The solution to this problem is to set the survey equipment to 'lock' on to the required HSPA channel, if such as feature is available on the survey device. A 'cell lock' or 'channel lock' of this kind overrides the phone's usual selection activities and forces it to camp onto a specific carrier. The surveyor must remember to cancel the lock and return the survey equipment to free running mode following the survey or risk compromising the accuracy of further surveys.

5.2.6   3G UMTS Handover – Soft Handover

In WCDMA, where all cells in an SFN layer will use the same radio channel, a mobile device in Connected Mode operating at or near the edge of a cell may cause interference to (and receive interference from) services in neighbouring cells. To avoid this situation causing noise control issues in the neighbouring cell, a Soft Handover (SHO) must be initiated, as described in Figure 5.19. The area of overlap and intersection between cells, where this adjacent cell interference must be controlled, is therefore known as the 'Soft Handover Region'.

The decision to invoke SHO is made by the network based on measurements supplied by the mobile device. If the current Serving RNC determines that an active mobile device has moved into the soft handover region between two cells, it forwards a copy of the connection(s) already being transmitted to the mobile device to the second Node B, which then begins to also transmits them on the downlink in its cell. The RNC will instruct the phone to search for the PSC of the new cell and begin receiving the additional downlink connection. It will also instruct the second base station to begin receiving the uplink signal from the phone and forwarding it to the RNC, which combines both copies of the uplink traffic (received via the two active base stations) before forwarding it to the core network.

SHO is triggered to ensure that the transmitted uplink signal from the mobile device is suitable for all cells affected by it, so the multiple base stations involved in the handover are able to send 'transmit power control' messages to the mobile device to make sure that its transmissions are kept below the level where they would 'drown out' signals from other phones that are sharing the same uplink radio channel in those cells.

If the mobile device travels past the Soft Handover Region into the new cell, meaning that the levels of interference caused and experienced by the phone will have

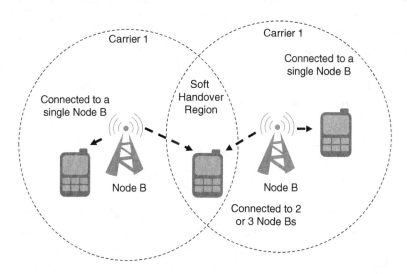

Figure 5.19   Soft Handover

dropped below threshold levels in the 'old' cell, the mobile device and the 'old' Node B will be instructed to drop their connections. The mobile device will no longer be in SHO and will be served by the 'new' Node B only.

If an active mobile device roams into an area where three cells intersect, a three-way SHO can take place, where the mobile device is in contact with three Node Bs simultaneously. The group of cells currently serving a mobile device in SHO are known as its 'active set'. A 3G mobile device in Connected Mode will be instructed to add and drop SHO connections to/from its active set over time and, in areas where there are several cells that offer similar signal strengths, this could happen with great rapidity.

For a variety of reasons, the 'soft handover candidate' cells employed during 3G handover are sometimes drawn from a set of neighbours that do not show up in Idle Mode surveys, meaning that details of those cells would not be captured if Idle Mode surveying alone was employed.

Scenarios in which unexpected neighbour cells might be used as SHO cells are related to received power levels in the mobile device. One scenario could be that the network advertises an incomplete set of local cells in Idle Mode NCL and the mobile device detects an unlisted cell as a SHO candidate when in Connected Mode. Although the net effect of SHO is that a call is handed over from one cell to another, the main motivation for SHO is the need to maintain power control in a SFN, so decisions about SHO are based on measurements of cells that were actually detected by the mobile device rather than being limited to only using cells that were contained in an NCL.

Forensic surveys of 3G networks need to take the potential for SHO into account. Important serving cell information may not be captured if surveys are undertaken in Idle Mode only, given the possibility that some SHO candidates may not show up in Idle Mode surveys. A commonly accepted way of capturing Connected Mode details during forensic surveys is to make a series of 'test calls', noting the cell(s) that are used to carry each call.

It should be noted that some types of survey equipment might only provide details of the 'primary' connected cell in an active set and may not provide details of other members of the survey equipment's current active set. This could limit the effectiveness of surveys undertaken using such equipment.

5.2.7   Other Kinds of 3G UMTS Handover

SHO is not the only form of handover employed in UMTS; examples of other types are shown in Figure 5.20.

SHO occurs when the old (source) and new (target) cells belong to different base stations, meaning that the RNC must get involved in mediating between the handover between the Node Bs, but intra-carrier handover can also take place between cells (sectors) belonging to the same Node B. Although the mechanism and the outcome

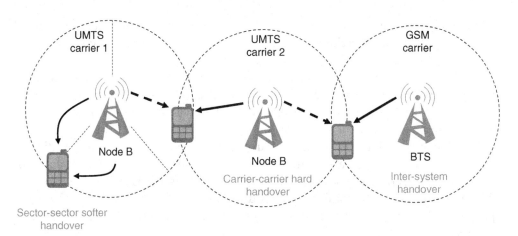

Figure 5.20 Softer, hard and inter-system handover

are the same, this form of intra-Node B handover is known as a SrHO (Softer Handover) and can be managed by the Node B without recourse to the RNC.

SHO can be characterised as 'make before break' handover, as a connection to a new base station is established before releasing the connection to the old base station. This is only possible if the old and new cells are on the same frequency, so SHO is really a function of the SFN environment.

Handovers between cells that are on different carriers cannot be managed using SHO techniques and must instead use much the same 'hard handover' techniques as employed in multi-frequency systems such as GSM.

Hard Handover (HHO) can be characterised as 'break before make' handovers and involve the mobile device releasing the old connection and retuning before picking up a new connection.

HHO is also sometimes employed if the source and target cells belong to different RNCs, even if the cells are on the same frequency layer, although inter-RNC SHO can be achieved if the RNCs are connected to each other.

Inter-system Handover (ISHO) is also supported in UMTS and mobile devices are able, in theory, to handover between 3G cells and those belonging to 2G or 4G networks. ISHO is only possible for 'multi-RAT' devices that support more than one radio access technology, which the majority of modern mobile devices are.

5.2.8 Cell Breathing

As has already been noted, in CDMA-based systems all mobile devices in the same cell share the same uplink and downlink radio channels, meaning that the base station receives signals from all devices mixed together on the same carrier. The uplink signals from each phone will have been created using a different scrambling code and each code produces a signal with identifiably distinct characteristics. The base station employs a set of parallel 'signal processors', all looking at the same uplink channel and each tasked with searching for the characteristics of a particular code and each therefore dedicated to serving a specific active phone.

For this 'code division' approach to work, however, all of the signals from active phones must reach the base station at roughly the same power level; if one signal is significantly more powerful than the others it will drown them out; if one signal is too quiet it will be drowned out by the others.

If a Connected Mode phone moves further away from its serving base station, it will be instructed to increase its transmit power in order to maintain the quality of its uplink connection and to keep the signal received at the base station within the detection window so that it does not get drowned out. The phone adjusts it uplink power level in line with 'transmit power control' messages sent to it from the base station.

Active phones will also be instructed to increase their uplink transmit power if there is an increase in 'background noise' in a cell; the primary cause of background noise is transmissions from other phones in the cell that are sharing the same uplink carrier.

There is a maximum permitted uplink transmit power level in each cell, which is advertised on the BCCH, and this has an effect on the usable size of a cell. The 'planned' size of a cell can be determined as the point at which a phone, that is transmitting at maximum permitted uplink power, sends a signal that stays within the detection window and can still be received at the base station. Any further away and the phone's signal will drop below the level of the other received signals at the base station and will no longer be detectable.

In very busy periods, when lots of background noise is being generated by other connections, a phone may find itself in this situation even though it is some way inside the planned area of the cell; the usable area of the cell will therefore have 'shrunk', meaning that closer phones will be able to connect to the base station but more distant phones will not.

Therefore, in CDMA networks the effective coverage area of the cell is not fixed but will change continuously depending on the traffic load, and on interference from other mobile devices and adjacent Node Bs. This phenomenon is depicted in Figure 5.21 and is known as 'cell breathing' due to the expansion of effective cell coverage during quiet periods (like the cell is breathing out) and the contraction of coverage (breathing in) during busier times. Cell breathing is an unavoidable consequence of the use of CDMA in a SFN configuration. This means that the coverage, capacity and interference parameters of each cell are interdependent on each other and consequently that a change in any one of those parameters has an effect on the others.

It is important to recognise that cell breathing is a condition of the uplink direction only. The effects of cell breathing are usually avoided by planning a cell's coverage area to conform to the 'breathed in' size. In any event, cell breathing is usually assumed to change the usable cell radius by no more than 5–10%.

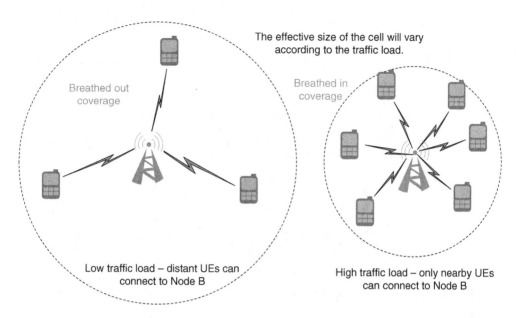

Figure 5.21   Cell breathing

5.2.9   3G UMTS Radio Measurements

3G devices are provided with a broadcast resource in each cell that acts as a known quantity for them to measure. This resource is known as the CPICH (Common Pilot Channel).

The CPICH uses a fixed channelisation code, which is the same in every cell of every UMTS network. The CPICH performs the same 'known measurable' service for UMTS cells as that provided by the BCCH Carrier in GSM.

As depicted in Figure 5.22, a 3G mobile device is able to employ the CPICH to calculate a number of measurable parameters [16]:

RSCP (Received Signal Code Power) – Measures the received energy of the CPICH code channel. This value is the 'wanted signal' and is represented as Ec in the Ec/No parameter mentioned below. Measured in dBm.

RSSI – Measures the raw received energy of an unprocessed downlink signal. It can be thought of as the overall 'noise' on the channel, which is made up of general radio interference plus the co-channel interference caused of signals generated by other mobile devices and base stations in local cells. Measured in dBm.

Ec/No (Energy [of wanted signal] over Noise) – This is the primary measurement in UMTS, on which most reselection and handover decisions are based. It is a measure of 'signal to noise ratio' that compares the RSCP received power of the 'wanted' signal (the measured cell's CPICH) to the RSSI power of the overall noise on the channel (which includes interference from neighbouring SFN cells and other noise). Ec/No is a comparative value and is measured in dB, but because the

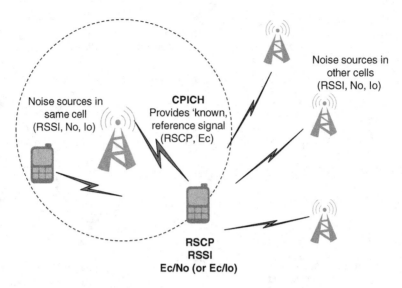

Figure 5.22   Measurement parameters

power of the background noise is generally much higher than the power of the
wanted signal, this value is always negative, for example –6 dB.

Typical ranges for RSCP are around –80 (very strong) to –120 dBm (very weak) and
RSSI ranges from around –75 to –100 dBm.

Typical ranges for Ec/No are from –3 dB (very strong) to –24 dB (very weak).

An Ec/No of –6 dB means that the wanted signal's power is one-quarter of the
strength of the overall noise, whilst an Ec/No of –20 dB means that the wanted signal
level is just 1/100 of the power of the background noise. Signal detection in these
circumstances is only possible because each transmission is coded using a known
scrambling code; meaning that a CDMA receiver already knows the characteristics of
the signal it is searching for. Without the 'known code' element it would be virtually
impossible for a receiver to distinguish a very weak wanted signal from the much
more powerful background noise.

Ec/No is the most useful measurement for UMTS. Knowing only RSCP is less useful and requires a degree of guesswork as, without knowing the RSSI noise level with
which the RSCP is competing it is difficult to accurately predict whether the signal-
noise ratio is usable or not.

Both RSCP and RSSI are measured in dBm, which is a logarithmic (rather than a
linear) value, so although Ec/No is the ratio of the two values (Ec [RSCP] divided by
No [RSSI]) the calculation used is really Ec (RSCP) – No (RSSI). This follows the
'law of powers' discussed previously (e.g. $x^3/x^2 = x^{3-2}$) and is used because the dBm
values that describe RSCP and RSSI are logarithmic values.

For example: if a mobile device measures the RSCP of a cell's CPICH as –100
dBm and measures the power of the RSSI background at –90 dBm, then the resulting

Ec/No value will be calculated as −100 (RSCP) − −90 dBm (RSSI). −100 − −90 = −100 + 90 = −10 dB (Ec/No).

Some forensic radio survey devices, CSurv for example, use an alternative form of Ec/No, known as Ec/Io (energy of the wanted signal over interference) and provides two measurements: Total Ec/Io (TotECIO) is the ratio of the wanted signal against all noise on the channel (from the current cell, from neighbour cells and from other sources), whereas ECIO is the ratio of the wanted signal against just the current cell's noise.

5.2.10   3G UMTS Channel Numbering

Within the spectrum allocated for cellular mobile communications, the radio channels are identified by an ARFCN [17].

UARFCNs (UMTS ARFCNs) are numbered for each 200 kHz step between 0 and 3276.6 MHz (16383 steps means 16383 UARFCNs).

The radio bands that have been defined for UMTS and their UARFCN ranges are numbered as shown in Table 5.4.

Table 5.4   UMTS ARFCN assignments.

Band	Uplink range	Downlink range	DL UARFCN		Band name	Deployment
			Low	High		
I	1920–1980	2110–2170	10 562	10 838	UMTS2100	Global
II	1850–1910	1930–1990	9662	9938	UMTS1900	Americas
III	1710–1785	1805–1880	1162	1513	UMTS1800	—
IV	1710–1755	2110–2155	1537	1738	UMTS1700	Americas
V	824–849	869–894	4357	4458	UMTS850	Global
VI	830–840	875–885	4387	4413	UMTS800	Japan
VII	2500–2570	2620–2690	2237	2563	Used by LTE	Global
VIII	880–915	925–960	2937	3088	UMTS900	Global
IX	1749.9–1784.9	1844.9–1879.9	9237	9387	UMTS1700	Japan
X	1710–1770	2110–2170	3112	3388	UMTS1700	Americas
XI	1427.9–1447.9	1475.9–1495.9	3712	3787	UMTS1500	Japan
XII	699–716	729–746	3842	3903	UMTS700	Americas
XIII	777–787	746–756	4017	4043	UMTS700	Americas
XIV	788–798	758–768	4117	4143	UMTS700	Americas
XIX	830–845	875–890	5112	5413	UMTS800	Japan
XX	832–862	791–821	5762	5913	Used by LTE	Europe
XXI	1447.9–1462.9	1495.9–1510.9	10 562	10 838	Used by LTE	Japan
XXII	3410–3490	3510–3590	9662	9938	Not yet used	—
XXV	1850–1915	1930–1995	1162	1513	Not yet used	—
XXVI	814–849	859–894	1537	1738	Not yet used	—

Based on 3GPP TS 25.104:5.2, reproduced with permission from 3GPP.

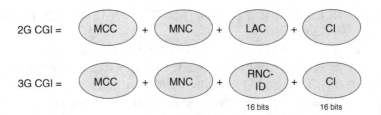

Figure 5.23 Comparison of 2G and 3G CIs

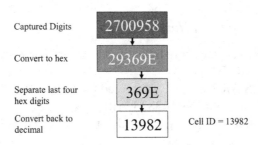

Figure 5.24 Determining 3G CIs

GSM ARFCNs are just index numbers that have been assigned to channel pairs and have no significance beyond that.

UARFCNs are calculated based on the centre frequency of the carrier being described. The formula followed is:

$$UARFCN = 5 \times [Carrier\ Frequency - Band\ Offset]$$

The band offsets align the equation to the specific frequency ranges in each radio band and are listed in 3GPP TS 25.104.

An example of UARFCN calculation is as follows:

UMTS2100 has a band offset of 0, so a 3G downlink channel with a centre frequency of 2124.8 MHz would have a UARFCN of 10 624 as:

$$5 \times [2124.8 - 0] = 10\,624.$$

5.2.11 3G UMTS Cell Identifiers

Ultimately, UMTS cells are uniquely identified by their CI [18].

3G CIs differ slightly in format from those employed by 2G cells in that they incorporate an RNC-ID in place of the LAC carried in the 2G version, as shown in Figure 5.23.

The raw data gathered by some forensic survey devices presents the RNC-ID and CI portion of the CGI as one number in decimal format. The CI part of this number can be gained by following the process outlined in Figure 5.24.

Service Area Codes

UMTS call records (known as CDR or Call Detail Records) technically do not capture the CI of the cells used to carry calls and data connections. They instead capture a value known as the SAC (Service Area Code), which is outlined in Figure 5.25.

SACs were originally created to allow operators to group cells that provide the same service in an area – so, cells that offer voice services at a site or those that serve the same sector could be grouped by SAC.

The UMTS CDR reporting format, which carries details of calls to the billing system, only records details of the used SAC. This means that UMTS call records do not provide details of the *cell* that was used but of the *SAC* to which the used cell belongs. This is not generally an issue as most networks assign SACs to cover individual cells, so the SAC and the CI are the same.

There are some operators, for example O2 UK, who elected to amalgamate co-sector cells into a combined SAC – meaning that use of any of the cells on a sector will be reported in CDRs using the same SAC. In these cases it is not possible to tell which particular cell carried a call, only that one of the cells within the SAC did so. This is further complicated if operators assign SACs using a different numbering scheme than that used to assign CIs, in which case it will not be possible to extrapolate details of the used CI from a knowledge of the SAC that the used cell belongs to.

This causes problems for forensic radio surveys, as the SAC is not broadcast on a cell's BCCH, which only shows the CI. If the SAC captured in call records does not relate to a cell's CI then it becomes difficult to correlate the information provided in CDRs to cell coverage details captured in radio surveys. In these circumstances, cell site analysts are often required to apply to the network operator in question to obtain a list of the CIs associated with each SAC, but even in these situations it will not be possible to determine exactly which cell was used to carry each individual call.

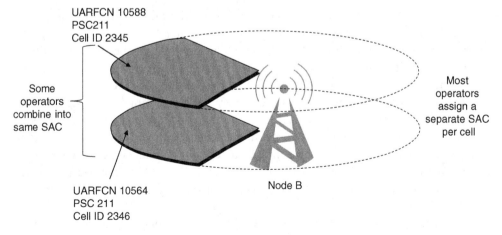

Figure 5.25 Service Area Codes

Physical Layer Identities

A mobile device can only collect the CI of a measured cell if it waits for the required information to cycle across the cell's BCCH. Depending upon where the cell is in its BCCH cycle when the mobile device first begins searching, it could take several tens of milliseconds for the CI to be captured. This could lead to unacceptable delays when taking neighbour signal strength measurements and during handovers, so UMTS provides a set of physical layer identifiers that are quicker and simpler for mobile devices to detect.

UMTS mobile devices can discriminate between cells locally at the physical layer by using a combination of their UARFCN and the PSC physical layer ID. If the network has been properly planned there should only be one cell per area using a given combination of UARFCN and PSC.

If the network has not been properly planned or if unexpected propagation conditions cause cells to extend beyond their planned size, then there is a possibility that a mobile device could detect multiple cells in an area that share a UARFCN/PSC pair. In these cases, users of the network could experience patchy connectivity and poor call quality and forensic surveys could have difficulty taking measurements of each individual cell that is sharing the same settings.

5.2.12   3G UMTS Cell Configurations

It is possible to increase the capacity of a UMTS site by deploying additional cells to it, either by adding extra sectors (and consequently, by reducing the coverage arc of any existing sectors to accommodate the new ones) or by adding extra radio carriers to existing sectors.

Given the SFN techniques that are usually employed in UMTS networks, the addition of extra carriers to existing sites is generally undertaken as part of the wider rollout of an additional frequency layer across all or part of the network.

The constraints imposed by the need to manage scrambling code reuse only apply to neighbouring cells that are on the same carrier (e.g. neighbouring cells on the same carrier cannot use the same scrambling code); they do not apply between cells on different carriers (e.g. neighbouring cells on different carriers are permitted to use the same scrambling code). This also applies to overlapping or 'stacked' cells on the same sector that are deployed on different carriers. It is common (although not universal) practice therefore to replicate the scrambling code usage pattern employed on the 'primary' carrier layer on any subsequent carriers deployed on the same sector.

If, for example, the original cell deployed to a sector operated on carrier 10564 and used scrambling code 211, a second cell deployed on carrier 10588 on the same sector could also use scrambling code 211 without causing any interference. Figure 5.25 illustrates these concepts.

UMTS network operators often employ additional carriers as a result of deploying HSPA services. In these scenarios they will reserve one carrier for voice/SMS services and a different carrier for HSPA.

It is possible to dynamically alter the use to which a carrier is put, so a two-carrier cell (that might usually have a 'voice' channel and an 'HSPA' channel) that is experiencing high demand for voice services but low demand for data could decide to use some of its 'HSPA' channel capacity to carry voice calls instead and vice versa.

5.2.13   3G UMTS Cell Discrimination

UMTS networks employ single frequency cell deployment techniques – this means that neighbouring cells and sectors all use the same radio channels and cannot avoid causing each other interference. Cells in a region of the network must all therefore use different downlink PSCs to ensure that their signals are separately distinguishable.

Figure 5.26 provides an example of a typical UMTS channel deployment.

All cells in this example use UMTS2100 UARFCN 10661.

Cell A1 uses PSC 122.

Cells A2 and A3, deployed to the same base station on carrier 10661 cannot also use PSC 122, to avoid co-code interference issues. Cell A2 uses PSC 123 and Cell A3 uses PSC 124.

Cells belonging to neighbouring base stations B, C and D also use UARFCN 10611 but all use different downlink PSCs.

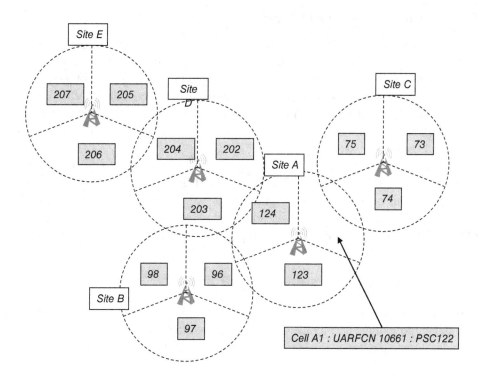

Figure 5.26   UMTS cell deployment

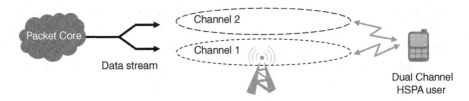

Figure 5.27   Multi-carrier HSPA

As there are 512 different PSC sequences available for the downlink there can be large geographical separations between base stations that use the same PSC. Coupled with the fact that in a UMTS SFN all but the strongest neighbour cell signals are overwhelmed by the local dominant active cell, this means that the chances of co-PSC interference being experienced by a mobile device are very low indeed.

In the unlikely event that a PSC has been reused on a frequency layer with too little separation between the base stations, a mobile device or survey device may find it impossible to distinguish between the transmissions of the two cells, especially if the two cells' signals are received with roughly equivalent signal strength at the mobile device. In this scenario the survey device will typically fail to capture CI details for either cell, and will merely record that the UARFCN and PSC were observed to be transmitting.

5.2.14   Multi-carrier UMTS Services

In the original iteration of UMTS, a mobile device would connect to a cell (or to some cells, if SHO was in progress) on only one single frequency layer at a time.

The evolved HSPA+ variant of UMTS introduced the ability for mobile devices to be allocated and use resources on more than one carrier concurrently. This 'multi-carrier' operation is designed to increase the potential data rates available to users, by allowing traffic to flow to their devices via more than one cell at a time. This is illustrated in Figure 5.27.

The first version of this scheme allowed mobile devices to access cells on up to two carriers at the same time and was known as DC-HSPA (Dual Carrier HSPA). Further advances in this concept, generally grouped as MC-HSPA (Multiple Carrier HSPA), have allowed four carrier, six carrier and eight carrier options to be developed, each of which introduces a further increase in potential user data rates.

5.3   4G LTE Networks

3GPP developed a successor to 3G UMTS known as LTE. The acronym reflects the original purpose of the development project that was started within 3GPP and was tasked with developing the 'long term evolution' of the 3GPP air interface, hence the name LTE

(Long Term Evolution). A parallel development project had the aim of developing the evolved 4G core network and was known as SAE (System Architecture Evolution).

In reality, this 4G system should more accurately be called EPS (Evolved Packet System), but the original development project name, LTE (or sometimes LTE/SAE), stuck and is more commonly used to describe this network type [19].

LTE shares many administrative and architectural elements with its predecessors, GSM and UMTS, and this allows operators to optionally run 'multi-RAT' 2G, 3G and 4G services in parallel. This in turn offers the opportunity to provide 'inter-system' services that allow mobile devices to roam across all three types of access network, depending upon which provides the best service at each location.

Although initially developed as one 4G solution amongst several, LTE has become the dominant global 4G technology, as the UWB (Ultra Wide Band) system that was planned as the upgrade to CDMA2000 was abandoned and as WIMAX operators began to switch over to LTE.

5.3.1   4G LTE Network Architecture

Like most cellular systems, an EPS is divided into separate RAN and core network environments [20].

The LTE radio access network is known as the EUTRAN (Evolved Universal Terrestrial Radio Access Network) and hosts the base stations and the transmission equipment that connects users with core network services. The core network is known as the EPC (Evolved Packet Core), it hosts the 'user plane' elements that convey user traffic across the network and the 'control plane' elements that provide session management, security and other administrative features.

EUTRAN

The main element in the EUTRAN, as shown in Figure 5.28, is the LTE base station, known as the eNB (EUTRAN Node B) [21].

The EUTRAN design does not incorporate an access network controller, such as the BSC or RNC found in GSM and UMTS access networks respectively. The access control functions of those devices have instead been devolved down to the individual eNBs to perform. The only other element found in the EUTRAN is the 'backhaul' network that carries traffic between the base stations and the core network.

The EUTRAN is organised into administrative areas known as Tracking Areas (TAs), which are detailed in Figure 5.29 and can be as small as just one cell or can be as large an operator's entire EUTRAN. A typical TA consists of cells belonging to a small local 'cluster' of base stations and might cover an area the size of a small town.

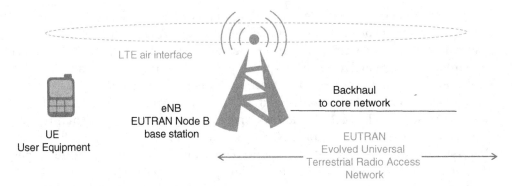

Figure 5.28   Evolved Universal Terrestrial Radio Access Network

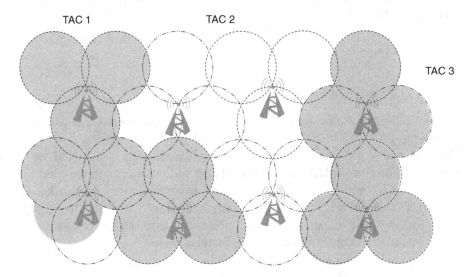

Figure 5.29   Evolved Universal Terrestrial Radio Access Network Tracking Areas

5.3.2   4G LTE Radio Interface

The LTE radio interface employs a multiple access technology known as OFDMA (Orthogonal Frequency Division Multiple Access), which is used on the downlink. An adapted version of this technology, known as SC-FDMA (Single Carrier Frequency Division Multiple Access) is used on the LTE uplink [21].

Legacy radio systems, such as those employed by GSM and UMTS, work on 'single carrier' principles, meaning that each downlink or uplink signal consists of a single stream of data transmitted across the channel's single 'carrier signal'.

OFDMA, in contrast, works by transmitting air interface traffic across multiple parallel low data-rate radio connections known as 'subcarriers'. Each wideband OFDMA radio channel consists of dozens or even hundreds of these parallel connections, each carrying a small part of the overall stream of traffic being conveyed by the radio channel. Basic OFDMA characteristics are outlined in Figure 5.30.

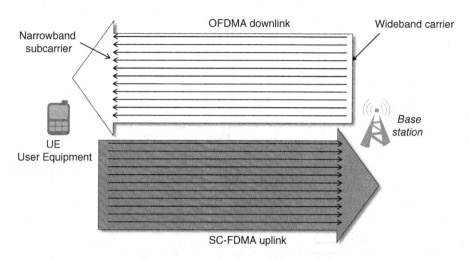

Figure 5.30 Orthogonal Frequency Division Multiple Access

A simple analogy might serve to explain these concepts: a traditional single carrier system can be thought of as being like a serial cable, over which the data to be transmitted is sent one bit at a time in series. The multi-carrier OFDMA system, on the other hand, is more like a ribbon cable, where data is transmitted in parallel streams.

This parallel, multi-carrier transmission technique has a number of benefits in terms of overall data throughput, mostly because the multiple carriers can carry a large aggregated traffic flow. Benefits are also gained in terms of resilience and quality, as each individual subcarrier can carry data at a comparatively low bit rate with a resulting long 'symbol period', thus giving receivers a long period over which to measure each bit of received data and make a decision on whether it should be interpreted as 1 or 0.

SC-FDMA, employed on the LTE uplink, can be thought of as a more power-efficient version of OFDMA that is more suited to the restricted-power environment of a battery-powered device like a mobile phone.

Unlike 2G GSM and 3G UMTS, which use fixed-width radio carriers (200 kHz wide for GSM, 5 MHz wide for UMTS), LTE has been designed to use a set of possible channel bandwidths, which range from just 1.4 MHz wide up to 20 MHz wide (the six supported bandwidth options are 1.4, 3, 5, 10, 15 and 20 MHz).

On both the downlink and the uplink, sets of adjacent subcarriers are grouped into units known as Resource Block s (RB), which are the basic unit of capacity allocation. There are 12 subcarriers per RB and the number of RBs available in a cell is determined by the size of the radio carrier used in that cell. A 1.4 MHz wide carrier supports six RBs (72 subcarriers), whereas a 20 MHz carrier supports 100 RBs (1200 subcarriers).

The minimum unit of capacity allocation for a user is 1 RB for one allocation period (1 ms).

Channel bandwidth and RB concepts are illustrated in Figure 5.31.

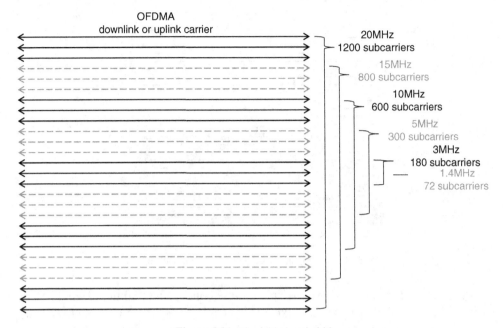

Figure 5.31 Scalable bandwidth

5.3.3 *4G LTE Cell Selection and Reselection*

LTE mobile devices employ the S algorithm to perform initial cell selection. This is very similar to the algorithm employed by UMTS but uses different measured values. It is detailed in 3GPP Specification 36.304 [22].

The S algorithm essentially instructs the mobile device to select the cell that offers the highest quality, although quality is actually determined by signal strength.

The S criteria are as follows:

$$S_{qual} = Q_{qualmeas} - \left(Q_{qualmin} + Q_{qualminoffset} \right)$$

$$S_{rxlev} - Q_{rxlevmeas} - \left(Q_{rxlevmin} + Q_{rxlevminoffset} \right) - P_{compensation}.$$

The requirements of the S algorithm are deemed to be met – that is a cell can be selected – if:

$$S_{qual} > 0 \text{ and } S_{rxlev} > 0.$$

The procedure states that the mobile device will search for the strongest cell and will then select it, which implies that if signals from more than one cell are detected during the process the cell with the highest value of S_{qual}/S_{rxlev} will be selected.

The parameters employed by the S process are as follows:

S_{qual} – cell selection quality value. Calculated by comparing a target cell's readings with a broadcast minimum value (and any offsets)

S_{rxlev} – cell selection RXLev value. Calculated by comparing a target cell's readings with a broadcast minimum value, any offsets and a value related to maximum permitted and possible transmit power capabilities.

$Q_{qualmeas}$ – measured quality value for a target cell. Gained from measuring the cell's RSRQ (Reference Signal Received Quality in 4G).

$Q_{rxlevmeas}$ – measured RXLev value for a target cell. Gained from measuring the RSCP of the cell's Reference Signals and expressed in dBm.

$Q_{qualmin}$ – minimum acceptable received quality level in target cell, advertised on the cell's BCCH.

$Q_{rxlevmin}$ – minimum acceptable received signal strength, expressed in dBm and advertised on the cell's BCCH.

$Q_{qualminoffset}$ and $Q_{rxlevminoffset}$ – are offsets that can be applied based on any network or channel priority instructions that have been given to the mobile device, for example a higher priority cell will have a positive offset applied to it to make it more likely to be selected.

$P_{compensation}$ – takes the maximum permitted transmit power on the target cell's uplink RACH (a BCCH value known as UE_TXPWR_MAX_RACH) and subtracts the mobile device's maximum physical transmit power level (P_MAX). The resultant value (or 0, whichever is stronger) is then subtracted from the other S_{rxlev} components.

Although expressed in fairly complex terms, the S criterion essentially works in the same way as the GSM C1 algorithm and allows the mobile device to select the local cell that provides the best RSRQ and RSRP values.

The selected cell is termed the serving cell.

Cell Reselection

Cell reselection analysis is performed according to the R criteria, which is again similar to the process performed in UMTS but uses different measured values.

The reselection process is only triggered if the measured signal provided by the current Active cell drops below a broadcast minimum value, known as Sx (which is based on the current S_{qual} value).

In normal cells, the R criteria are as follows:

$$R_s = Q_{meas,s} + Q_{hyst}$$

$$R_n = Q_{meas,n} - Q_{offset.}$$

The $Q_{meas,s}$ and $Q_{eas,n}$ parameters are related to the measured quality of the current serving cell and a candidate neighbour cell respectively and are based on the current S_{rxlev} values for those cells, whilst Q_{hyst} and Q_{offset} are offset parameters that can be broadcast in a cell to temporarily or permanently alter the perceived attractiveness of that cell or its neighbours as reselection candidates.

There is no direct equivalent of the GSM CELL_RESELECT_HYSTERESIS parameter that is used to control inter-LAC handovers. Where required such functions can be achieved by applying Q_{hyst} and Q_{offset} values to cells along TAC (Tracking Area Code) boundaries.

LTE supports a similar Idle Mode network and frequency prioritisation scheme as those used in 2G and 3G networks, so it is possible that forensic surveyors will have difficulty securing measurements of some channels without using 'cell lock' or 'channel lock' techniques.

5.3.4   4G LTE Handovers

4G networks can be deployed in single frequency or multi-frequency configurations but are typically deployed, like 3G networks, as a set of SFN layers each built on a different radio carrier. Handovers can be supported between cells on the same frequency layer (intra-carrier) or on different frequency layers (inter-carrier).

Additionally, it is possible to perform handovers between 4G cells (intra-LTE handovers) and from 4G cells to 'legacy' 2G or 3G cells – including CDMA2000 cells – and back again (inter-system handovers).

LTE must therefore employ a range of handover techniques, which are detailed in Figure 5.32, and include [23]:

- Intra-carrier, intra-LTE handover
- Inter-carrier, intra-LTE handover
- Inter-carrier, inter-system handover.

Intra-carrier, intra-LTE handovers use techniques that are similar in effect to the intra-carrier handover mechanism used in CDMA based networks, but which are very different in practice.

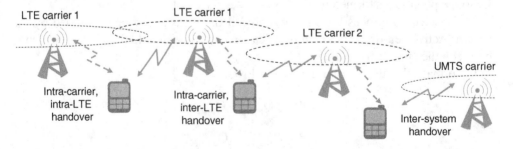

Figure 5.32   4G handovers

As LTE does not use code division, the swap from one intra-carrier cell to another is effected by moving the phone from a RB assigned by the source base station to a RB assigned by the target base station. In an LTE SFN, neighbouring base stations are configured to use different sets of RB on a shared channel in cell boundary areas – for example, on a cell boundary between Base Station A and Base Station B, where both nodes share Carrier 1, the first base station may be configured to use RB from the top half of the channel, while its neighbour uses the RB in the lower half. Interference between users of the two cells is therefore avoided by directing them to different parts of the shared radio carrier. Away from cell boundary areas, both base stations are free to use the entire set of RB on the shared carrier.

The lack of code division in LTE also means that there is no necessity for LTE to support the 'two way' or 'three way' aspects of SOH used in 3G; an LTE phone will be connected to only one cell at a time and, therefore, in reality, LTE always uses HHO techniques.

Inter-carrier, inter-LTE handover is always managed using HHO techniques – a mobile device will be instructed to release the RB allocated to it on one radio carrier and will be directed to a new set of RB on a different radio carrier.

Inter-system handover also uses HHO techniques and involves not only a change in radio carrier (as different network types are typically allocated different areas of spectrum) but also a change of radio interface technology, from the OFDMA techniques used for LTE to the TDMA or CDMA techniques used in legacy networks.

5.3.5 4G LTE Cell Configurations

3GPP has defined a large number of possible frequency bands in which LTE can be deployed.

Most countries and regions have adopted several LTE bands for use by operators, and most operators will have acquired licences that cover several radio bands.

Whether an operator elects to deploy a multi-frequency network or a multi-carrier SFN, they are likely to use resources spread across two or more radio bands, as illustrated in Figure 5.33.

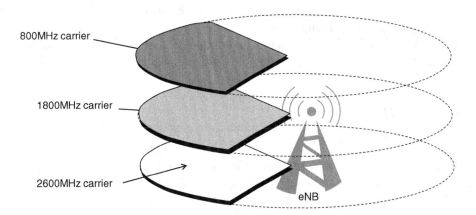

Figure 5.33   Multi-band operation

In the United Kingdom, for example, LTE operators have acquired licences that cover the 800 (Band 20), 1800 (Band 3) and 2600 MHz (Band 7) FDD radio bands, with some operators also gaining spectrum in the 2600 MHz TDD band (Band 38).

These operators have deployed networks that combine channels from different bands in each base station. So a given base station may generate an 800, 1800 and 2600 MHz cell on each sector to increase the overall amount of capacity available from that site.

In this scenario, each carrier is typically deployed as a SFN layer and will be shared with all neighbouring sites.

5.3.6   4G LTE Radio Bands and Channel Numbering

3GPP has designed LTE to fit in with the radio channel allocation regulations in force in different parts of the world [24].

Each national and regional government is responsible for determining the set of radio applications that will be permitted within their jurisdiction and for defining the spectrum to be used by each service.

Although a great deal of work has been done to harmonise spectrum allocation, the radio bands assigned for use by cellular mobile services can differ markedly around the world.

3GPP has therefore been forced to define multiple different radio bands to be used by LTE in various regions. As LTE supports both an FDD option and a TDD (Time Division Duplex) option it has also been necessary to separately define sets of FDD and TDD resource.

The full set of radio bands currently defined for LTE is shown in Table 5.5.

Individual LTE radio channels are identified by a channel number; the EARFCN (EUTRAN ARFCN).

Each EARFCN identifies a different frequency in a frequency band. The frequencies are placed at regular intervals of 100 kHz across each radio band. Operators are free (within the limits of their allocated spectrum) to place their channels wherever they think they will be most effective. The centre frequency of each channel must align with one of the EARFCN frequencies. The EARFCN therefore identifies the centre frequency of an LTE uplink or downlink channel.

Like UMTS UARFCNs, LTE channel numbers are assigned based on a calculation that takes the channel centre frequency plus a band-specific offset into account.

5.3.7   4G LTE Cell Identifiers

Physical Layer Cell Identities

LTE cells are provided with a simple method of identifying themselves at the physical layer, using a PCI. Each cell in a given geographical area should be assigned a different PCI, providing mobile devices with a simple way of differentiating between them

Table 5.5 3GPP defined operating bands for LTE (as of Release 12, Spring 2014) and typical deployment regions.

Band	Uplink range (MHz)	Downlink range (MHz)	Duplex mode	Naming and deployment
1	1920–1980	2110–2170	FDD	LTE2100 (global)
2	1850–1910	1930–1990	FDD	LTE1900 (Americas)
3	1710–1785	1805–1880	FDD	LTE1800 (global)
4	1710–1755	2110–2155	FDD	LTE1700/2100 (Americas)
5	824–849	869–894	FDD	LTE850 (Americas, Africa and Asia Pacific)
6	830–840	875–885	FDD	LTE850 (Japan)
7	2500–2570	2620–2690	FDD	LTE2600 (global, not United States)
8	880–915	925–960	FDD	LTE900 (global, not United States)
9	1749.9–1784.9	1844.9–1879.9	FDD	LTE1700 (Japan)
10	1710–1770	2110–2170	FDD	LTE1700 (Americas and Asia Pacific)
11	1427.9–1447.9	1475.9–1495.9	FDD	LTE1500 (Japan)
12	699–716	729–746	FDD	LTE700 (Americas and Asia Pacific)
13	777–787	746–756	FDD	LTE700 (Americas)
14	788–798	758–768	FDD	LTE700 (Americas)
15	Reserved		—	—
16	Reserved		—	—
17	704–716	734–746	FDD	LTE700 (Americas)
18	815–830	860–875	FDD	LTE850 (global, not Europe)
19	830–845	875–890	FDD	LTE850 (global, not Europe)
20	832–862	791–821	FDD	LTE800 (Eastern Europe, Africa and Middle East)
21	1447.9–1462.9	1495.9–1510.9	FDD	LTE1500 (Japan)
22	3410–3490	3510–3590	FDD	LTE3500 (global)
23	2000–2020	2180–2200	FDD	LTE2000 (global S-band)
24	1626.5–1660.5	1525–1559	FDD	LTE1600 (global L-band)
25	1850–1915	1930–1995	FDD	LTE1900 (Americas)
26	814–849	859–894	FDD	LTE850 (Americas)
27	807–824	852–869	FDD	LTE850 (Americas)
28	703–748	758–803	FDD	LTE700 (Asia Pacific)
29	No uplink	717–728	FDD	LTE700 (Americas)
30	2305–2315	2350–2360	FDD	LTE2300 (global)
31	452.5–457.5	462.5–467.5	FDD	LTE450 (Americas, Eastern Europe and Asia)
32	Not used		—	—
33	1900–1920		TDD	LTE2100 TDD (global)
34	2010–2025		TDD	LTE2100 TDD (global)
35	1850–1910		TDD	LTE1900 TDD (Americas)
36	1930–1990		TDD	LTE1900 TDD (Americas)
37	1910–1930		TDD	LTE1900 TDD (Americas)
38	2570–2620		TDD	LTE2600 TDD (global)
39	1880–1920		TDD	LTE1900 TDD (China)
40	2300–2400		TDD	LTE 2300 TDD (global)
41	2496–2690		TDD	LTE2500 TDD (Americas)
42	3400–3600		TDD	LTE3500 TDD (global)
43	3600–3800		TDD	LTE3700 TDD (Americas)
44	703–803		TDD	LTE700 TDD (Asia Pacific)

Based on 3GPP TS 36.104:5.2, used with permission from 3GPP.

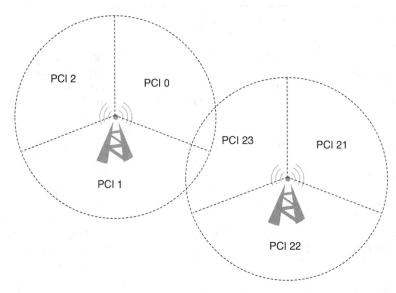

Figure 5.34   Physical cell IDs

without the need to access and decode each cell's BCCH (which carry the cell's full ECGI – EUTRAN Cell Global Identifier).

The PCI is divided into two parts, as shown in Figure 5.34: the Physical Layer Group ID, of which there are 168; the Physical Cell ID, of which there are three per group [25].

Typically, Base stations that have been configured following the traditional three-sector model will be assigned a PCI Group and each of the three cells will use a different PCI from within that group. Base stations that have more than three sectors will need to be assigned more than one PCI Group.

The PCI is carried by each cell's PSS (Primary Synchronisation Signal) and the PCI Group ID is carried by the cell's SSS (Secondary Synchronisation Signal).

As with legacy PCI – the BSIC used in 2G GSM and the PSC used in 3G UMTS – PCIs should be unique within a limited geographical area, especially in LTE networks that use an SFN configuration. There are 504 different PCIs available (168 groups × 3 cells per group), so it should be possible for network operators to ensure that PCIs are not reused within the same area.

Cell Identities

The PCI is designed to allow cells to be discriminated locally at the physical layer. Network-wide or global cell discrimination is provided by the ECGI (EUTRAN Cell Global Identifier) [26].

Each LTE cell has an ECGI, which is globally unique and can therefore be used to identify a mobile device's location down to the cell level in any network in the world.

Like 2G and 3G CGI formats, the 4G ECGI starts with the host network's PLMN (Public Land Mobile Network) ID, consisting of MCC and MNC. The remainder of the ECGI consists of an eNB ID (unique within the PLMN) and a CI (unique within the eNB).

An example of an LTE ECGI is as follows: 2349003250690

- MCC = 234 (United Kingdom)
- MNC = 90 (unassigned MNC, used as an example)
- eNB + CI = 03250690.

The eNB + CI portion of the ECGI is in reality a 28-bit number, the interpretation of which, according to 3GPP, is that the first (most significant) 20 bits are the eNB ID and the last (least significant) eight bits are the CI.

It should be noted that the 20 + 8 interpretation of the eNB + CI component is only the recommended partitioning method, network operators are free to partition the 28-bit field in any way they wish. Most forensic radio survey devices therefore just present a decimal version of the entire 28-bit number, the range of which is from 0 to 268435455.

To recover the individual eNB and CI components (assuming the 20 + 8 split) it is necessary to convert the decimal number to hexadecimal, then separately to convert the first five digits (including any leading zeroes) and the last two digits back to decimal, as shown:

03250690 (decimal) = 0319a02 (hex)
eNB ID = 0319a (hex) = 12698 (decimal)
CI = 02 (hex) = 02 (decimal).

So, ECGI 2349003250690 decodes as: United Kingdom, <network>, eNB 12698, CI 2.

5.3.8   4G LTE Cell Discrimination

An LTE cell is identified by a combination of EARFCN and PCI.

LTE networks can employ single frequency cell deployment techniques, which means that neighbouring cells and sectors all use the same radio channels and cannot avoid causing each other interference. Cells on the same frequency layer in a region of the network should all therefore use different downlink PCIs to ensure that their signals are separately distinguishable.

Figure 5.35 provides an example of a typical LTE channel deployment.

All cells in this example use LTE1800 EARFCN 1617.

Cell A1 uses PCI 30.

Cells A2 and A3, deployed to the same base station cannot also use PCI 30, to avoid PCI interference issues. Cell A2 uses PCI 31 and Cell A3 uses PCI 32.

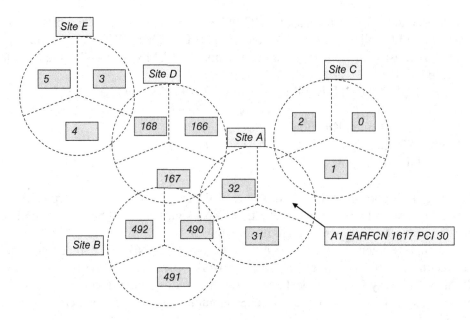

Figure 5.35   LTE cell deployment

Cells belong to neighbouring base stations B, C, D and E also use EARFCN 1617 but all use different PCIs.

As there are 504 different PCI arranged within 168 PCI Groups there can be large geographical separations between base stations that use the same PCI Group. Coupled with the fact that in an LTE SFN all but the strongest neighbour cell signals are often overwhelmed by the local dominant serving cell, this means that the chances of co-PCI interference being experienced by a mobile device in a properly planned network should be reasonably low.

In the unlikely event that a PCI has been reused on the same frequency layer with too little separation between the base stations, a mobile device or survey device may find it impossible to distinguish between the transmissions of the two cells, especially if the two cells' signals are received with roughly equivalent signal strength at the mobile device. In this scenario the survey device will usually fail to capture CI details for either cell and will merely record that the EARFCN and PCI were observed to be transmitting.

5.3.9   4G LTE Radio Measurements

Measurements of the LTE air interface are taken from radio channel elements known as RSs. RSs are transmissions of a known type and power that are embedded in both downlink and uplink RBs and provide a 'known quantity' against which measurements can be taken.

Each Connected Mode mobile device will transmit measurement reports back to its serving eNB base station at regular intervals, indicating the current quality of the downlink service the mobile device is experiencing. The base station uses the feedback reports to decide whether and when to adjust the parameters of the downlink and uplink service provided to each mobile device. For example, in situations where a mobile device reports good quality, the base station might decide to reduce the amount of error protection that is employed on the radio link for that device, which has the effect of increasing the data rate of the service provided to that mobile device.

There are three main measurement types employed by LTE [27]:

RSRP (Reference Signal Received Power) – this is the average amount of power (in dBm) measured across the RS received during one measurement period. RSRP is a measurement of the wanted 'signal' of a specific transmission on the air interface.

RSSI (Received Signal Strength Indicator) – this is a measurement of the total radio power (in dBm) received across all subcarriers (not just those carrying RS) during a measurement period. RSSI is a measurement of unwanted 'noise' in the cell and is made up of interference, thermal noise in the receiver, co- and adjacent-channel interference leaking in from other cells and other sources.

RSRQ (Reference Signal Received Quality) – this is the comparative ratio of RSRP to RSSI (expressed in dB) and provides an indication of the strength of the wanted 'signal' against that of the unwanted 'noise'. RSRQ is the 'signal to noise ratio' of the measured channel during a single OFDMA Symbol period.

Typical Values

Typical ranges for RSRP are from around –80 dBm (very strong) to around –140 dBm (very weak).

Typical ranges for RSSI are from around –60 dBm (very strong) to around –110 dBm (very weak).

Typical ranges for RSRQ are from –3 dB (very strong) to –30 dB (very weak).

RSRQ is often considered to be the most useful measurement for LTE, as it offers details of the signal to noise ratio. RSRP is also quite commonly used but can be of less use as, without knowing the RSSI noise level it is difficult to tell whether the signal to noise ratio makes the cell usable or not.

In practice, decisions related to air interface quality and performance are made in relation to measured or reported SINR (Signal to Interference Noise Ratio), but these values are not typically captured or reported by forensic survey devices. This is because the measurements captured by survey devices usually relate to an instant in time – one symbol period or subframe, for example – and as SINR is a long-term average these values are outside of the scope that the survey device is designed to monitor.

5.3.10   4G LTE Variants – TDD and FDD

LTE was designed to be as widely applicable as possible and therefore has variants that support the two main cellular radio configuration options: FDD and TDD.

LTE-FDD operates using paired radio carriers in each cell, although, unlike in legacy FDD system such as GSM or cdmaOne, uplink and carriers do not need to be symmetrical (allowing, for example, a 20 MHz downlink to be paired with a 5 MHz uplink).

3GPP has defined over 30 different potential allocations of spectrum for FDD services (see Table 5.5) and networks based on this variant are widely deployed.

LTE-TDD (or TD-LTE as it is also known) operates using a single unpaired channel per cell, which is dynamically split using time division principles between uplink and downlink transmission periods.

UMTS had two TDD variants which were widely deployed in China and other Asia-Pacific areas (UMTS-TDD and TD-SCDMA), but were less widely adopted elsewhere. TD-LTE was designed as the 4G evolution for existing UMTS TDD operators and has also been widely deployed in the regions that previously favoured TDD technologies.

TD-LTE has also found a market in regions that were previously uninterested in TDD technologies, like Europe and North America. This is partly a reflection of the need for operators to meet the demand for LTE capacity in any way that they can, but is also related to the growing acceptance of TDD-based technologies as being an inherently more economical way of utilising cellular spectrum resources.

5.3.11   4G LTE Voice Call Options

4G LTE networks were designed to carry data services only; there is no option to carry 'circuit switched' services, like traditional voice telephony connections, over LTE. This is not to say that voice is not supported by LTE, however, it is just that the provision of this most basic of mobile services is more complicated than in older network types.

Voice calling can be supported in one of two ways in LTE:

1. Calls can be carried using Internet telephony type techniques via the LTE network's IP-based data connections using a service known as VoLTE (Voice over LTE).
2. Operators who have access to legacy 2G or 3G networks can elect to instruct LTE mobile devices to make calls via legacy network facilities in a service known as CSFB (Circuit Switched Fallback).

The two methods, VoLTE and CSFB, are not mutually exclusive; networks can opt to support both, as illustrated in Figure 5.36.

Voice over LTE

With VoLTE services, the 4G phone stays connected to the LTE network and transmits voice traffic using VoIP (Voice over IP) techniques. The network will capture CDR call records for the call and will list CIs of 4G cells for the calls. VoLTE calls

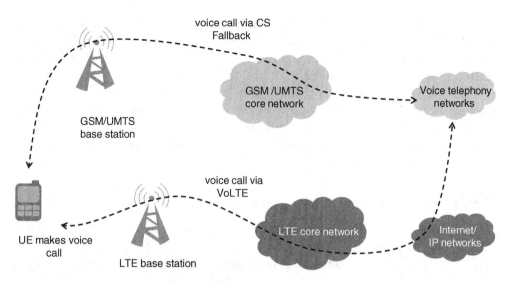

Figure 5.36 LTE voice options

can be made via 3G HSPA networks and can be handed over between 3G and 4G cells but cannot be carried by 2G networks. If a CDR shows a 4G CI for a call, that call must have been carried by VoLTE.

Although LTE networks began to be deployed in 2010, handset and network equipment vendors struggled to get VoLTE working reliably and it was not until 2014 that voice over IP services began to be offered in any volume. The majority of LTE networks initially offered only CSFB to support voice and most networks continue to use this as their primary 4G voice option.

The interoperability of VoLTE is defined by the GSMA in their IR.92 specification [28].

Circuit Switched Fallback

If CSFB is used, when a 4G-connected phone requires a voice connection it is instructed to 'fall back' to a legacy 2G or 3G cell and make the call from there. The fallback is managed as a reselection to a 2G or 3G cell if the mobile device was in Idle Mode at the time or is managed as a reselection or an inter-system handover if the device was in Connected Mode. Calls made using CSFB will have 2G or 3G CIs listed in the CDR [29].

SMS text messaging can be carried by 4G cells, even without VoLTE being deployed, and do not require a fallback to take place, so text events may be listed in CDRs with 4G CIs.

In RF survey terms, the potential use of CSFB means that it might still be necessary to take surveys of 2G and 3G cells in relation to a case even if the target phone in question was an LTE phone.

The decision about which resources to survey will typically be based on the content of the individual CDRs – if a 4G CDR contains 2G or 3G CIs, surveys of those access network types will also be required. If the CDR shows only 4G CIs then, potentially, only 4G surveys may be required.

Until VoLTE is more widely rolled out (and more importantly until it is supported on forensic radio survey equipment), any test calls made as part of an LTE forensic radio survey will be likely to use CSFB and will connect via 2G or, more likely, 3G cells. The use of 2G or 3G resources should be clearly signalled in the captured survey data, so that analysts are aware that non-4G cells were used.

5.3.12   LTE Evolution – LTE -Advanced and Beyond

3GPP has released specifications for the next stage beyond LTE, which is known as LTE-Advanced (LTE-A).

LTE-A uses much the same set of principles of standard LTE but adds a concept known as 'carrier aggregation'. This is similar to the multi-carrier services offered by more advanced HSPA + networks and allows a mobile device to connect to and exchange traffic over multiple cells concurrently.

LTE-A, in theory, allows a mobile device to be scheduled traffic capacity on up to five concurrent cells, which boosts the data rates that are potentially available to network users quite considerably.

Signalling and control activities all take place via one cell, which is known as the PCell (Primary Cell), additional traffic connections can be established via up to four SCells (Secondary Cells). The PCell and all SCells in use at any one time by a mobile device must belong to the same eNB base station.

LTE-A can aggregate carriers that belong to the same radio band (known as intra-band aggregation) or can aggregate carriers that belong to two or even three different radio bands. For example, a mobile device that supports carrier aggregation could find itself concurrently exchanging traffic via an LTE800 cell, an LTE1800 cell and an LTE2600 cell, with all cells and connections controlled by the same eNB.

Plans for further evolutions of LTE are laid out in 3GPP specifications and early work has also begun on 5G (Fifth Generation) mobile cellular services. 5G, in whatever form it eventually takes, is not likely to be deployed commercially before 2020.

References

[1] European Telecoms Standards Institute (2014) *Mobile Technologies GSM*, http://www.etsi.org/index.php/technologies-clusters/technologies/mobile/gsm (accessed 1 June 2014).
[2] Third Generation Partnership Project (2014) *GPRS and EDGE*, http://www.3gpp.org/technologies/keywords-acronyms/102-gprs-edge (accessed 1 June 2014).
[3] 3GPP Technical Specification (2013) *Network Architecture*, TS 23.002 v11.6.0 Section 4.2, www.3gpp.org (accessed 26 July 2014).
[4] 3GPP Technical Specification (2013) *Physical Layer on the Radio Path: General Description*, TS 45.001 v11.0.0, www.3gpp.org (accessed 26 July 2014).

[1] 3GPP Technical Specification (2013) *Functions related to Mobile Station (MS) in Idle Mode and Group Receive Mode*, TS 43.022 v11.0.0, www.3gpp.org (accessed 26 July 2014).

[2] 3GPP Technical Specification (2013) *Radio Subsystem Link Control*, TS 45.008 v11.5.0 Section 6.4, www.3gpp.org (accessed 26 July 2014).

[3] 3GPP Technical Specification (2013) *Radio Transmission and Reception*, TS 45.005 v11.3.0 Section 2, www.3gpp.org (accessed 26 July 2014).

[4] 3GPP Technical Specification (2013) *Numbering, Addressing and Identification*, TS 23.003 v11.6.0 Section 4.3, www.3gpp.org (accessed 26 July 2014).

[5] 3GPP Technical Specification (2013) *Radio Subsystem Link Control*, TS 45.008 v11.5.0 Section 7.2, www.3gpp.org (accessed 26 July 2014).

[6] 3GPP Technical Specification (2013) *Radio Subsystem Link Control*, TS 45.008 v11.5.0 Section 8.1, www.3gpp.org (accessed 26 July 2014).

[7] 3GPP Technical Specification (2013) *Radio Subsystem Link Control*, TS 45.008 v11.5.0 Section 8.2, www.3gpp.org (accessed 26 July 2014).

[8] Third Generation Partnership Project (2014) UMTS, http://www.3gpp.org/technologies/keywords-acronyms/103-umts (accessed 1 June 2014).

[9] 3GPP Technical Specification (2013) *UTRAN Overall Description*, TS 25.401, www.3gpp.org (accessed 26 July 2014).

[10] 3GPP Technical Specification (2013) *User Equipment (UE) Procedures in Idle Mode and Procedures for Cell Reselection in Connected Mode*, TS 25.304 v11.4.0 Section 5.2.3, www.3gpp.org (accessed 26 July 2014).

[11] 3GPP Technical Specification (2013) *User Equipment (UE) Procedures in Idle Mode and Procedures for Cell Reselection in Connected Mode*, TS 25.304 v11.4.0 Section 5.2.6, www.3gpp.org (accessed 26 July 2014).

[12] 3GPP Technical Specification (2013) *Physical Layer; Measurements*, TS 25.215 v11.0.0, www.3gpp.org (accessed 26 July 2014).

[13] 3GPP Technical Specification (2013) *Base Station (BS) Radio Transmission and Reception (FDD)*, TS 25.104 v11.7.0 Section 5, www.3gpp.org (accessed 26 July 2014).

[14] 3GPP Technical Specification (2013) *Numbering, Addressing and Identification*, TS 23.003 v11.6.0 Section 12, www.3gpp.org (accessed 26 July 2014).

[15] Third Generation Partnership Project (2014) LTE, http://www.3gpp.org/technologies/keywords-acronyms/98-lte (accessed 1 June 2014).

[16] 3GPP Technical Specification (2013) *General Packet Radio Service (GPRS) Enhancements for Evolved Universal Terrestrial Radio Access Network (EUTRAN) Access*, TS 23.401 v11.7.0 Section 4, www.3gpp.org (accessed 26 July 2014).

[17] 3GPP Technical Specification (2013) *Evolved Universal Terrestrial Radio Access (E-UTRA) and Evolved Universal Radio Access Network (EUTRAN); Overall Description*, TS 36.300 v11.6.0, www.3gpp.org (accessed 26 July 2014).

[18] 3GPP Technical Specification (2013), *Evolved Universal Terrestrial Radio Access Network (EUTRAN); User Equipment (UE) Procedures in Idle Mode*, TS 36.304 v11.5.0 Section 5.2, www.3gpp.org (accessed 26 July 2014).

[19] 3GPP Technical Specification (2013) *General Packet Radio Service (GPRS) Enhancements for Evolved Universal Terrestrial Radio Access Network (EUTRAN) Access*, TS 23.401 v11.7.0 Section 5.5, www.3gpp.org (accessed 26 July 2014).

[20] 3GPP Technical Specification (2013) *Evolved Universal Terrestrial Radio Access (E-UTRA); Base Station (BS) Radio Transmission and Reception*, TS 36.104 v11.6.0 Section 5.2, www.3gpp.org (accessed 26 July 2014).

[21] 3GPP Technical Specification (2013) *Evolved Universal Radio Access (E-UTRA); Physical Channels and Modulation*, TS 36.211 v11.4.0 Section 6.11, www.3gpp.org (accessed 26 July 2014).

[22] 3GPP Technical Specification (2013) *Numbering, Addressing and Identification*, TS 23.003 v11.6.0 Section 19.6, www.3gpp.org (accessed 26 July 2014).

[23] 3GPP Technical Specification (2013) *Evolved Universal Terrestrial Radio Access Network (EUTRAN); Physical Layer Measurements*, TS 36.214 v11.1.0, www.3gpp.org (accessed 26 July 2014).

[24] GSM Association (2013) Newsroom content uploads, http://www.gsma.com/newsroom/wp-content/uploads/2013/04/IR.92-v7.0.pdf (accessed 1 June 2014).

[25] 3GPP Technical Specification (2013) Circuit Switched Fallback (CSFB) in Evolved Packet System (EPS); Stage 2, TS 23.272 v11.6.0, www.3gpp.org (accessed 26 July 2014).

6

Other Cellular Network Types

The Third Generation Partnership Project 2 (3GPP2) inherited responsibility for CDMA2000 from Qualcomm and is now the umbrella organisation covering cdmaOne (IS-95) and the various iterations of CDMA2000 (1x RTT, EV-DO) [1].

The following sections provide outline details of cdmaOne and more in depth information regarding CDMA2000.

6.1 2G IS-95/cdmaOne

The 2G cdmaOne network type was originally developed by Qualcomm as a proprietary system ('cdmaOne' is a trademarked brand name belonging to Qualcomm), but was standardised by the United States industry body the TIA/EIA (Telecoms Industry Association/Electronic Industries Alliance) as TIA-EIA-95; it was better known as Interim Standard (IS)-95. A large amount of development and standardisation for cdmaOne and later systems was also undertaken under the auspices of the CDG (CDMA Development Group) [2].

cdmaOne, as its name suggests, employed CDMA (Code Division Multiple Access) radio technologies and was the first commercially available system to be based on these techniques – CDMA had previously been used mainly for military communications systems.

The first cdmaOne systems began to go into service in the mid-1990s. As with other 2G systems, cdmaOne employed digital transmission techniques that offered

Forensic Radio Survey Techniques for Cell Site Analysis, First Edition. Joseph Hoy.
© 2015 John Wiley & Sons, Ltd. Published 2015 by John Wiley & Sons, Ltd.

consistent quality, improved security and higher capacity than the analogue 1G systems that it replaced.

The original IS-95 specifications were published in 1993, but the first deployed networks were based on a later revision of the standard known as IS-95A. IS-95A networks offered digital voice and dial-up (circuit switched) data services with data rates of up to 14.4 kbps. IS-95B upgrades began to be deployed from 1999 and are classified as being 2.5G networks, as they offer greater data rates than IS-95A (up to 64 kbps or even 115.2kbps) and also offer a packet switched data service similar to that offered by other 2.5G systems such as the General Packet Radio Service.

IS-95/cdmaOne networks were initially deployed in the United States but spread to other parts of the Americas and to parts of Asia, the Middle East, Africa and Eastern Europe. Many cdmaOne deployments have evolved into CDMA2000 1x RTT networks, but the CDG still (as of mid-2014) lists around 90 operational networks in a variety of countries around the world.

From a forensic survey point of view, IS-95/cdmaOne networks are functionally similar to CDMA2000 1x RTT networks, so details of network structure, operation and measurement will be given in the 3G CDMA2000 1x RTT section.

6.2   3G IS-2000/CDMA2000 1xRTT

CDMA2000, also known as IS2000, was originally developed by ANSI and Qualcomm as a 3G evolution of their 2G cdmaOne (IS95) technology. Responsibility for CDMA2000 development was later passed to 3GPP2. A considerable amount of CDMA2000 development has also been provided by the CDG.

CDMA2000-based networks were originally deployed in the United States and the wider Americas and have since spread to other areas and there are examples of these network type in most regions of the world. The CDG website (as of mid-2014) lists CDMA2000 deployments of various generations in over 120 countries or territories.

6.2.1   CDMA2000 Generations and Variants

The original IS-2000/CDMA2000 standard offered an evolution path for existing 2G IS-95/cdmaOne networks which, initially, reused the same channel structures as the 2G system.

IS-95/cdmaOne systems employ radio channels with a bandwidth of 1.25 MHz. The first versions of CDMA2000 were also based on 1.25 MHz wide channels, but later versions were planned to concatenate multiple channels to increase available data rates. The initial version was therefore known as '1x', as it used one radio channel per cell, and later iterations were to be known as '3x' (three concatenated 1.25 MHz channels per cell), '6x', '9x' up to a planned maximum of '12x'. The term RTT (Radio Transmission Technology) was also used, so the earliest iterations of the system were known as CDMA2000 1x RTT. Aspects of the planned '3x' version have

Table 6.1 cdmaOne and CDMA2000 generations and variants.

Variant	Generation	Date	Comments
IS-95	2G	1993	Initial description
IS-95A	2G	1995	Voice, dial-up data
IS-95B	2.5G	1999	Voice, up to 64 kbps PS data
CDMA2000 1x	3G	2000	Voice, 150 kbps PS data, 1.25 MHz FDD carrier
1x Advanced	3G	2001	Voice, 300 kbps PS data, 1.25 MHz FDD carrier
1x EV-DO Rev. 0	3.5G	2002	PS data only, 2.4Mbps (DL), 150 kbps (UL) 1.25 MHz FDD carrier
EV-DO Rev. A	3.5G	2006	3.1Mbps (DL), 1.8Mbps (UL) 1.25 MHz FDD carrier
EV-DO Rev. B	3.5G	2010	14.7Mbps (DL), 5.4Mbps (UL), 5 MHz FDD carrier

Based on information from www.cdg.org, used with permission from CDG.

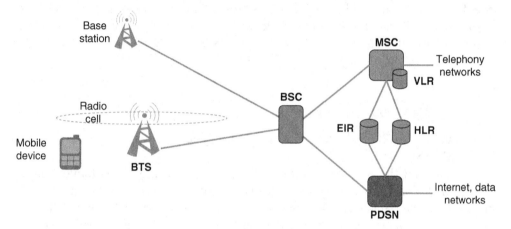

Figure 6.1 CDMA2000 1x network architecture

been rolled into the EV-DO (Evolution – Data Only or Data Optimised) concept, which provides a 3.5G upgrade to CDMA2000, but the 6x, 9x and 12x options seem to have been abandoned.

If the predecessor 2G and successor 3.5G systems are also taken into account, the full list of versions and revisions for CDMA2000 is as shown in Table 6.1.

6.2.2   CDMA2000 Network Architecture

The basic architecture of a cdmaOne/CDMA2000 1x network is outlined in Figure 6.1 [3].

3GPP2 made a conscious decision to reuse much of the terminology related to 3GPP 2G and 3G network types [GSM (Global System for Mobile), UMTS (Universal Mobile Telecommunications System)], so many of the elements in Figure 6.1 will be familiar from previous chapters.

- BTS – the Base Transceiver Stations manage individual cells and offer radio transmit and receive services for users.
- BSC – as in GSM networks, the Base Station Controller manages the activities of a group of base stations and handles radio resource management and handoff activities.
- MSC – again, as in legacy GSM, the MSC (Mobile Services Switching Centre) manages voice and messaging services and also handles the interconnects to the wider telephony network.
- HLR – the Home Location Register is the central repository of subscriber data and authentication information.
- VLR – the Visitor Location Register is co-located with an MSC and contains a local copy of the subscription profile, downloaded from the HLR, of each user currently served by the MSC.
- EIR – the Equipment Identity Register stores details of registered mobile devices.
- PDSN – the Packet Data Service Node provides the management and interconnection of packet data services between the BSC and an external PSDN (Packet Switched Data Network) such as the Internet.

6.2.3   CDMA2000 Network Structure

Individual CDMA2000 networks are classed as 'systems' and each is identified by a 15-bit SID (System ID).

The need for the SID, in addition to the more common MCC (Mobile Country Code)/MNC (Mobile Network Code) network identifiers, can be explained by looking at the United States cellular market for which it was designed. Cellular licences in the United States are assigned based on 'market' boundaries, where services in different markets can be provided by a variety of operators. It was anticipated that this type of environment would lead to a great deal of competition and would work against the formation of large nationwide networks. The expectation was that cellular coverage would be provided by a patchwork of different operators and that, even if one company managed to obtain large numbers of licences, they would be likely to have to deploy a non-contiguous service. The SID can therefore be used to identify not only each operator, but also each separate area of deployment for each operator across the country.

The allocation of SIDs is managed by IFAST (International Forum on ANSI-41 Standards Technology), where 'ANSI-41' refers to the signalling standard employed within most CDMA2000 networks and some other network types [4].

Each System is sub-divided into a set of Networks, each of which has its own NID (Network ID). The NID is 16 bits long, allowing up to 65 536 Networks per System, although NID 0 and NID 65 535 are reserved.

A SID/NID pair, as shown in Figure 6.2, identifies a collection of cells within a specific network and is used to control roaming, both within a System and between

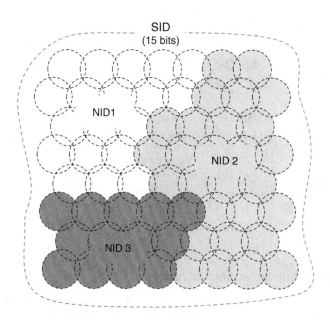

Figure 6.2   CDMA2000 SID and NID

Systems. Each mobile device will be provided with a 'Home SID' setting, which allows it to determine whether it is being served by its 'home' system or not. Mobile devices may also be provided with a PRL (Preferred Roaming List), which provides details of the SID/NID combinations in which the device is able to roam outside of its home system.

The NID can be used to control area roaming within a SID. Devices can be provided with a Home NID and also with allowed and forbidden NIDs. This allows networks to determine areas where specific classes of user are (and are not) permitted to roam. Many operators simply define a single NID (NID1) in each SID and do not bother to define any further internal structure.

The SID/NID pair has some forensic relevance as it identifies a subset of network coverage, rather like the MCC/MNC/LAC does in GSM and UMTS. However, the SID/NID does not typically identify an individual cell; that is the job of the Base Station ID (BSID).

6.2.4   3G CDMA2000 User and Device Identities

In addition to sharing architectural terminology with GSM networks, the developers of CDMA2000 also tried to harmonise much of the administrative terminology, especially where user and device identifiers were concerned, although there are also some 3GPP2-specific terms [5].

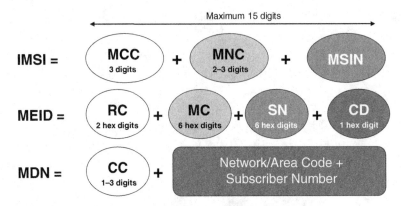

Figure 6.3   CDMA2000 identifiers

International Mobile Subscriber Identity

The International Mobile Subscriber Identity (IMSI) in 3GPP2 performs the same function as in 3GPP networks; it provides a unique identifier for each subscription. As detailed in Figure 6.3, the 3GPP2 IMSI is up to 15 digits long, with an MCC and MNC followed by an MSIN (Mobile Station Identification Number).

Unlike the 3GPP IMSI, which is always allocated to a SIM card, the 3GPP2 IMSI can be associated with a mobile device. cdmaOne did not support the SIM concept, identifying subscriptions instead by the Electronic Serial Number (ESN) of the mobile device, and CDMA2000 deployments can choose to use hardware-based identification modules, which include an IMSI, in place of SIMs.

Temporary Mobile Subscriber Identifier

A Temporary Mobile Subscriber Identifier (TMSI) is allocated to a subscriber by the serving MSC/VLR following a successful registration and authentication event. As in 3GPP networks, the allocation of a TMSI allows the network and the mobile device to exchange signalling and control information without needing to continually transmit the subscription's sensitive IMSI over the radio path.

Electronic Serial Number

The ESN uniquely identifies each mobile device and is 'burnt in' to the device during manufacture and is the equivalent of the 3GPP IMEI (International Mobile equipment Identifier).

The original ESN was 32 bits long and consisted of an 8-bit manufacturer ID, followed by a 24-bit device serial number. This configuration was superseded by a version that used 14 bits for manufacturer and 18 bits for device serial number. This is turn was superseded by the current scheme, operated by 3GPP2, which assigns an MEID (Mobile Equipment ID) to each CDMA device [6].

Mobile Equipment ID

The MEID is 56 bits long and is typically represented as a 14- or 15-digit hexadecimal number consisting of a two-digit RC (Regional Code), followed a six-digit MC (Manufacturer Code), then a six-digit SN (Serial Number) and lastly a one-digit CD (Check Digit), all as detailed in Figure 6.3.

Mobile Directory Number

The Mobile Directory Number (MDN) is the phone number associated with a subscription and is the equivalent of the 3GPP MS-ISDN. Each subscription will have at least one MDN associated with it and is used to allow calls to be made to and from a mobile device.

The MDN contains up to 15 digits and is outlined in Figure 6.3.

6.2.5   3G CDMA2000 Radio Interface

cdmaOne and CDMA2000 use a form of CDMA on the radio interface which is similar to that used by 3G UMTS but with a few key differences [7]. Some basic features of 3GPP2 CDMA are illustrated in Figure 6.4.

The 3GPP2 version of CDMA was originally based on comparatively narrow radio channels with a bandwidth of 1.25 MHz (as opposed to the 5 MHz channels employed in UMTS). The Wideband Code Division Multiple Access technique used by UMTS employs two levels of CDMA coding – scrambling codes, which are used to differentiate between different base stations and mobile devices, and

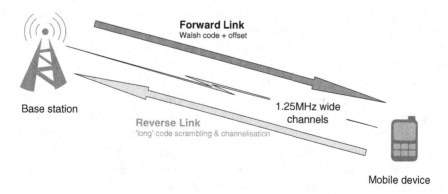

Figure 6.4   3GPP2 CDMA concepts

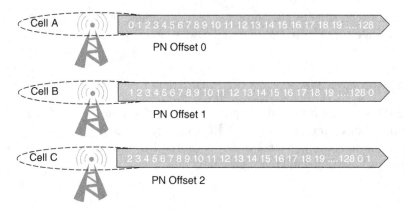

Figure 6.5   3GPP2 CDMA concepts

channelisation codes that are used to differentiate between different logical channels – the 3GPP2 version employs Walsh Codes (or PN [Pseudo Noise] codes), which are used in both 'short' and 'long' code formats that are variously used for channelisation and scrambling purposes.

In 3GPP2 CDMA, cells are deployed in single frequency networks and all cells use the same set of codes for all transmissions. To allow traffic belonging to different cells to be distinguished, each cell in an area applies a specific 'offset' to the PN codes they use (known as a 'PN Offset') – this means that although all cells use the same set of codes, they will not all be transmitting the same part of the code at the same time. Walsh codes are constructed so that they are 'orthogonal' to each other (meaning that the signals being created by different versions of the same code are distinctly different to each other) as long as there is a sufficient offset (e.g. time difference) between neighbouring versions of the code. This concept is illustrated in Figure 6.5.

The degree of synchronisation involved in ensuring that each cell transmits its codes with the exact amount of offset generally requires each site to obtain a synchronisation reference from an extremely accurate source; this is typically achieved by ensuring that each site has a GPS receiver and obtains its sync from that.

All 3GPP2 systems use FDD (Frequency Division Duplex) and there are no TDD (Time Division Duplex) options. In place of the 'downlink' and 'uplink' terminology employed by 3GPP, the CDMA2000 family usually employ the terms 'forward link' (base station to mobile device) and 'reverse link' (mobile device to base station).

cdmaOne and CDMA2000 1x employ symmetrical forward/reverse link allocations, with a 1.25 MHz channel being used in each direction. Some configurations of EV-DO are able to employ 'multi-carrier' forward channels that can concatenate three or four 1.25 MHz channels together to provide a higher bandwidth downlink service and use either a 1x 1.25 MHz reverse link or a 3.75 MHz link that offers higher capacity.

Each cell transmits a 'pilot' channel, which mobile devices take measurements of, and a number of administrative logical channels plus a set of user traffic channels. cdmaOne cells support a maximum of 64 code channels that are shared between administrative and user traffic functions, CDMA2000 employs a wider set of up to 256 code channels.

The main logical control channels employed in CDMA2000 include:

- F-PICH – Forward Pilot Channel, uses a known Walsh code, Walsh 0, which is transmitted with a specific offset to differentiate each local cell. It provides a known quantity for mobile devices to search for and measure. All other code channels in a cell follow the offset of the pilot channel.
- F-SYNC – Forward Synchronisation Channel, provides time synchronisation services to mobile devices in a cell and also provides details of the Walsh codes assigned to key control channels, such as the F-BCCH, in the cell.
- F-BCCH – Forward Broadcast Control Channel, carries cell wide control and identification information which is generically known as 'overhead' messaging.
- F-PCH – Forward Paging Channel, carries paging messages to idle devices.

The F-BCCH provides details of cell access parameters CDMA, network identification (SID, NID, BSID) and other system parameters.

In cdmaOne networks the Sync and Paging channels carry the cell 'overhead' control messaging, the BCCH was only introduced with CDMA2000.

6.2.6   3G CDMA2000 Cell Selection

3GPP2 Cell selection and reselection are known as 'acquisition' and 'idle handoff'.

As with the C1 and C2 algorithms employed in GSM (and the S and R algorithms employed in UMTS and LTE), 3GPP2 systems take measurements of the pilot channels of the cells they can detect locally and compile a list known as the 'pilot set'.

When in Idle Mode, the cell with the strongest pilot in the pilot set is selected as the serving cell and the mobile device will monitor the F-BCCH channel in that cell. Figure 6.6 illustrates some CDMA2000 handoff concepts.

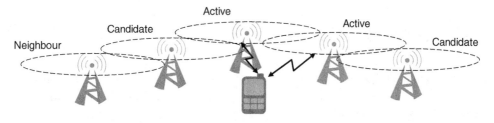

Figure 6.6   3GPP2 Handoff concepts

6.2.7 3G CDMA2000 Handover

As CDMA2000 employs code division techniques in a single frequency network environment, it has access to the same range of generic handoff/handover types as are available to UMTS, these include:

- Softer handoff – this is inter-sector handoff between sectors of the same base station on the same frequency.
- Soft handoff – this is cell to cell handoff between different base stations on the same frequency.
- Softer–Soft handoff – which is when a three-way handoff is set up that includes softer handover between two sectors of the same site and soft handoff with a cell belonging to a different site, all on the same frequency.
- Hard handoff – this is handoff between cells that are deployed on different frequency layers or handoff between a CDMA cell and a cell belonging to another RAT (Radio Access Technology), such as 4G LTE.

CDMA2000 mobile devices take measurements of each detected cell's pilot channel to enable them to make selection and reselection decisions, so information related to detected cells is stored in a group of Pilot Sets, which for soft handoff includes:

Active Set – this is the set of cells to which the mobile device is currently connected and is able to exchange traffic with. As with UMTS, CDMA2000 allows a maximum of three pilots to be in an Active Set concurrently.

Candidate set – this is the set of 'best neighbours' that could join the Active Set but whose received pilot strength hasn't quite reached the required threshold.

Neighbour Set – this is the set of other detected pilots that are not sufficiently strong to join the Candidate Set and are unlikely to become handoff candidates in the near future.

Remaining Set – this is the set of potential pilots available in the network or radio band being employed but which are not currently detected.

Each cell broadcasts a set of handoff management parameters that allow the mobile device to judge when to add pilots to the Active Set and when to drop them back out again.

6.2.8 3G CDMA2000 Radio Measurements

As with other CDMA-based systems, the single frequency environment of a CDMA2000 network requires that an absolute received signal strength measurement of each pilot is captured and that a comparison of each pilot signal against background noise is then made.

Like in UMTS, the main measurement taken in CDMA2000 is RSCP (Received Signal Code Power), which is an absolute measurement of received F-PICH power for a given PN pilot code. RSCP is measured in dBm.

Mobile devices also measure the absolute value of received channel noise and interference to calculate a value known as Io, which is also measured in dBm.

The usability of a cell/pilot can best be judged by comparing the power of the received pilot (RSCP, also known for the purposes of this calculation as Ec) with the power of the channel noise and interference (Io) – this provides a comparison of the signal to noise ratio known as Ec/Io, which is measured in dB.

RSCP has typical reported range of approximately −40 dBm down to −120 dBm.

Ec/Io has a typical reported range of −5 dB down to −30 dB.

6.2.9   3G CDMA2000 Band Classes and Channel Numbering

IS95/cdmaOne was originally specified to operate in the United States PCS 1900 MHz band but as its popularity began to spread, further bands were defined for its use.

cdmaOne and CDMA2000 now have a set of Band Classes defined which specify the frequency resources allocated to the system in different parts of the world [8].

Individual channel centre frequency allocations are marked out in 0.05 MHz (50 kHz) intervals within each band, which provides a degree of flexibility as to where each 1.25 MHz wide channel is sited. Each Band Class operates across a specific range of frequencies that are of differing bandwidths, so some bands support large numbers of channels while others support smaller allocations.

Channel numbers are assigned sequentially within each Band Class, generally starting from either 0 or 1 depending upon the band, so there is no overall channel numbering scheme.

Table 6.2 provides an overview of the current Band Class and Channel Numbering arrangements for CDMA2000 systems.

Some CDMA2000 Band Classes are further subdivided into frequency allocation 'Blocks', which aid the process of licensing spectrum to individual operators. Band Class 1 (US PCS 1900 band), for example, is subdivided into six allocations blocks (Blocks A–F).

Not all of the Band Classes that have been defined have actually been used.

6.2.10   3G CDMA2000 Cell Identifiers

Physical Layer Identifiers

Fundamentally, each cell/sector in an area of a CDMA2000 access network is identified by the pilot channel's PN Offset.

The pilot channels of all CDMA2000 cells use the same code sequence (Walsh Code 0) but by carefully synchronising the operation of neighbouring

Table 6.2   CDMA2000 band classes and channel numbering.

Band class	Forward link frequencies	Reverse link frequencies	Channel numbers	Description
0	860–894	815–849	1–1323	800 MHz band
1	1930–1990	1850–1910	0–1199	1.8–2.0 GHz PCS band
2	917–960	872–915	0–2108	872–960 MHz TACS band
3	832–870	887–925	1–1600	832–925 MHz JTACS band
4	1840–1870	1750–1780	0–599	1.75–1.87 Korean TACS band
5	420–493	410–483	1–2108	450 MHz NMT band
6	2110–2170	1920–1980	0–1199	2GHz IMT2000 band
7	746–758	776–788	0–240	Upper 700 MHz band
8	1805–1880	1710–1785	0–1499	1800 MHz band
9	925–960	880–915	0–699	900 MHz band
10	851–940	806–901	0–919	Secondary 800 MHz band
11	420–493	410–483	0–2016	400 MHz European PAMR band
12	915–921	870–876	0–239	800 MHz PAMR band
13	2620–2690	2500–2570	0–1399	2.5 GHz IMT2000 band
14	1930–1995	1850–1925	0–1299	United States 1.9 GHz PCS band
15	2110–2155	1710–1755	0–899	AWS band
16	2624–2690	2502–2568	140–1459	United States 2.5 GHz band
17	Not specified			United States 2.5 GHz forward link only band
18	757–769	787–799	0–240	700 MHz public safety band
19	728–746	698–716	0–360	Lower 700 MHz band
20	1525–1559	1626.5–1660.5	0–680	L-band
21	2180–2200	2000–2020	0–399	S-Band

Based on 3GPP2 Specification C.S0057-E, used with permission from 3GPP2.

base stations and by offsetting the code sequence that each of them generates, it is possible to ensure that neighbour cells operate 'orthogonally' to each other. This is achieved by having each cell start transmitting the pilot at a different and specific point in the code sequence, as long as the offset between neighbouring pilots is sufficiently long (in time) it should be possible for mobile devices to distinguish between the pilot signals being sent by different local cells. Figure 6.7 illustrates this concept.

There are 512 possible PN Offsets available in CDMA2000, so it should be possible to assign a locally unique offset to each cell and only reuse offsets in cells that are physically distant from each other. If an offset is reused in cells that are in too close a proximity to each other then it will typically result in the inability of mobile devices to accurately measure the pilot signals of either cell.

In addition to the PN Offset, cells are differentiated by their channel number, but as CDMA2000 channel numbers are non-unique between bands it is also desirable to know the cell's Band Class.

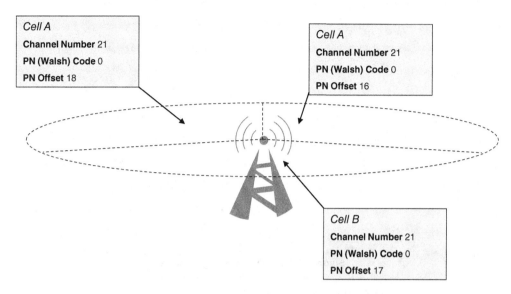

Figure 6.7   CDMA2000 physical layer cell discrimination

In summary, the physical layer cell differentiators for CDMA2000 are a combination of Band Class/Channel Number/PN Offset.

Cell Identifiers

There is a degree of variability in the way that different CDMA2000 operators choose to identify individual cells at the 'logical' level.

Some operators identify individual cells for administrative purposes using a CGI (Cell Global Identifier) that is superficially the same as the GSM CGI, namely MCC/MNC/LAC/CI. However, the CDMA2000 air interface overhead channels do not carry this data structure, so it is not possible to use it for forensic radio survey purposes.

Using only the information that is available over the air interface signalling channels, CDMA2000 cells can be identified by a combination of MCC/MNC/SID/NID and BSID.

The BSID is carried in the System Parameters message on the F-BCCH in CDMA2000 cells and should be available to most forensic survey devices [9].

The BSID is a 16-bit value that is typically used to identify a base station within a SID/NID area and in most cases is structured to identify both the site and each individual sector transmitted by the site. For example, some operators choose to represent the BSID in hexadecimal and use the first (most significant) three hex digits to identify the base station and the last digit to indicate the sector – so base station ab8 might have three sectors labelled 1, 2 and 3, meaning that the site would generate three different BSIDs – ab81, ab82, ab83.

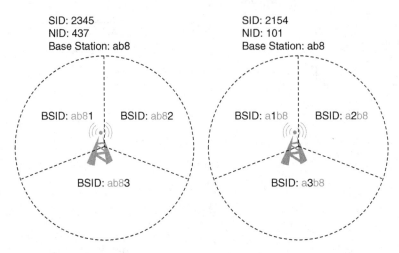

Figure 6.8   BSID numbering schemes

Other operators place the sector identity portion of the BSID in the second most significant digit, so base station ab8 and sectors 1, 2 and 3 would be represented as a1b8, a2b8 and a3b8. Examples of this numbering system are shown in Figure 6.8.

Apart from specifying the length of the BSID (16 bits) and its method of transmission (System Parameters), 3GPP2 is content to leave it to individual networks to decide how best to structure and use the identifier.

3GPP2 has incorporated one more method of allowing cells to be individually distinguished. The signalling overhead channels in each cell have space to carry details of the base station's geographical location, in the form of a GPS latitude/longitude reference. Not all networks transmit this information, but knowledge of a site's location can further help to identify it, but only if access to a site location database is available.

6.2.11   3G CDMA2000 Cell Configurations

CDMA2000 cells can be deployed following the same options that are available to other cellular systems: omnidirectional or sectorised. Sectorised cells can be configured using either the OTSR (Omnidirectional Transmit Sectorised Receive) or STSR (Sectorised transmit Sectorised Receive) models. The sectorised concepts are recapped in Figure 6.9.

Given the large geographical areas that CDMA2000 networks are required to serve in countries such as the United States, the OTSR model is one that has proven to be attractive to operators. From a forensic survey point of view, all of the forward link sectors on an OTSR site will share the same PN Offset (as they are all generated from the same radio transmitter) and will all share the same BSID.

In fully sectorised sites, each sector will have its own PN Offset and will typically also have its own BSID.

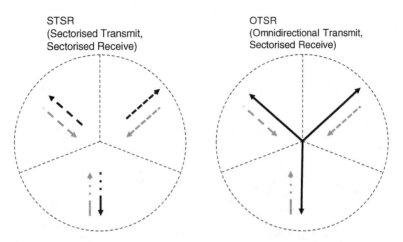

Figure 6.9   Sector Transmit/Receive Configurations

Table 6.3   EV-DO variants.

Variant	Generation	Date	Comments
EV-DO Rev. 0	3.5G	2002	PS data only, 2.4 Mbps (DL), 150 kbps (UL) 1.25 MHz FDD carrier
EV-DO Rev. A	3.5G	2006	3.1 Mbps (DL), 1.8 Mbps (UL) 1.25 MHz FDD carrier
EV-DO Rev. B	3.5G	2010	14.7 Mbps (DL), 5.4 Mbps (UL), 5 MHz FDD carrier

Based on information from www.cdg.org, used with permission from CDG.

6.3   3G CDMA2000 EV-DO

The EV-DO (Evolution – Data Optimised) variant of CDMA2000, as the name suggests, offers a service optimised to carry packet data rather than voice – when subscribers wish to make or receive voice calls, their user devices are required to fallback to a 1x RTT cell. Some device vendors produce terminals with dual radio interfaces that allow devices to simultaneously attach to a 1x RTT cell for voice and an EV-DO cell for data services. This combination of EV-DO for data and CDMA2000 1x for voice is sometimes classified as SVDO (Simultaneous Voice and Data).

Each iteration of EV-DO supports the use of the same 1.25 MHz wide radio carriers as are used by cdmaOne and CDMA2000 1x, but employ more advanced radio transmission techniques to increase the carrying capacity of those radio channels.

EV-DO also supports higher data rates by permitting cell deployments in wider carriers, either by aggregating 3x 1.25 MHz carriers or by sending a signal that directly occupies a 5 MHz channel.

There is also a version of EV-DO that allows 4x 1.25 MHz carriers to be aggregated to support higher bandwidth services.

The set of EV-DO technologies is recapped in Table 6.3.

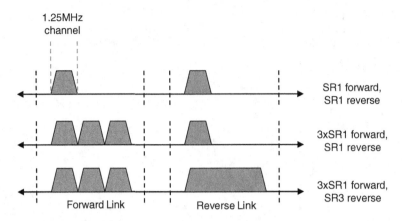

Figure 6.10   EV-DO carrier configuration options

EV-DO Revision (Rev) 0 initially provided services within the 1x 1.25 MHz structure employed by CDMA2000 1x, but offered higher data rates through the use of
more complex radio transmission techniques. 1x-compatible services are generically
known in EV-DO as SR1 (Spreading Rate 1).

The term 'spreading rate' refers to the rate at which the CDMA code transmitted in
a cell is generated; the higher the code rate, the wider the bandwidth of the radio
channel required to carry it. An SR1 signal fits within a standard 1x 1.25 MHz radio
channel. EV-DO also supports SR3 (Spreading Rate 3), where signals are transmitted
that require a 3x (3x 1.25 MHz or 3.75 MHz) channel to carry them. Although the
direct SR3 signal occupies 3.75 MHz, each carrier is deployed in a bandwidth of 5
MHz to ensure that there is a 'guard band' around it to minimise the amount of adjacent channel interference caused to neighbouring carriers. There is also a 'multicarrier' option, in which downlink/forward link transmissions are split and transferred
across three separate, but bound, SR1 1x 1.25 MHz channels.

The general set of transmission options for EV-DO cells are highlighted in Figure 6.10
and use a mix SR1 (1x) and SR3 (3x) techniques in the following combinations:

- SR1 (1x 1.25 MHz) forward link, SR1 (1x 1.25 MHz) reverse link
- 3x SR1 (3x 1.25 MHz) FL, SR1 (1x 1.25 MHz) RL
- 3x SR1 (3x 1.25 MHz) FL, SR3 (1x 3.75 MHz) RL.

6.3.1   EV-DO Forensic Surveys

The measurements for EV-DO are essentially the same as for CDMA2000 1x.
   The physical layer cell identifiers are:

- Band Class
- Channel Number

Table 6.4   3GPP2 technical specifications.

Reference letter	Subject area	Key specifications
A	Access networks	A.S0011
C	Radio interface	C.S0001
N	Core network (pre-December 2012)	N.S0005
P	Packet network	P.S0001
S	Services	S.R0005
X	Core network (current)	X.S0011

From www.3gpp2.com, used with permission of 3GPP2.

- PN Offset
- EV-DO additionally defines a Sector ID that can individually identify each sector in a network.

The main radio measurements are:

- RSCP
- Ec/Io.

In cells that support 3x SR1 multicarrier operation, identification and measurements are taken of the nominated primary carrier in the set.

6.3.2   CDMA2000 Specifications

3GPP2 publish freely available versions of their specifications on their website (www.3gpp2.com/Public_html/specs/index.cfm).

Their specification set is divided into subsections known as Technical Specification Groups, examples of which are listed in Table 6.4.

Details related to commercial and interoperability aspects of the 3GPP2 family of networks are handled by the industry body CDG [2].

6.4   Surveying Other Technologies

There are many types of cellular network in use around the world. The network types detailed in this and the previous chapter account for well over 90% of current active user connections and will therefore represent the bulk of network types that will be surveyed for forensic purposes, but there is always a chance that a surveyor will be required to take measurements of less common network types.

Details of the types of information likely to be captured during surveys of some of these less common network types are presented below. It should be noted that most of the technologies listed below would not be accessible to standard forensic survey devices. Surveys of these network types would therefore need to use either a specialised survey device or would need to use the 'engineering mode' function of an applicable handset, if such a facility exists.

6.4.1   2G iDEN

iDEN (Integrated Digital Enhanced Network) networks were developed by Motorola in the 1990s and were deployed in North and South America, Israel, Jordan, Singapore and other areas. The most significant deployment was by Nextel in the United States, which was later acquired by Sprint. This network was decommissioned in mid-2013.

The system was designed to offer a mix of traditional one to one telephony and group-call 'press to talk' services.

iDEN networks typically operate in the 800, 900 or 1500 MHz frequency bands. The radio transmission technique employs a very basic form of Orthogonal Frequency Division Multiple Access (in 4G), with four subcarriers transmitted in each channel, which is then shared between subscribers on a Time Division Multiple Access (TDMA) basis.

iDEN cells are discriminated at the physical layer using a combination of channel number and CC (Colour Code). The CC is a four-bit identifier, which allows for very basic local cell discrimination to take place with 15 different possible values.

Network wide cell discrimination is provided by the Cell ID, which is a 16-bit field offering 65 535 cell IDs per network.

The main measurement taken of iDEN cells is RXLev (received signal strength), which is measured, is dBm and has a typical reporting range of −30 to −130 dBm.

iDEN is a proprietary solution developed by Motorola, as such there are no openly available standards to Ref. [10].

6.4.2   2G Personal Digital Cellular (Japan)

PDC (Personal Digital Cellular) was a 2G network type developed for deployment in Japan, it was used by the NTT DoCoMo, Softbank and KDDI networks. It operated in the 800 and 1500 MHz bands and used TDMA-based transmission techniques.

The last PDC network service was switched off (by NTT) in 2012, so radio survey-ors will not encounter this network type.

PDC was standardised by the Japanese telecoms standards agency ARIB (Association of Radio Industry Businesses), ARIB standard RCR STD-27 [11] provides a general introduction to the system.

6.4.3   2G TETRA

TETRA (Terrestrial Trunked Radio) is designed to offer a mix of voice, messaging and low-rate data services and is mainly aimed at professional and public safety users.

TETRA is a heavily adapted version of GSM and employs a form of TDMA in which each radio channel supports four timeslots.

Each TETRA cell is identified at the physical layer by its channel number, which is calculated from a combination of channel frequency, band specific offset and channel bandwidth. The main TETRA bands operate between 390 and 400 MHz, with ARFCNs (Absolute Radio Frequency Channel Number in 2G) that run between 3600 and 3999 and a channel spacing of 25 kHz.

Above the physical layer, each cell is uniquely identified by its LAC (Location Area Code), which is the equivalent of the GSM cell ID. Cells are made globally unique by the addition of the MCC and MNC values that relate to the network – for example, the Airwave public safety in the United Kingdom uses MCC 234 and MNC 78.

The typical measurements captured for TETRA include:

- Cell identity – MCC, MNC, LAC
- Carrier frequency in MHz
- ARFCN (between 0 and 4096)
- RSSI – received signal strength between −10 and −111 dBm, with typical values of −80 dBm (strong) to −105 dBm (weak)
- C1 (for serving cell) and C2 (for neighbour cells).

TETRA's main distinguishing feature is that it offers PTT (Press to Talk) group calling services in which user terminals are members of 'talk groups' – when one group member keys their PTT 'presell' a voice connection is established to all of the other group members who are connected and available. TETRA also offers 'one to one' calling services, which are essentially mobile phone calls, text messaging and data services. If a TETRA gateway is installed in a network then one to one calls can also be established with phones connected to external networks.

TETRA terminals are able to operate in two different modes: 'trunked' mode, in which calls are connected via a base station and can be connected to terminals in the target talk group wherever they happen to be in the network; 'direct' mode, which allows terminals to connect directly to each other, without going via a base station. DMO (Direct Mode Operation) is also sometimes known as 'back to back' mode, and is used to allow groups to maintain communication with each other even if they are out of network coverage. A specific subset of TETRA channels is made available to support DMO. It is possible to allow devices that have network connectivity to act as bridges for local DMO devices that do not; this can allow a DMO device to maintain contact with the wider network even if they have no network connectivity themselves.

TETRA networks have been extensively deployed in Europe and in most other regions.

TETRA is specified by ETSI (European Telecoms Standards Institute), which was the organisation originally responsible for developing GSM and which is a contributing member of 3GPP [12]. Commercial and interoperability issues related to the system are dealt with by T&CCA (TETRA + Critical Communications Association) [13].

6.4.4   2G GSM for Railways

GSM for Railways (GSM-R) is an adaptation of 2G GSM900 that was created to serve the needs of railway operators.

In addition to the standard GSM voice and messaging features, GSM-R also supports 'press to talk' group calling functions to allow members of the same team to maintain 'all informed' communications.

As GSM-R is simply a re-purposed version of GSM900, the cell identifiers, physical layer identifiers, channel numbering and radio measurement types are all identical to those employed by standard GSM.

Most radio survey devices do not support GSM-R, however, so forensic radio surveys of this network type would need to be undertaken using appropriately capable equipment.

GSM-R specifications are mainly based on ETSI/3GPP documents [14] but much of the 'railway-specific' content is defined by the UIC (International Union of Railways – www.uic.org) as part of their EIRENE programme [15].

6.4.5   3G TD-SCDMA (China)

UMTS is available in both FDD and TDD versions, with the FDD variant being by far the most widely deployed. The TDD UMTS services are offered in two main variants, known as UMTS-TDD$_{HCR}$ (High Chip Rate) and UMTS-TDD$_{LCR}$ (Low Chip Rate).

UMTS-TDD$_{HCR}$ is essentially just a single carrier TDD version of 'standard' FDD UMTS but UMTS-TDD$_{LCR}$ employs a very different air interface technology known as TD-CDMA (Time Division – Synchronous Code Division Multiple Access), was developed in China to meet the perceived needs of the Chinese market.

The 'chip rate' of a UMTS system is similar to the 'spreading rate' referenced in CDMA2000 and influences the bandwidth of the radio channel occupied by the coded signal. TD-SDCMA transmits a 'low' code rate signal of 1.28 Mcps (which stands for 'mega-chips per second, where a 'chip' is a component of the code being transmitted), which requires a 1.6 MHz wide channel to carry it. Standard UMTS FDD and UMTS-TDD$_{HCR}$ transmit a code signal with a chip rate of 3.84 Mcps and require a wider 5 MHz channel to carry them.

All three systems, standard UMTS FDD, UMTS-TDD$_{HCR}$ and UMTS-TDD$_{LCR}$, share the same core network design and use the same signalling and control mechanisms, with the main differences being reserved for the operation of some aspects of the air interface technology.

TD-SCDMA is a TDD technology, so uplink and downlink traffic share the same radio channel. The resources of the shared channel are divided into seven timeslots, which can be dynamically configured to be used to carry either uplink or downlink

Table 6.5 TD-SCDMA radio survey parameters.

Value	Range	Comment
UARFCN Band 34	9400–9600	Unpaired TDD carriers have
UARFCN Band 39	10050–10125	only one channel number
UARFCN Band 40	11500–12000	
RSCP	−20 to −116 dBm	Wanted pilot strength
RSSI	0 to −140 dBm	Channel noise

Based on 3GPP TS 25.105:5.2, used with permission from 3GPP, reporting values based on Anite NEMO Handy capabilities.

traffic. If a base station determines that there is more demand for downlink capacity in a cell it could, for example, elect to set five of the seven timeslots to carry downlink traffic and leave just two to carry uplink. Alternatively, if the uplink load was higher at a point in time, there could be more timeslots allocated to uplink than to downlink.

TD-SCDMA employs 1.6 MHz wide radio channels and, in China, has been assigned radio capacity in three unpaired TDD radio bands:

- Band 34 – 2010–2025 MHz
- Band 39 – 880–1920 MHz
- Band 40 – 2300–2400 MHz.

TD-SCDMA services in China are operated by China Mobile, which indicates on its website [16] that services are currently only deployed in bands 34 and 39. China Mobile also indicates that the data rates currently achievable on its network are around 1.6 Mbps on the downlink and 0.5 Mbps on the uplink.

The measurements captured for TD-SDCMA surveys are similar to those for UMTS-FDD, as are the cell selection/reselection principles. A set of key TD-SCDMA survey parameters is summarised in Table 6.5.

TD-SCDMA does not employ soft handover and uses strict techniques to ensure that each mobile device maintains synchronisation with its current serving base station. The lack of co-channel, soft handover operation means that TD-SCDMA devices are not required to calculate the comparative Ec/No signal to noise ratio value that is so important in 'standard' UMTS.

Physical layer cell identification in TD-SCDMA is achieved by a combination of channel UMTS ARFCN (UARFCN) and downlink scrambling code, as it is in UMTS-FDD, this therefore allows the system to make use of single frequency network techniques if required. The method by which mobile devices are informed of the scrambling code to use for TD-SCDMA cell differs from the method employed in UMTS-FDD, as TD-SCDMA cells advertise a value known as the Cell Parameters ID.

There are 128 Cell Parameters ID values (in the range 0–127), each cell on the same frequency layer in an area will be assigned a different ID, which should ensure that no two cells in the same area use the same downlink scrambling code.

TD-SCDMA employs the same logical level cell identification as standard UMTS, namely MCC/MNC/RNC ID/CI and cells will also advertise their LAC.

TD-SDCMA is specified in a series of 3GPP Technical Standards and is usually dealt with alongside 'standard' UMTS topics. 3GPP refers to TD-SCDMA as UMTS$_{LCR}$ and as 1.28 Mcps TDD. Examples of 3GPP standards that provide details of the TD-SCDMA air interface are TS 25.102 [17], TS 25.105 [18] and TS 25.304 [19].

6.4.6   4G WIMAX

WIMAX (Worldwide Interoperability for Microwave Access) was originally developed to support FWA (Fixed Wireless Access) services, which provide data services to fixed locations but deliver them via a radio connection rather than via a telephone line or cable.

Standardisation work on WIMAX was coordinated by the IEEE (Institute of Electrical and Electronics Engineers), the body also responsible for coordinating development of WiFi (Wireless Fidelity).

WIMAX standards are published under the collective reference number 802.16 and most specifications are freely available from the IEEE website (www.ieee.org). IEEE 802.16 specifications deal solely with the operation of the radio air interface.

The name 'WIMAX' was created by a separate body, the WIMAX Forum, which coordinates commercial and interoperability aspects of the standard, WIMAX Forum specifications also detail the wider network architecture beyond the air interface.

Early versions of WIMAX concentrated on FWA services, but the 802.16e update issued in 2005 introduced a number of enhancements that were collectively known as Mobile WIMAX. Mobile WIMAX was further enhanced in 802.16 m (also known as WIMAX Release 2) in 2009.

Mobile WIMAX offers a cellular service that is mainly optimised for packet data but that is capable of carrying voice (as VoIP, Voice over Internet Protocol) traffic as well.

In general, WIMAX is designed to operate within a frequency range of 2–66 GHz; Mobile WIMAX typically operates in licensed frequency bands of 2.3, 2.5, 2.8, 3.3 and 3.5 GHz with some deployments also made in unlicensed spectra in the 5.8 GHz range.

WIMAX services of some kind, either fixed or mobile, have been deployed in most regions and territories. All of the licensed bands in the above list are heavily used for other services in most countries and there has been little or no coordination of the allocation or deployment of WIMAX services, meaning that frequency band usage around the world is something of a 'patchwork' of coverage.

The main measurements that are taken of WIMAX cells are:

- RSSI (Received Signal Strength Indicator), which has a reporting range of −40 to −120 dBm (although according to the IEEE specification, the lowest value that can be measured is −103.75 dBm.

- CINR (Carrier to Interference and Noise Ratio), which compares the wanted signal to the channel noise and has a typical reporting range (depending upon the survey equipment) of +40 dB to −32 dB.

WIMAX networks are typically deployed using frequency reuse techniques, so neighbouring cells will use different carriers. Differentiation between cells at the physical layer will therefore usually be based on carrier frequency – there is no defined channel numbering scheme in WIMAX, base stations report their cell carrier frequencies in MHz.

Cell differentiation at the logical level is performed using a BSID. This is a 48-bit identity, which is usually presented as six groups of two hex digits separated by colons – e.g. 12:34:56:ab:cd:99 – and is essentially identical to the MAC (Medium Access Control) addressing scheme used by Ethernet local area networks.

One method of providing hierarchical cell identification is to use the first 24 bits of the BS ID to identify the network operator and the remaining bits to identify individual base stations.

Current versions of the IEEE 802.16 family of specifications are available from the IEEE website [20] and copies of the commercial and interoperability standards published by the WIMAX Forum are available from their website [21].

6.4.7   Non-Cellular: WiFi

WiFi is a generic name for a set of WLAN (Wireless Local Area Network) technologies, which were developed under the coordination of the IEEE using the standards reference number 802.11 [22].

The name 'WiFi' (which stands for Wireless Fidelity, although the full version of the name is rarely used) was coined by an industry body, the WiFi Alliance (www.wi-fi.org), which coordinates commercial and interoperability aspects of WiFi deployment.

The first IEEE 802.11 standards were published in the late 1990s and have been regularly updated with new versions and capabilities since then.

The majority of residential broadband routers offer a WiFi-based interface to connect to the user's computers, mobile phones and other devices and numerous operators provide public access WiFi 'hotspot' services.

802.11 WiFi services are typically offered in a 'nomadic' fashion rather than being truly mobile. 'Nomadic' in this sense means that a user can travel to an area covered by a WiFi hotspot but will then disconnect from that hotpot if they move out of range. The basic WiFi standards do not provide a generally available method of supporting handover between hotspots, so the type of seamless mobile connectivity offered by cellular networks is not typically provided by WiFi.

The base unit that provides WiFi connectivity is known as an Access Point and each is typically capable of serving 16, 32 or more users simultaneously. Each Access

Table 6.6   802.11 WiFi variants and frequency bands.

Variant	Frequency band (GHz)	Number of channels	Comments
802.11a	5.8	42	Rarely used
802.11b	2.4	14	Commonly used
802.11g	2.4	14	
802.11n	2.4	13	
	5.8		
802.11ac	5.8	42	Recently defined
802.11ad	60	4	

Point will identify itself by broadcasting an SSID (Service Set Identifier), which will be picked up and reported by mobile devices.

LAN (Local Area Network) devices are locally identified within their network using a MAC address. This is a 48-bit identity, which is usually presented as six groups of two hex digits separated by colons – e.g. 12:34:56:ab:cd:99 – which is allocated to the device during manufacture and should be globally unique. Both Access Points and client devices (such as laptops and mobile phones) will have been assigned unique MAC addresses during manufacture.

The SSID of an access point can usually be changed by the owner via the unit's configuration settings, but the MAC address should remain constant. Additionally, Access Points belonging to the same 'hotspot' provider generally all transmit the same SSID, meaning that differentiation of different APs must usually be based on a combination of SSID and MAC address.

Different 802.11 variants and generations have been designed to make use of different frequency bands, as outlined in Table 6.6.

Simple WiFi access points must be manually configured to use a particular band and channel, but more advanced devices support a 'frequency hopping' scheme that allows them to automatically scan for the 'cleanest' local channel. This facility seeks to ensure that users get the best quality services, but also means that access points and hotspots might jump between channels over time.

Measurements of WiFi access points are generally limited to capturing the RSSI of the wanted signal, with typically reported measurement values of +20 to –110 dBm. The range of real world values typically captured for WiFi coverage range from around –30 dBm (very strong) to –90 dBm (very weak).

Most forms of 802.11 WiFi and most types of access point are designed to provide radio coverage across a very limited area, with maximum distances of 10–20 m being common. Surveys of specific WiFi hotspots should therefore be undertaken as close as possible to the access point's location. Surveys of general WiFi hotspot coverage can be expected to capture details of many access points, especially if conducted in urban areas.

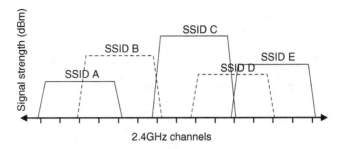

Figure 6.11 WiFi surveys

WiFi hotspot/access point differentiation is based on the following elements:

- SSID – hotspot name
- MAC Address – permanently assigned to the access point
- Channel Number – which can change over time.

WiFi surveys can be conducted using a variety of survey devices and applications. Forensic survey devices such as CSurv, NEMO, TEMS and CSU-4L all capture WiFi data and there are many dedicated WiFi survey devices such as the Wi-Spy dongle and Metageek InSSIDer analysis software.

The type of graphical output typically provided by WiFi survey devices is shown in Figure 6.11.

References

[1] Third Generation Partnership Project 2 (2014) Home Page, http://www.3gpp2.com (accessed 1 June 2014).

[2] CDMA Development Group (2014), Home Page, https://www.cdg.org (accessed on 1st June 2014)

[3] 3GPP2 Technical Specification (2009) *cdma2000 Wireless IP Network Standard: Introduction*, X.S0011-001-E v1.0, www.3gpp2.com (accessed 26 July 2014).

[4] International Forum on ANSI-41 Standards Technology (2007) *System Identification Number (SID) Assignment Guidelines and Procedures v2.2*, http://ifast.org/files/GuidelinesMay2007/ IFAST%20SID%20Guidelines%20r2.2.pdf (accessed 26 July 2014).

[5] 3GPP2 Technical Specification (2009) *Mobile Application Part (MAP) Introduction*, X.S0004-000-E v9.0, www.3gpp2.com (accessed 26 July 2014).

[6] 3GPP2 Technical Specification (2013) *MAP Support for the Mobile Equipment Identity (MEID)*, X.20008-A v1.0, www.3gpp2.com (accessed 26 July 2014).

[7] 3GPP2 Technical Specification (2012) *Introduction to cdma2000 Standards for Spread Spectrum Systems*, C.S0001-F v1.0, www.3gpp2.com (accessed 26 July 2014).

[8] 3GPP2 Technical Specification (2010) *Band Class Specification for cdma2000 Spread Spectrum Systems Revision E*, C.S0057-E v1.0, www.3gpp2.com (accessed 26 July 2014).

[9] 3GPP2 Technical Specification (2012) *Upper Layer (Layer 3) Signaling Standard for cdma2000 Spread Spectrum Systems*, C.S0005-F v1.0 section 3.7.2.3.2.1, www.3gpp2.com (accessed 26 July 2014).

[10] Motorola Solutions (2014) *iDEN – Motorola Product Page*, http://www.motorolasolutions.com/US-EN/Product+Lines/iDEN+Networks (accessed 2 June 2014).

[11] Association of Radio Industries and Businesses (2005) *Personal Digital Cellular Telecommunications System Specifications –STD-27 (fascicles 1-3) Rev. L-E (English Translations)*, http://www.arib.or.jp/english/html/overview/st_ej.html (accessed 2 June 2014).

[12] European Telecoms Standards Institute (2014) *TETRA*, http://www.etsi.org/technologies-clusters/technologies/tetra (accessed 30 May 2014).

[13] TETRA + Critical Communication Association (2014) Home Page, http://www.tandcca.com (accessed 2 June 2014).

[14] European Telecoms Standards Institute (2013) *GSM-R Technical Specification ETSI TS 102 281*, http://webapp.etsi.org/workprogram/Report_WorkItem.asp?WKI_ID=41948 (accessed 2 June 2014).

[15] UIC – International Union of railways (2013) *UIC ERIENE/GSM-R Specifications, ERIENE — System Requirements Specification*, http://www.uic.org/spip.php?article676 (accessed 2 June 2014).

[16] China Mobile Ltd (2104) *Networks and Technologies*, http://www.chinamobileltd.com/en/business/networks.php, (accessed 2 June 2014).

[17] 3GPP Technical Specification (2013) *User Equipment (UE) radio transmission and reception (TDD)*, TS 25.102 v11.5.0, www.3gpp.org (accessed 26 July 2014).

[18] 3GPP Technical Specification (2013) *Base Station (BS) Radio Transmission and Reception (TDD)*, TS 25.105 v11.5.0, www.3gpp.org (accessed 26 July 2014).

[19] 3GPP Technical Specification (2013) *User Equipment (UE) Procedures in Idle Mode and Procedures for Cell Reselection in Connected Mode*, TS 25.304 v11.4.0, www.3gpp.org (accessed 26 July 2014).

[20] Institute of Electrical and Electronic Engineers (2014) *IEEE 802.16: Broadband Wireless Metropolitan Area Networks (MANs)*, http://standards.ieee.org/about/get/802/802.16.html (accessed 30 May 2014).

[21] WIMAX Forum (2014) *WIMAX Forum standards – Releases*, http://resources.wimaxforum.org/resources/documents/technical/release (accessed 2 June 2014).

[22] Institute of Electrical and Electronic Engineers (2014) *IEEE 802.11: Wireless LANs*, http://standards.ieee.org/about/get/802/802.11.html (accessed 2 June 2014).

7

Forensic Radio Surveys

7.1 Forensic Radio Survey Objectives

Forensic radio surveys are used in support of a cell site analysis (or cell tower tracking) investigation and are usually undertaken for one of three reasons:

1. To determine the set of cells that serve or provide coverage at or around a location.
2. To determine the extent of serving coverage of a given cell.
3. To determine serving coverage of a set of cells along a given route.

Forensic radio surveys are most often performed in support of historical cell site analysis (and so are intended to determine where a mobile device could have been when certain calls were made), but may also be performed to gather intelligence as part of 'live' events such as kidnaps (and so are intended to determine the current location of a mobile device).

Details of cell coverage can also be obtained from each cellular network operator; this type of information is discussed in Chapter 8.

7.2 Forensic Radio Survey Terminology

The following descriptions of survey methods and techniques make use of some specific terminology that is commonly employed to describe survey concepts. An understanding or clarification of the meaning of some of the key terms used may be beneficial before commencing the descriptions.

Forensic Radio Survey Techniques for Cell Site Analysis, First Edition. Joseph Hoy.
© 2015 John Wiley & Sons, Ltd. Published 2015 by John Wiley & Sons, Ltd.

'*Target phone*' – this is a mobile/cell phone that is relevant to the investigation. Typically, the investigators in a case have obtained copies of the call records for each target phone and are able to determine the set of cells used by them at significant times (the term 'Subject Phone' is sometimes preferred).

'*Attribution*' – this is evidence that a particular mobile phone belongs (or was used by) a specific individual. If there is no solid attribution evidence linking an individual to a target phone then there is less chance of being able to infer the movements of that individual based on the target phone's call records.

'*Survey results*' – these are the set of radio measurements captured during a forensic radio survey at a location, which provide details of the set of cells that serve or provide coverage there.

'*Serving cell*' – this is a cell that a mobile/cell phone would elect to use to establish a connection at a given location. The equivalent term in CDMA-based networks is 'active cell' or 'active set'. A forensic radio survey report typically indicates the set of cells that were detected at a location and will highlight those that were observed as serving cells during the survey.

'*Provide coverage*' – most locations will be covered by signals from multiple cells per network, especially in urban areas. Cells that can be detected at a location but are not observed as being serving cells can be described as 'providing coverage' there. A more technically accurate description might be that they provide 'non-serving' coverage at the surveyed location.

'*Network*' – the cellular service provided by one telecoms provider. Most countries support multiple networks and most surveyed locations will be covered by cells belonging to multiple networks. During a forensic radio survey it is usually necessary to separately survey the sets of cells belonging to each network operator at a location.

'*Technology*' – 2G GSM, 3G UMTS, 3G CDMA2000 and 4G LTE are all examples of different cellular technologies. During a forensic radio survey it is usually necessary to separately survey the sets of cells of each transmitted technology belonging to each network operator at a location.

'*Survey device*' – this is the radio equipment used to conduct forensic radio surveys and capture cell site measurements. There are several types of device commonly in use by forensic radio surveyors.

'*RFPS*' – Radio Frequency Propagation Survey, is the term used to describe this discipline in some countries.

7.3 Forensic Radio Survey Types and Techniques

A number of techniques are employed by forensic radio surveyors; the specific techniques to be employed on each survey job are usually determined by the characteristics of the associated case and by the assertions that the cell site analysts are attempting to prove or disprove.

The range of typical cell site techniques includes the following.

7.3.1   Spot/Location Surveys

'Spot' or location surveys are designed to capture details of the set of serving and non-serving cells that provide coverage at or near a given location. Generally the locality chosen for the survey is the address where an incident has occurred or where a person of interest in an investigation lives or works. The typical type of question that a spot or location survey would be commissioned to answer would be 'could the user of target phone A have been at location B when call C was made?'

The basic concepts related to spot and location surveys are illustrated in Figure 7.1.

Spot/location surveys generally work best when they are captured over an extended period; a typical ideal value for this would be between 5 and 10 min.

The journey of a radio signal from base station to mobile phone rarely follows a direct line-of-sight path; signal propagation is aided by reflection, deflection or diffraction and the signal as received by a mobile phone often consists of components that have travelled over numerous multipath routes.

The dynamic combining of multipath effects, coupled with other factors, causes the strength of received radio signals to fluctuate over very short periods of time. These effects are known as 'fades' and can sometimes be abrupt and deep – fades of the order of 30 dB (or –1000×) are not unknown, although these are usually associated with a phone moving into a radio shadow or some similar form of attenuation.

The signal provided by one cell may be very strong at one moment but may then suffer from the negative reinforcement of two or more multipaths and offer only a moderate or weak signal in the next instant. This means that the ranking of the set of cells that can be detected at a location can vary constantly and, in areas of non-dominance, may even mean that different cells compete to be the serving cell at that location.

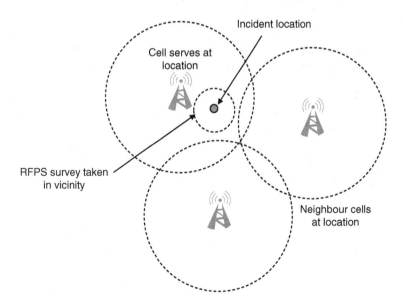

Figure 7.1   Spot/location survey

The benefit of surveys taken over an extended period of time is therefore that the effects of fading and variable signal strength can be 'averaged out' and a more consistent picture of the average coverage provided to the location can be gathered.

Spot surveys are usually employed to capture a set of average coverage measurements for a location.

Traditionally, 'spot' readings were just that – a set of readings taken at a static spot location, with the surveyor siting in a car parked outside a suspect's home address, for example.

In recent years the idea of the 'static' spot survey has been deprecated in favour of a more mobile 'location' or 'local coverage' survey, in which the surveyor walks or drives around the periphery of the address or site being surveyed whilst capturing survey measurements. The reasons for the deprecation of the static spot survey include the variability of coverage that can be experienced over relatively short distances and the shadowing effects of buildings, both of which can mean that the measurements obtained at one spot may not be representative of the measurements that could be captured just a few metres away. The difficulty with spot surveys is therefore related to the surveyor's ability to choose the optimum 'spot' at which to capture survey data.

Multipath effects play an important role in the propagation of most mobile phone signals, as do geography, antenna height, building types and density and a number of other factors. All of these can conspire to concentrate radio signals towards particular areas and may also conspire to direct them away from others.

The signal-blocking effects of buildings also play a part in the variability of coverage. Take, for example, a hypothetical survey undertaken outside a typical residential address, as shown in Figure 7.2.

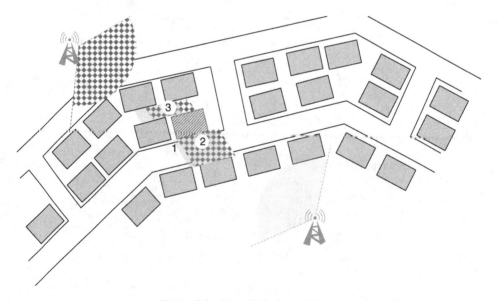

Figure 7.2   Hypothetical spot survey

In this example the address is expected to be served by two cell sites; one of which is located 1.0 km to the rear of the address and the other 0.5 km to the front.

Network readings could be taken at any number of points around the property, but three representative possible locations are shown in the diagram: Spots 1 and 2 are at the front of the property; Spot 3 is at the rear.

In this example, Spot 1 would appear to be the ideal point at which to undertake the survey as it is clear of the radio shadow created by the target address in relation to the rear-facing cell and is also clear of the radio shadow caused by the property on the other side of the radio in relation to the front-facing cell.

Surveys taken at Spots 2 or 3 might result in readings whose accuracy has been compromised by the locations at which they were taken, as each of them is in the radio shadow of one or other of the surrounding buildings.

7.3.2   Location Surveys

The choice of 'spot' is a perennial problem when dealing with spot surveys; however one potential method of dealing with it is to employ the concept of the 'location' or 'local coverage' survey.

A location survey is a walk or drive survey conducted in the immediate 'vicinity' of the target location. There is no agreed definition of the term 'vicinity' in relation to cell site analysis, with different experts employing their own preferred value, but a reasonably representative definition could be that the 'vicinity' of a location includes areas up to a distance of maybe 100 m from the ideal spot location, as shown in Figure 7.3.

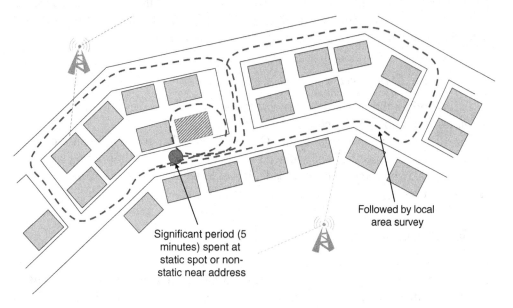

Figure 7.3   Local coverage survey

A location survey involves a continuous set of readings taken whilst the surveyor is travelling around the local vicinity of a significant address. Location survey results should be able to provide a set of average readings not just for the significant location but also for its immediate vicinity. If the survey is taken with the surveyor moving relatively slowly (walking or driving slowly), then the survey should also be able to capture some of the long term variability in coverage in the area, as multipath and other effects cause signal strengths to fluctuate.

The conclusions reached in cell site reports usually refer not just to the significant location but also to the 'vicinity' of that location – for example, a cell site report will often reach conclusions that are worded in a similar way to 'the data is consistent with the target phone being at or in the vicinity of the location'. A location survey can be beneficial in this regard as it provides an indication of the coverage provided both 'at' the location and 'in the vicinity' of the location, as mentioned in the conclusion. This type of survey also has the benefit of being able to sample the cellular coverage provided to all sides of a building or site and should therefore average out the effects of the radio shadows cast by the building and its neighbours in relation to the local cell sites.

In some cases, a non-static location survey may not be possible, due to geography, the type of building or location (e.g. a flat on the 20th floor of a tower block) or concerns for the safety of the surveyor, so an alternative technique is to take a series of static surveys in different locations around the target address and combine the results later.

Another emerging trend in location surveys is to use multiple survey devices (or to use a device like the CSurv MuNST that incorporates multiple survey elements) connected to the same network/technology, so that multiple captures are taken from the same network. This has the effect of multiplying the amount of data captured about a location and also provides an in-built 'sanity check' of captured data, in the sense that the results of multiple captures could identify a set of results that offered anomalous data, that would not be available if only one device were employed.

7.3.3   Static Spot Surveys

Notwithstanding the discussion above regarding locality surveys, there may still be scenarios in which a static spot survey is required.

For example, if an investigation has CCTV (Closed Circuit TV) evidence of a suspect standing at a specific spot and seen in the video to be on their phone at around the time that a significant call was made, it would be beneficial to undertake a static survey at that specific spot to see if the cell used for the call serves there.

7.3.4   Indoor Surveys

Forensic radio surveys are generally undertaken outdoors but may occasionally be requested for indoor locations – for example, if a suspect's alibi places them at home when a significant call was made it would be instructive to determine if the cell used for that call serves inside their address.

It is often possible to get a good impression of the cellular resources that would be available inside an address by taking survey measurements outside, especially if the locality survey technique is employed and the surveyor is able to walk around the periphery of the property.

In some circumstances, though, this may not be possible. If, for example, the suspect lives in a 20th floor apartment it might only be possible to undertake a survey on the public landing at the front of the apartment; this would generally not give a particularly accurate impression of coverage in the rear rooms of the property. Similarly, surveys taken at ground level at the rear of the apartment building may not adequately reflect the difference in coverage experienced in the less cluttered radio environment of the 20th floor.

In these scenarios it is advantageous to gain access to the property to undertake surveys inside the address.

This may not always be possible without a court order or search warrant so, in situations in which an indoor survey *should* have been undertaken but was not permitted, the forensic radio engineer should make it clear in their report that the resulting survey may provide sub-optimal results.

7.3.5   All-Network Profiles

Spot/location surveys are typically undertaken to gather evidence related to a specific set of target phones and are therefore often conducted on just a subset of networks or technologies at a time.

All-network profiles are usually undertaken on all networks and all technologies at a spot or location and can be thought of as a linked set of location surveys, as shown in Table 7.1.

There is an investigative benefit in undertaking forensic radio surveys at key locations as close in time as possible to the events that are under investigation. Prompt and timely surveys can ensure that details related to the state of each network at or near the time of those events are as accurate as possible.

Network configurations change over time; new cells can be added, old cells can be decommissioned, the antennas on a cell site can be 'reorientated' to point in different directions and all of these changes have an effect on the observable cellular coverage at a location. The longer that investigators wait before commissioning a forensic radio survey at a significant location, the greater the potential for network coverage to have changed.

All-network profiles are often undertaken immediately after an investigation commences, sometimes within hours or days of the events to be investigated and often before any suspects have been identified or any call records have been seized. An early all-networks profile ensures that investigators have an accurate 'snapshot' of cellular coverage at key locations in their case, ready to be used if and when specific target phones are identified.

Table 7.1 Example of an all-network profile.

	Network 1				Network 2				Network 3		
	Channel/PCI	Cell ID (status)	Average signal		Channel/PCI	Cell ID (status)	Average signal		Channel/PCI	Cell ID (status)	Average signal
2G	45/27	20456 (s)	−82.34	2G	23/12	234 (s)	−75.12	2G	787/34	45398 (s)	−82.04
	47/29	30456 (s)	−87.23		67/13	345 (n)	−98.12		821/42	2123 (n)	−87.38
	105/23	10456 (n)	−93.87		97/16	12765 (n)	−100.67		698/31	54901 (s)	−88.13
	51/23	38765 (n)	−99.56		—	—	—		670/32	19801 (n)	−90.87
	—	—	—		—	—	—		701/36	20700 (n)	−91.45
3G	10637/145	10987 (s)	−4.56	3G	10712/87	32154 (s)	−7.38	3G	10836/198	56901 (n)	−5.78
	10637/146	20987 (n)	−14.56		—	—	—		10811/198	55901 (s)	−6.87
	—	—	—		—	—	—		10811/199	55902 (s)	−9.21
	—	—	—		—	—	—		10836/198	56903 (n)	−5.78
	—	—	—		—	—	—		10836/199	56902 (n)	−12.72
	—	—	—		—	—	—		10761/14	8790 (n)	−16.87
4G	6400/31	10976542 (s)	−8.32	4G	6300/121	435213 (s)	−7.02	4G	1617/341	62541 (s)	−8.3
	6400/32	10976543 (n)	−14.34		6300/122	435212 (s)	−11.59		—	—	—
	—	—	—		2850/21	143256764 (n)	−18.34		—	—	—
	—	—	—		—	—	—		—	—	—
	—	—	—		—	—	—		—	—	—

The potential disadvantage of speculative all-network profiles, as outlined above, is one of cost; all-network profiles can take a long time to complete and can be expensive. Investigators could commission an all-network profile at a location in the days after an incident, only to find later that all of the target phones identified in the case belonged to just one network – the money spent surveying the other networks could therefore be considered as wasted.

As with most investigative tools and techniques, investigators need to balance the cost and benefit of each technique and make a judgement on a case by case basis. Many investigators consider the benefits of early all-network profiles to outweigh the potential disadvantages.

7.3.6   Cell Coverage Surveys

Cell coverage surveys are intended to determine the extent of serving coverage of a particular cell in a way that allows the approximate 'footprint' of the cell to be mapped. This type of survey is generally performed as a drive survey and the results provide a snapshot of cell coverage at the time the survey was taken. The typical type of question that a cell coverage survey would be commissioned to answer would be 'what is the extent of the area that phone A could have been in when call C was made?'

Given the variability of cellular coverage, a cell coverage survey can only hope to provide an indicative representation of the area served by a cell and it should always be borne in mind that there could be outlying areas of coverage that are not shown on the resulting coverage map but which may also be served by the cell.

The typical 'product' of a cell coverage survey is a map that shows the approximate serving footprint of the target cell.

The coloured dots in Figure 7.4 indicate the location (obtained from a GPS fix) where a target cell was selected as the serving or active cell by the survey device. Cell coverage maps often also show a background trace of smaller coloured dots that indicate the path driven during the survey.

Cell site analysis is not an exact science; in all but a few limited scenarios, cell site analysis cannot definitively state that a target phone must have been at one location and no other location at the time that a call was made. The best that cell site analysis can generally say is that a target phone was somewhere within the serving coverage of the cell used to carry a call at the time that the call started. A cell coverage survey seeks to set some approximate limits on the size of that area, by providing evidence of the area within which the target cell serves.

Derived Service Area

The United Kingdom FSS (Forensic Science Service) developed a concept that they call the 'DSA' (Derived Service Area) of a cell, which is based on the results of a cell coverage survey.

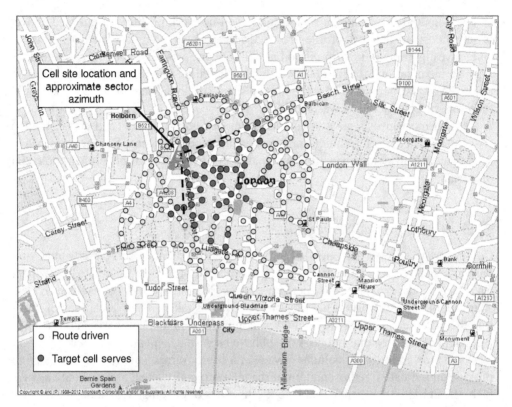

Figure 7.4 Typical cell coverage profile. Source: Microsoft AutoRoute map reproduced with permission from Microsoft

The DSA is the cell's serving footprint, as shown in a cell coverage survey, but is extended to cover some of the gaps that may appear in contiguous cell coverage in such a survey map as well as being extended slightly beyond the surveyed boundary of the cell to accommodate the variability of cell coverage.

When presented in a report, a cell's DSA is identified by a line drawn around the areas within which it is reasonable to assume that the cell has the potential to serve, even if it was not specifically measured as a serving cell in all of those areas during a coverage survey. The DSA therefore also sets the boundary of the area within which it is reasonable to assume that the target phone must have been located when calls using the target cell were made. A typical DSA map is shown in Figure 7.5.

The DSA concept is a useful tool but is arguably partly subjective in nature, as it requires the surveyor to make a judgement about which areas to include with the DSA boundary, and is therefore only currently employed by a minority of cell site experts.

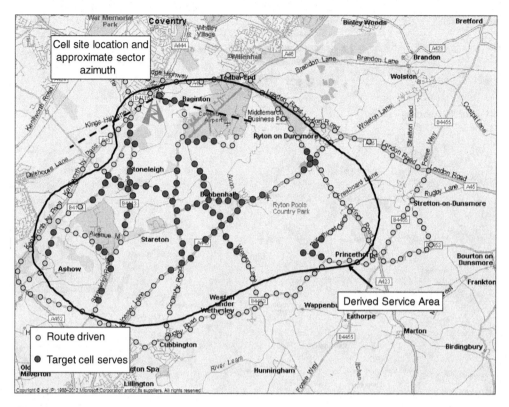

Figure 7.5   Derived Service Area map. Source: Microsoft AutoRoute map reproduced with permission from Microsoft

7.3.7   Route Profile Surveys

A route profile survey is an example of a 'scenario testing' survey that employs similar methods to a cell coverage survey but, whereas a coverage survey seeks to determine the area served by a single cell, a route profile attempts to represent the progression of cells that serve along a given route. The type of question that a route profile survey would be commissioned to answer would be 'could call C have been made by target phone A somewhere along this route?'

This technique can also be used to infer a direction of travel during an alleged journey on the basis of cell handover. A handover from Cell A to Cell B, for example, could indicate the direction that the target phone was travelling in at the time.

The route specified for a route profile is usually dictated by the circumstances of the case being investigated. For example, if the getaway car carrying a gang of armed robbers was observed to travel along a particular route following a bank raid, then the route driven by a forensic radio surveyor would match the path taken by the gang.

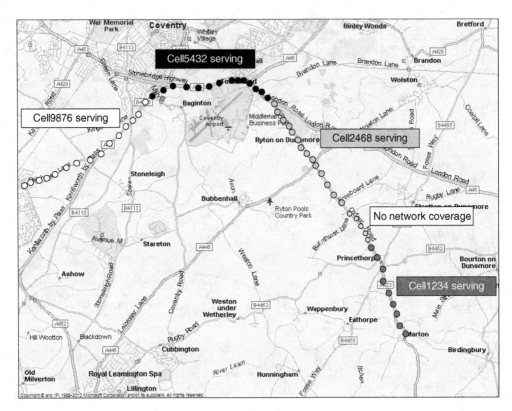

Figure 7.6   Typical route profile. Source: Microsoft AutoRoute map reproduced with permission from Microsoft

In cases where a specific route cannot be determined, the method sometimes employed is to separately drive each of the potential routes that could have been taken by a target phone's user to see if there is a match with the target phone's progression of cells.

The usual method employed when undertaking a route profile is a drive survey, although if the route to be surveyed is not accessible by car – for example, a footpath or canal towpath – or if the significant individuals involved in the events under investigation were thought to have been walking or cycling when the calls of interest were made, a walk survey may be required instead. There have also been cases where route profiles of train journeys and even of boat or ferry trips have been required.

An example of a typical route profile map is shown in Figure 7.6. Such maps are usually colour-coded to indicate the extent of serving coverage of each of the cells detected along the survey path.

As with cell coverage surveys, the results of route profiles should be viewed as indicative rather than absolute as they can only show coverage along the surveyed route at the time that the survey was undertaken – with the survey device typically moving quickly through each surveyed point along the route – and cannot take account of long term variations in signal strength over time.

If a route profile survey fails to capture the expected data, if for example the cell used by a target phone during the time that the user of the target phone is suspected of having been travelling along the surveyed route, then further surveys could be undertaken of the same route. Usually, if additional surveys are required the forensic radio surveyor will drive or walk the survey route at a much slower pace than on the first attempt, possibly stopping for a few minutes every few hundred metres, depending on the length of the route. This method gives the survey device a much better chance of picking up any long term variations in serving cell coverage along a route and may allow more than one serving cell to be detected in each subsection of the route.

Another technique that can be employed is to check the neighbour information captured during a route survey. If the expected cell was not detected as serving along the route but was detected as a strong neighbour cell, then that may provide an indication of points along the survey route where the surveyor could stop and spend time to see if the expected cell is selected as an occasional serving cell.

7.4   Idle Mode versus Connected Mode Surveys

Mobile phones that are powered on and attached to a valid network will generally be in one of two modes:

1. Idle Mode, where they are not connected but are available to make calls.
2. Connected (or Dedicated) Mode, where they are connected to a base station and are able to actively exchange user traffic with the network.

There are also a number of 'limited service' modes, which are typically employed when normal connections are not available (such as 'emergency calls only' mode).

Most forensic radio survey devices are able to capture radio measurements in both Idle Mode and Connected Mode and there has been an on-going debate within the cell site community regarding the appropriate mode(s) in which to capture different types of surveys.

7.4.1   Idle Mode

Mobile devices are typically battery-powered, which requires system designers to incorporate ways of saving power during periods when the device is not actively connected to the network.

The Idle Mode concept employed in all types of cellular network has been designed to serve this purpose.

When a device is powered on but is not engaged in a call, exchanging text messages or actively exchanging data with a data service it will generally drop into Idle Mode to preserve power.

The generic activities undertaken by a 3GPP mobile device in Idle Mode are set out in Ref. [1]; these are generically representative of the activities required in other types of network and include:

- Network selection
- CSG (Closed Subscriber Group) selection
- Cell selection and reselection
- Location registration.

7.4.2   Connected Mode

When a mobile device is required to establish a radio connection to its serving cell it moves from Idle Mode into Dedicated or Connected Mode – Dedicated mode was the term originally used in 2G GSM networks, Connected Mode is the term used in 3G and 4G networks – both terms mean essentially the same thing. The most recent versions of GSM specifications now also use the term Connected Mode – it is used in this text to ensure consistency.

The generic functions performed by a device when in Connected Mode include:

- Transfer of user traffic and control information to/from the network
- Taking serving and neighbour cell measurements
- Reporting radio measurements and traffic quality information to the serving base station
- Performing handovers to neighbour cells if instructed to do so by the network.

7.4.3   Transition from Idle Mode to Connected Mode

One of the key functions of cell site analysis is to attempt to determine the choices that a mobile device would make when instructed to transition from Idle Mode to Connected Mode at a certain location. This is one of the main reasons why forensic radio surveys are undertaken; to determine which cells serve at a location and therefore which cells a mobile device is likely to choose to use there.

When an Idle Mode device is required to establish a connection it will typically attempt to use the cell it is currently camped on. That is one of the main reasons why mobile devices perform reselections – to ensure that they are camped on the 'best' available local cell in preparation for moving to Connected Mode.

7.4.4   Idle Mode Surveys

Idle Mode surveys attempt to capture the radio environment as observed by a mobile device 'at rest'.

This means that the survey is essentially a passive sampling of the locally available radio channels during which, barring Location Updates, the survey device would not be expected to actively transmit.

A survey device will typically mimic the activities of a 'real' mobile device by processing the neighbour cell list (e.g. the BA List employed by 2G networks or the NCL used by 3G and 4G networks) provided by the current serving cell, searching for cells that are members of the list and measuring their received signal strengths. Unlike a 'real' mobile device, however, a forensic survey device will display and record the readings that have been captured for later analysis.

An Idle Mode survey is typically conducted by ensuring that the survey device has been fitted with a SIM card belonging to the appropriate network (or has had a network lock put in place), ensuring that is in Idle Mode, starting the data recording and leaving the equipment alone for the duration of the survey. Some survey equipment allows the surveyor to 'lock' it to a specific technology, band, channel or Cell ID and Idle Mode surveys are usually conducted with the equipment locked to the required technology (e.g. 2G, 3G or 4G). Technology and band locks also overcome some of the issues related to any Idle Mode behaviour settings the operator may have in force, which may attempt to force the device to camp on cells belonging to a specific technology or band when idle.

Idle Mode surveys conducted with a channel or cell lock in place are often undertaken to allow the capture of measurements of a specific resource – for example, if the surveyor has been instructed to obtain measurements of only the 3G channel an operator uses to carry data services.

Benefits

Idle Mode surveys are useful in the sense that they replicate the state that most target phones will be in most of the time – that is switched on and waiting to be used.

The objective of most forensic radio surveys is to attempt to determine which cell(s) a phone might choose to use if it were instructed to connect to the network from a particular location. As mobile devices usually attempt to use the cell on which they are currently camped when they are required to connect, it is advantageous to know which cell(s) a device will select and reselect in Idle Mode at that location.

Idle Mode surveys provide information about the relative likelihood of a cell being selected as serving at a location or in an area, but to be generally representative of the coverage at a location they typically need to be conducted over a relatively extended period. Between 5 and 10 min is a common duration range for Idle Mode surveys.

Disadvantages

From the description provided above, an Idle Mode survey would appear to be the ideal mode in which to undertake surveys. There are, however, a number of aspects of mobile operation that either cannot be tested in Idle Mode or that interfere with the decision-making process employed by idle mobile devices.

The first of these issues relates to the fact that a mobile device 'attempts' to use the cell on which it is currently camped, it cannot be guaranteed that any connection attempt will succeed. An Idle Mode survey will only indicate that a cell was selected as serving based on its signal strength, it will not indicate how busy that cell is or how likely it is that it would be able to accept a call set up request. Additionally, most types of survey equipment do not capture details about the quality of the signal from a cell, only its strength.

For surveys undertaken in very busy locations (city centres, shopping malls, airports, etc.) there exists the possibility that, when a mobile attempts a call set up via the locally strongest cell, that cell might be too busy to accept the connection, meaning that either the call will be rejected or that Directed Retry will be invoked. Directed Retry is a network technique that, as the name suggests, directs a mobile device to retry its connection attempt via a specific neighbouring cell in the event that the selected cell is too busy. The strongest cell might not, therefore, be the most commonly used at that location.

Another example of a potential issue, particularly in 2G scenarios, is where the selected cell is using a channel that is suffering co-channel (another nearby cell using the same channel) or adjacent channel (a cell nearby using a channel that is adjacent to the target cell in spectrum terms) interference. Sources of interference might be more periodic and might, possibly, be associated only with particular times of the day or be experienced only when the network is particularly busy. In both of these scenarios the mobile device might measure a strong signal from the target cell, which actually turns out to have unusably poor quality.

In reality, neither of the scenarios outlined above is particularly likely or commonly encountered, and as the issues that might cause them could be linked to a specific level of traffic or to a specific time period, they would be very difficult to recognise and to accommodate in a survey schedule.

A more pertinent issue associated with Idle Mode surveys is related to the ability of network operators to impose 'Idle Mode behaviour' on attached devices. Idle Mode behaviour settings can be used to make particular cells, channels or even whole technologies more or less attractive to idle devices. This can either be done implicitly, by adjusting the reselection parameters advertised by cells or explicitly by providing mobile devices with prioritisation instructions when they leave Connected Mode.

The net effect of both of these prioritisation schemes is that a mobile device may not necessarily select a local cell as serving even if it has the highest measured signal strength.

An even more likely issue related to Idle Mode surveys is that a device in Idle Mode (provided with an Idle Mode BA List or Neighbour Cell list by its current serving cell) might not have been provided with details of all local cells or channels, meaning that it might not know to take measurements of them. This scenario is often referred to a 'missing neighbour' and the net effect of it is that, if a handset is not informed of the presence of a local cell, it will not try to measure its signal strength and will not report this to the network. When the network looks for handover opportunities the 'missing cell' will not form part of any decision-making process and thus it could lead to calls getting dropped unnecessarily, which is a network performance issue.

The points made in this section serve to highlight the notion that an Idle Mode survey alone might not capture all relevant information about the coverage provided at a location.

Conclusion

Idle Mode surveys are generally simple to undertake and can provide a sufficiently representative view of the cellular coverage provided at a location or in a locality.

Idle Mode surveys are generally recommended for spot/locality surveys and for cell coverage surveys. They are also useful for route profile surveys. However, Idle Mode surveys only capture information about the strength of a cell's signals at a location, they do not capture information about the usability of that cell, nor do they capture serving cell information about cells that have been deliberately configured not to be selected as Idle Mode serving cells.

Surveying only in Idle Mode could potentially lead to relevant information about the coverage at a location being missed.

7.4.5   Connected Mode Surveys

A Connected Mode survey is usually undertaken in one of two ways – a constant connection or intermittent test calls.

In both of these methods the survey device is instructed to make a call (or receive a call, send/receive a text message or connect to the Internet) whilst the survey application is capturing measurement data.

The justification put forward for performing Connected Mode surveys is that they can answer the question that Idle Mode surveys cannot; namely, 'which cell would a phone actually choose to be connected to if it made a call from here?'

Connected Mode surveys, like Idle Mode ones, are often conducted with at least a technology lock in place. This is to ensure that only measurements of the required technology are captured, although 'free running' surveys (with no technology lock in place) can also be performed if the propensity for inter-system reselection/handover between 2G, 3G and/or 4G needs to be tested as well.

Benefits

The main benefit of a Connected Mode survey is that it provides unarguable evidence of the cell(s) that would actually be selected by a mobile device to initiate a call and by the network to maintain a call (via handover). The results of test calls should not invalidate the results of Idle Mode surveys, however, they simply provide additional evidence regarding the set of serving cells.

This potentially means that any issues related to the availability of a cell, the quality of the connection to a cell or the 'attractiveness' of a cell are nullified and a more complete picture of the usable resources at a location can be gained.

The use of Connected Mode test calls has benefits for 2G surveys, especially surveys of networks that deliberately configure some cells not to be selected as serving cells in Idle Mode.

An example of this can be seen in 2G networks that to deploy some cells in the E-GSM900 and GSM1800 bands as 'capacity layer' or 'capacity extension' cells. These cells have their advertised C2 reselection parameters configured with a permanent negative offset so that Idle Mode devices, after applying the offset, measure their signal strengths as lower than other local cells and therefore do not select them. Networks do this to ensure that the capacity of these cells is reserved exclusively to provide TCH (Traffic Channel) capacity to serve Connected Mode devices. A phone may therefore initially establish a cell using a P-GSM900 cell but then be quickly handed over to an E-GSM900 or GSM1800 capacity layer cell to continue the call. The P-GSM900 cell would show up on a target phone's call records as the start cell for calls and the capacity layer cell would show up as the end cell for those calls. A survey undertaken in Idle Mode only would measure the capacity layer cell as a strong neighbour but would not list it as serving, requiring the additional test calls to attempt to complete the coverage picture in that cell.

There are also particular benefits associated with the use of Connected Mode test calls in relation to 3G surveys. CDMA-based 3G (UMTS and CDMA2000) networks are typically deployed in 'single frequency network' configurations. This means that all cells in a network (or in a particular 'layer' of a network like the voice layer or the data layer) will be using exactly the same radio channel as their neighbours.

Handover in a single frequency environment is therefore not as simple as it is in the 'multi-frequency' environment employed by, for example, GSM (where each cell uses different channels to those allocated to its neighbours). For power control and connection management reasons, single frequency CDMA-based networks must use a handover technique known as 'soft handover' in which the connection to the 'new' cell is established *before* the connection to the 'old' cell is released. This means that, for a brief period of time during a handover, a 3G device will be connected to two (and sometimes, three) cells simultaneously.

For a variety of technical reasons, the 'soft handover candidate' cells employed during 3G handover are sometimes drawn from a set of neighbours that do not show up in Idle Mode surveys, meaning that details of those cells would not be captured if Idle Mode surveying alone was employed. One reason for this could be that the network advertises an abbreviated set of local cells in Idle Mode neighbour cell lists that do not include more distant neighbour cells that might only occasionally provide signals that are strong enough to serve.

Disadvantages

There are several disadvantages to Connected Mode surveying that are generally put forward by critics of the technique.

The first relates to the additional cost and activity overheads that are associated with Connected Mode surveys. Additional costs may be incurred by making the calls that move the survey equipment into Connected Mode in the first place. Short duration test calls can usually be managed by dialling a 'free' phone number, such as the operator's voicemail or customer services number. Longer duration or constant connection surveys would usually require the surveyor to dial a 'real' phone number, which may incur charges.

There are potentially also costs related to the additional time it takes to conduct the surveys (making the calls, redialling if calls drop and recording additional call-related information).

The second objection to Connected Mode surveys is more contentious and relates to the possibility of (intentional or unintentional) bias that can be applied by the surveyor.

The argument put forward here is that during an Idle Mode survey there is very little that the surveyor can do to influence the results obtained (short of setting a lock on the survey device, placing it inside a Faraday bag or lying about the location of the test). During a Connected Mode survey, especially one that involves intermittent test calls, a surveyor could influence the choice of cell used to carry a test call by only dialling a number when the cell they favour is shown as the current serving cell. Whether consciously or not, the surveyor's influence could promote a cell that sometimes serves in Connected Mode into being the cell recorded as the only cell that serves in Connected Mode.

Of course, one way of avoiding this scenario is to use a survey device that automates and randomises the test call process, such as CSurv MuNST. In any event, if the results of Idle Mode surveys at the location are also taken into account (and are given equal weight), they will show the uninfluenced set of Idle Mode servers and this information can be combined with the Connected Mode results to provide a balanced view of coverage.

There is one further point to make in relation to Connected Mode surveys, it is more of a comment than a disadvantage, and that is that surveys involving femtocells are unlikely to be able to be handled in Connected Mode. Most femtocells are deployed in the 'closed' state, meaning that they will only accept connection requests from phones that registered as belonging to the cell's CSG. Phones that are not registered with the CSG will not be able to make calls via the femtocell, apart from emergency calls, and this includes forensic radio survey test calls. Femtocell surveys will almost always have to be conducted in Idle Mode only.

Conclusion

Given the potential for information related to the usability of cells to be missed if surveys are conducted only in Idle Mode, the use of Connected Mode would seem to be beneficial. However, there are two methods for undertaking surveys in Connected Mode; constant connections and intermittent test calls.

It can be argued that maintaining constant connections during surveys offers the ability to test the propensity for a cell to be selected in a network-controlled handover, rather than testing the likelihood of it being used for initial call set up. As discussed elsewhere in this chapter, constant connection surveying also introduces the potential for 'cell dragging', especially for cell coverage surveys.

Conversely, the intermittent test call option would seem to offer the ability to test the selection of a serving cell during the transition from Idle Mode to Connected Mode. This method seems to offer the most appropriate way of obtaining the very information that a cell site survey is intended to discover.

Using only Connected Mode for spot/location surveys would therefore not be recommended, but the use of intermittent test calls during these types of survey would.

Connected Mode surveying of either type (constant or intermittent) for cell coverage or route profile purposes would not be recommended, as the main purpose of these types of survey is to determine the extent of the area within which a particular cell would be selected as serving or to determine the areas within a route at which a particular cell would be selected for call setup. There are a small minority of potential scenarios in which Connected Mode test calls might be used during a route profile survey, but these would all relate to attempts to determine the areas in which a certain cell might be used as an 'end cell' during calls.

7.4.6   Mixed Mode Surveys

A typical mixed survey would incorporate periods of Idle Mode surveying augmented with a series of Connected Mode test calls.

This type of mixed survey could be handled in a number of different ways:

- The engineer could complete an Idle Mode survey and then make a separate series of test calls.
- The engineer could begin capturing an Idle Mode survey and then make a series of test calls using the same survey device during the Idle Mode survey.
- The engineer could capture an Idle Mode survey on one survey device and simultaneously make a series of test calls with a separate device and then combine the results obtained from both devices.

Mixed mode surveys are usually employed when undertaking spot/location surveys; the technique is less commonly employed for coverage or route profile surveys, for reasons that will discussed later in this chapter.

Benefits

Employing a mix of Idle and Connected Mode techniques during spot/location surveys allows forensic radio survey engineers to test the decisions made by mobile devices during the transition from idle to connected, whilst at the same time being able to capture long duration Idle Mode measurements. It also allows them to test the selection of cells that might not be seen to serve in Idle Mode due to network-defined offsets or other Idle Mode behaviour settings.

Disadvantages

The addition of test calls to an Idle Mode spot/location surveying schedule may increase the amount of time necessary to complete surveys at each location and the requirement to make calls may increase the overall cost of surveying.

The arguments against using Connected Mode test calls outlined in the Connected Mode section above may still be valid for mixed surveys. For example, the suggestion that the survey engineer could consciously or otherwise influence the outcome of a Connected Mode survey by selecting the time at which calls are made based on the current serving cell – but again, this objection can be countered by ensuring that the results of both the Idle Mode and Connected Mode surveys are combined and given equal weight.

Conclusion

Cell site analysis is mainly undertaken to determine whether particular cells could be used as 'start' cells at a given location. Therefore, the information gathered during a cell site survey would be expected to identify, from the pool of local cells, the set of Idle Mode serving cells on which devices are most likely to camp and the cells that would be selected by devices when they are required to transition from Idle Mode to Connected Mode. A mix of Idle Mode surveying supported by Connected Mode test calls would therefore seem to cover all requirements.

From a practical point of view, however, Connected Mode test calls may not always be necessary. Survey engineers are usually provided with a set of target cells for each location/area/route to be surveyed. If the expected target cells are detected as serving cells during the Idle Mode phase of a survey it may not be necessary to also undertake Connected Mode test calls – this may be especially true of route profile surveys.

7.4.7   Suggested Survey Modes

A summary of the suggested mode selection choices for forensic radio surveys follows:

- Employ Idle Mode for spot/location surveys, supplemented by intermittent Connected Mode test calls if expected serving cells are not detected in Idle Mode.
- Idle Mode spot/location surveys will be more representative if captured over a relatively long duration of 5–10 min.

- Employ intermittent Connected Mode test calls at locations to prove specific points, such as the use of cells as 'end cells' and the likelihood of them being used for handovers.
- Intermittent Connected Mode test calls are also useful in checking whether cells that have been optimised not to serve in Idle Mode might serve in Connected Mode.
- 3G test calls might be more useful if they last for up to 1 min (if call length is controllable) to provide a decent opportunity for any 'soft handover only' cells to be selected.
- Cell coverage surveys should be undertaken in Idle Mode only.
- Route Profile surveys should be undertaken in Idle Mode, supplemented by Connected Mode test calls in specific circumstances, such as cells that only appear as 'end cells'.

7.5 Additional Survey Techniques

7.5.1 Surveying using Multiple Devices

A single forensic radio survey device will capture details of the radio signals it can detect and the signal strengths that it measures.

All radio receivers, at least of the quality employed in commercial mobile phones, are constructed from elements that have comparatively wide variations in composition and tolerance and each device will measure received signal strengths slightly differently. It can be a little unsettling for forensic surveyors to realise that no two phones or survey devices will ever capture exactly the same set of measurements at a location, even if they are used side by side and capture their survey data simultaneously. In fact it is not that uncommon to view the output of side-by-side survey devices and see that they are measuring the signal from the same cell with a signal strength difference of 3 or 6 dB or even more.

This in turn means that the picture of radio coverage captured by a forensic survey device will not match exactly the conditions experienced by the target mobile phones that are the subject of cell site investigations.

One way of ensuring that cell site conclusions are based on a reasonably accurate understanding of the radio coverage at a location is to undertake 'long duration' spot/ location surveys of 5–10 min or more. The long capture durations help to create a more accurate view of the average conditions at a site by ensuring that a representative sample of the variations that occur there are recorded.

Another way of improving the accuracy of a survey is to use multiple survey devices simultaneously – or, in the case of devices like the CSurv MuNST, which incorporates 10 or 20 separate radio sub-units, survey using multiple sub-units set to the same network and generation simultaneously. If more than one device/sub-unit (with SIMs from the same provider and the same technology/band/channel locks in place) are used, the natural differences between the multiple receivers should lead to the capture of a wider and more diverse set of measurements. This is illustrated in Figure 7.7.

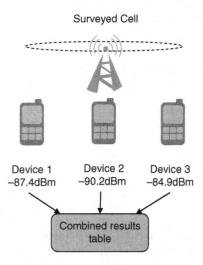

Figure 7.7 Multiple device surveys

If the results of these separate but simultaneous surveys are combined they can arguably provide a more comprehensive view of the coverage at the survey location.

Surveying using multiple devices can be beneficial and it is a practice that is enthusiastically endorsed by several cell site experts. Other experts argue just as strongly against this practice, however, due to the complications that can be introduced into a case.

The potential downsides of using multiple devices revolve around the fact that the devices will undoubtedly record a different set of measurements to each other. Combining these outputs into a single set of results can be useful, but if the individual results are viewed separately and are compared against each other they can be used to undermine the whole basis of the survey. This is a tactic that that is employed by some defence experts.

The narrative for this form of attack goes something like this: 'the prosecution has provided details from multiple survey devices which do not agree with each other. How can we therefore believe that the results of these surveys can accurately tell us about the choices that could have been made by the defendant's phone at the same location?'

Another line of attack, which is sometimes used if the results from two different survey devices vary significantly, is: 'how do we know that one or both of these devices are accurately calibrated? How do we know that all of the other results obtained using that equipment are not also flawed?'

The very diversity of information that multiple device surveys provide can be turned against the surveyor if there are significant differences between the sets of results that are obtained.

Multiple device surveying can therefore be seen as beneficial, as long as the potential downsides are understood and are taken into consideration.

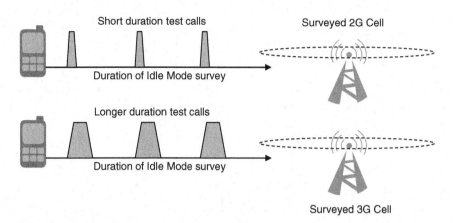

Figure 7.8 Test call patterns

7.5.2   Test Call Protocols

Connected Mode test calls are often made either during or after an Idle Mode survey, with the survey equipment capturing details of the cell(s) used during each call.

Some surveyors advocate making a series of several short duration calls (1–2s each), while others prefer to make just one or two longer duration calls (1 min or more). Longer duration calls are more beneficial during 3G surveys as they increase the potential for detecting soft handover candidates that may not appear during Idle Mode only surveys. These options are illustrated in Figure 7.8.

In some circumstances, surveyors elect to perform a 'constant connection survey', which employs just one long duration call that lasts for the entire survey period. Constant connection surveys are usually only undertaken in order to prove a very specific point, such as the potential for a cell to be used during soft handover when it is not detected in Idle Mode or the potential for a cell to be used as an 'end cell' when it is not shown to serve in Idle Mode.

Many forensic surveyors and cell site experts are wary of employing constant connection techniques, especially during cell coverage or route profile surveys, mainly due to the potential for triggering what is known as 'cell dragging'.

Most networks employ different parameters for determining and controlling Connected Mode handover than they do for controlling Idle Mode reselection. If a survey device has a test call setup it will use the cell that best satisfies the cell reselection criteria at that point in time. If the survey device is moving, as part of a coverage or route survey, it may move out of the area within which the connected cell might normally be expected to be the best reselection candidate and into an area where a neighbouring cell might be expected to serve.

If the handover parameters for the serving cell are set to values that discourage a handover from taking place it might be the case that the device 'drags' serving coverage from the current cell into an area where that cell would not usually be expected to be selected as an Idle Mode server and where that cell would not usually be used to set up a new call.

The effect of such cell dragging might therefore be to present a false impression of the area within which the connected cell naturally serves and could unintentionally mask areas in which other cells might be expected to be the main Idle Mode serving cell.

Constant connection test calls are therefore typically used only in scenarios where they can provide specific evidence.

4G LTE networks do not necessarily support voice calls. Most networks offer voice service using CSFB (Circuit Switched Fallback, which forces the mobile device to drop back to a 2G or 3G cell in order to establish a call. Some networks offer VoLTE (Voice over LTE), which carries voice calls within a data session using VoIP technologies, but support for VoLTE is far from universal among networks or survey equipment vendors.

In the case of 4G LTE surveys (and also in some circumstances, of 3G HSPA/ HSPA + surveys) it may be desirable to force the survey equipment to establish a data session, so that details of the Connected Mode serving cell for data services can be captured. One way of forcing a data session to be established is to use the survey device's browser to connect to the Internet; in theory, any website connection would serve this purpose but if the intention was to check the cells used for a fast data connection (or check for handover or 'end cell' usage during longer duration sessions) then surveyors may wish to connect to a service such as YouTube and stream a video for a few minutes. An alternative method, if the survey equipment can be appropriately configured, is to use an FTP (File Transfer Protocol) application to upload or download a fixed-size file to/from an FTP site. This has the benefit of performing a data session test with a known amount of data.

One of the objections to making test calls (and this applies to test data sessions as well) is that establishing connections costs money. Many surveyors use prepay 'pay as you go' SIMs in their survey equipment, especially if they are surveying using multiple devices or sub-units, and maintain the minimum amount of 'top up' credit on each SIM. To avoid using any of this credit, surveyors often make test calls to 'free' numbers, such as the operator's customer care or top up number.

The adoption of test data sessions for 3G and 4G surveys, using the methods suggested in this section, would inevitably lead to the consumption of data and the use of the SIM's prepay credit. One way of avoiding data charges is to ensure that the web pages requested during data sessions are 'free' resources from within the operator's Intranet. Examples of this could be the operator's WAP home page or a top up/credit check service page, although surveyors would need to experiment with test data sessions to find 'free to use' resources in the surveyed network that do not consume data credit when accessed.

Test calls and data sessions can provide useful additional information for surveys but, as outlined above, there are ways in which they can be undertaken without adding to the cost of those surveys.

7.5.3   Lock Files

A 'lock file' is created using a survey device's 'cell lock' feature – which is not supported by all devices on all technologies.

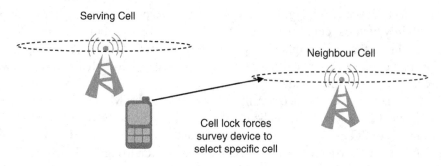

Figure 7.9 Cell locks

As the name suggests, a lock file is created by systematically working down the list of detected Cell IDs displayed on the survey equipment screen and applying a cell lock to each of them in turn. Surveyors sometimes also apply a lock to any Cell IDs listed on their target cell list that have not yet been detected, just to see if the cell is available even if it had not been listed on the survey screen.

The effect of the cell lock, as shown in Figure 7.9, is to override the normal cell selection/reselection routine in the survey device (which is essentially based on selecting the cell that is currently strongest) and forcing the device to camp on a cell that it might not necessarily choose to be served by at that point in time.

Lock files are beneficial because some survey devices do not routinely capture Cell IDs or other relevant details for neighbour cells; they only routinely capture full details of serving cells.

Survey results that indicate the presence of neighbour cells (by capturing details of their radio channel and physical layer ID) but do not identify them (via a Cell ID) are of limited use forensically, as call records generally do not provide details of a cell's radio channel or physical layer identity, so the act of forcing the survey device to camp on each detected cell in turn theoretically improves the chances of full details of all neighbour cells being captured.

Depending upon the type of survey device and the strength of the signal from the neighbour cell, the cell lock usually needs to be applied for a reasonable period of time – up to 10–20 s for very weak neighbours – and even then there is no guarantee that the required details will actually be captured if the signal is of poor quality.

7.5.4 Orbit Tests

An orbit test, also known as 'spinning a site', is undertaken to determine the true azimuths of the sectors on a site, which is sometimes necessary if the azimuth information provided by the network operator is out of date or inaccurate.

As illustrated in Figure 7.10, the orbit test involves walking or driving around a site, usually at a distance of 100 m or less, and noting when the received signal strength from each sector's Cell ID reaches its peak value on the survey device measurement

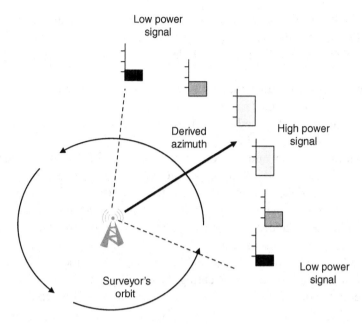

Figure 7.10 Orbit test

screen. Noting the location at which the peak signal is measured in relation to the site itself allows the surveyor to calculate a derived azimuth for each sector.

Obtaining accurate information about each significant cell's true azimuth is beneficial for two reasons.

First, it can be a useful indicator that some or all of the cells in an area have undergone 'reorientation', possibly as part of an optimisation programme, in the time since the call records for a target phone were produced. This in turn can provide the forensic surveyor and their instructing cell site analyst with an indication that the results obtained in the radio surveys might not be fully applicable to the calls made by the target phone – if the network has changed in the time between the calls being made and the survey being undertaken.

A second benefit is that knowledge of the accurate azimuth can allow the cell site analyst to mark the correct azimuths on any cell coverage maps created to support their cell site report, which ensures that any evidence put before a court is as accurate as possible.

7.6   Survey Preparation

7.6.1   Survey Specifications

Before undertaking a forensic radio survey it is considered good practice for the surveyor to have an indication of the cells that are expected to be found at each location.

A cell site analyst will use the call records for each target phone to draw up a list of the cells used by those phones at times when the users of those phones were suspected of being at locations of interest. They will use this information to draw up a target cell list of each location, area or route to be surveyed.

An example target cell list might include some of the following information:

- Location/cell coverage area/route to be surveyed.
- Networks and technologies to be surveyed.
- List of cells of interest (based on Cell ID) that are expected to be detected on each network/technology at that location.
- Additional information, such as 'cell X was only used for data sessions', 'end cell only' or 'cell Y was used extensively' can aid the survey engineer's understanding of the objectives of the survey and the relative importance of each cell.
- An indication of whether an attempt should be made to arrange for the survey to be conducted inside an address. This would be an issue in cases where the suspected user of a target phone was assumed to have been indoors when calls were made and is especially relevant to locations such as flats and apartments, where an outdoor survey would have difficulty replicating the coverage provided at altitude.
- Any potential dangers or sensitivities that should be taken into consideration by the surveyor.
- Cell address/location details for expected cells, including cell azimuth details.
- Cell and survey location map, which would usefully include details of the cell locations and azimuths in relation to the survey location.
- Providing extracts of the relevant CDR (Call Detail Records) that show the calls made using the cells to be surveyed can also be useful, this can add useful context for a surveyor by indicating, for example, the sequence and rapidity with which particular cells were used.

An example target cell list is shown in Figure 7.11 and the accompanying cell location map is shown in Figure 7.12.

7.6.2 Preparing Survey Devices

Each type of forensic radio survey has its own generic actions and activities, but the specific actions to be performed for each individual survey are typically dictated by the circumstances of the case, by the location at which the survey is to be based and by the surveyor's personal preferences.

It is therefore difficult to draw up a set of suggested activities and guidelines that are relevant in all scenarios and that match with every surveyor's preferences. The following suggested actions and activities should therefore be seen as at least partly subjective, as they are based on the author's experiences and preferences.

Before commencing each new survey, it is recommended that surveyors check the following:

- Make sure the survey device has sufficient charge for the predicted survey duration (if not on external power).

Case Name:	R v Moriarty	Case Ref:		CSAS-1-2014			

link to map

Location Surveys — Expected Cells

Location	C - 24 Lower St, Anytown	2G		3G		4G	
Significance	Robbery address	Cell ID	Comment	Cell ID	Comment	Cell ID	Comment
Vodafone	Surveyor Comments	6776	voice & data use	4144	data use only	not required	
		244	used once				
		245	end cell only				
O2	Surveyor Comments	20234	only used once	10298	data use only	not required	
		30234	end cell only	30298	end cell only		
		13423	used extensively				

Coverage Surveys
Survey 1

Network	O2	Technology	2G	Cell ID	13423	link to map
Site Name	Bubbenhall	Azimuth	20	Site Location	Bubbenhall Rd, Ryton - 52.45672, -1.09765	

Route Surveys

Route ID	1	Description	Marston to Kenilworth via A423, A45, B4115				
			Expected Cells				
Comments:		2G		3G		4G	
		Cell ID	Comment	Cell ID	Comment	Cell ID	Comment
O2	Surveyor Comments	free		free		not required	

Surveyor	Jim Smith	Organisation	RF Group Ltd	Comments:
Survey Date		Equipment		
Requested By	Jill Jones	Date	02/02/2014 15:12	

Figure 7.11 Target cell list. Source: CSAS target cell list reproduced with permission from Forensic Analytics

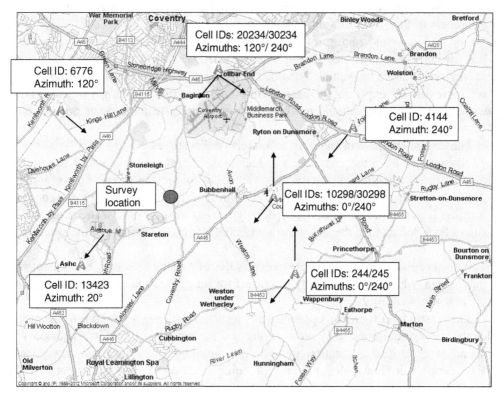

Figure 7.12 Cell location map. Source: Microsoft AutoRoute map reproduced with permission from Microsoft

- Make sure the survey device has sufficient free memory or enough spare data cards to store the expected survey data file(s).
- Make sure any previously applied technology, channel or cell locks have been reset if not required for the new survey.
- Make sure the required technology, channel or cell locks for the new survey are set.
- Make sure that the 'save as' filename for the survey is correct and reflects the survey being undertaken – this could include making sure that the correct location, network and technology are listed in the filename if a specific file naming convention is being employed.

Failure to perform these simple checks can lead to surveys failing to capture the required information, meaning that they may have to be rerun at extra cost.

7.6.3   Survey Safety

Forensic radio surveys can be hazardous for a number of reasons.

Spot and all-network surveys, for example, are generally undertaken outside, often close to roads and usually require the surveyor to concentrate on the information being displayed on the test equipment. Location surveys are often undertaken as walk surveys but also require the surveyor to devote at least part of their attention to the output displayed on the survey device's screen. Care should be taken to maintain an awareness of the surveyor's surroundings.

Cell coverage and route profile surveys are usually undertaken as drive surveys. It is strongly recommended that at least two people should be involved in a drive survey, one to drive the vehicle and one to operate the survey equipment. Lone working surveyors are exposing themselves and other road users to danger due to driving without due care and attention whilst also attempting to monitor or operate their survey equipment.

The circumstances surrounding a forensic radio survey should also be borne in mind: surveys are often undertaken at locations where traumatic events have taken place or near the addresses of witnesses, victims or suspects in a case or their families. A surveyor may unwittingly cause further distress if the reason for their presence at a location is guessed. Additionally, especially related to surveys near the addresses of suspects or defendants, there is a danger of attack from the suspect or their family or friends.

A further risk is related to the surveyor's use of expensive survey equipment in a public area, which might make them a target for mugging or might lead to their vehicle being broken into.

The reason for a surveyor's presence at a location is often guessed due to the nature of the survey equipment they are carrying and the amount of time it is necessary to spend at or around a location in order to capture spot or location survey data. Experience has shown that the risk of being discovered increases greatly if the surveyor elects to take a photo of the survey location to act as proof that they surveyed at the correct location. If a photo is required then it usually a good idea to take it after the survey has been completed.

Some surveyors decide to provide themselves with an 'alibi' for being at a location, sometimes using false identification credentials so that they can claim to be from a mobile phone company or a public utility. Some surveyors have gone to the length of acquiring false ID cards or even branded work clothing (such as a high visibility jacket) from the organisation they wish to claim to be working for.

To ensure the surveyor's safety during a survey it is recommended that they:

- Always let colleagues know where they are planning to survey and when.
- Consult the investigators in the case to determine if there are any specific risks or sensitivities related to any survey locations.
- Arrange a police escort for particularly dangerous or sensitive locations.
- Conduct a local coverage survey (where the surveyor moves around), which is less likely to attract attention than a static spot survey (where they stay still in one place).
- Keep survey equipment out of sight if at all possible.
- Only take site photos if they are sure it is safe.
- Maintain awareness of their location, especially if near a road or other hazards.
- Use a driver when conducting drive surveys.

7.7   Typical Survey Actions and Procedures

7.7.1   Spot/Location Surveys

This technique has been described as a spot/location survey to reflect the fact that there are two schools of thought regarding the ways in which it could be undertaken. Figure 7.13 illustrates some of the differences between these two concepts.

A traditional 'spot' survey was a static survey, with the surveyor typically sitting in a vehicle in a fixed spot near to the location or address to be surveyed, capturing radio measurements over a period of 5–10 min. This version of survey has the benefit that it allows the surveyor to capture the long term variations in serving cell and signal strengths at the location; the signal strength averages calculated from the survey results can therefore be deemed to provide a representative reflection of the total coverage at the site. Static spot locations, however, have the disadvantage that they do not reflect the variations in coverage across the wider area caused by things such as radio shadows. The argument often made against static spot surveys is that they are great for showing the coverage at that spot, but provide no detail of coverage 10 or 20 m away.

The 'location' survey is a more dynamic affair and usually involves the surveyor driving or walking around the perimeter of the location or address to be surveyed and also the surrounding area up to maybe 50 or 100 m in all directions.

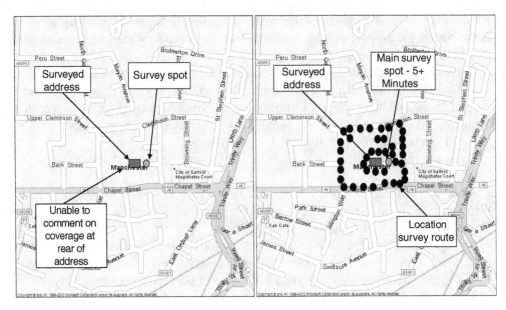

Figure 7.13   Spot/location surveys. Source: Microsoft AutoRoute map reproduced with permission from Microsoft

Some surveyors undertake a location survey as part of a more general drive survey of the surrounding area. This type of survey is generally acceptable as long as the survey involves a stop or series of stops for a moderate period of time (2–3 min per stop, for example) in the immediate vicinity of the target address. Simply driving past the address as part of a drive survey, without stopping, could prove problematic as such a survey may not have allowed sufficient time to capture the long-term variations in serving cells and signal strengths at the location. If this 'drive survey' technique is used, the recommendation is that the surveyor spends at least 5 min in the immediate vicinity of the surveyed location, possibly making a series of short stops at points around the address.

One further technique that is employed by some organisations is to undertake surveys of a series of locations during one long drive survey. The process for this kind of survey would be to start the survey equipment recording and then drive to and around the first location, then drive to the second location and so on, all with the survey equipment recording continuously. In principle this type of survey would capture essentially the same information at each location as separately captured surveys would. In general this technique is valid, as long as the surveyor accurately notes the time that they arrived at and left each location so that the overall survey dataset can be edited into batches of data that relate to each location.

There are benefits and disadvantages to both approaches – static spot surveys and non-static location surveys – but in general the non-static 'location' or 'local coverage' survey seems to be becoming the more popular option.

Spot/Location Survey Actions

The set of individual surveys to be conducted at a location will vary depending upon the networks/technologies to be surveyed, which is dictated by the networks and technologies used by the case's target phones as detailed in their call records, the geography of the location and the circumstances of the case, but the full set of captures that might be undertaken for spot or location surveys could include the following, per network:

- Free running 2G survey, with survey device technology locked to 2G
- Free running 3G survey, with survey device technology locked to 3G
- Additional 3G surveys with band or channel locks in place if the network employs Idle Mode behaviour settings that make surveys of HSPA 'data' channels difficult
- Lock-on survey to capture details of 3G neighbours if the survey device does not automatically capture neighbour Cell IDs
- Free running 4G survey, with survey device technology locked to 4G
- Additional 4G surveys with band or channel locks in place if the network employs Idle Mode behaviour settings that make surveys of some channels difficult
- Lock-on survey to capture details of 4G neighbours if the survey device does not automatically capture neighbour Cell IDs.

Multiple 3G and 4G surveys are suggested due to the possibility (indeed, the strong likelihood) that operators will employ some form of Idle Mode behaviour control to limit mobile devices to camping on a specific frequency layer when idle. Band lock surveys therefore allow the surveyor to guarantee that they capture cell coverage details of the non-camping bands.

The set of surveys suggested above would be conducted in Idle Mode but would typically also each include a set of Connected Mode test calls.

Obviously, such an extended set of surveys could take a considerable amount of time and surveyors may be able to reduce the intensity of surveys by conducting just 'free running' Idle Mode surveys on each network/technology first and then supplementing these with 'channel lock' or 'band lock' surveys and Connected Mode test calls if required.

The terms used in the above are explained as follows:

- A *'free running'* survey is one that is conducted with a technology lock in place but no other locks (no band, channel or cell locks), allowing the survey device to reselect freely between the available cells of the selected network/technology.
- A *'band lock'* locks the survey device to a specific frequency band – GSM1800 or LTE2600, for example. Bands locks are often required when the free running reselection between radio bands is curtailed by some form of Idle Mode behaviour control.
- A *'channel lock'* locks the survey device to a specific channel (ARFCN, UARFCN or EARFCN). It is also used in cases where free reselection between channels is hindered by an operator's Idle Mode behaviour settings.

Spot/Location Survey Procedures

The suggested set of procedures for a spot/location survey includes the following:

- Ensure that each survey device or submodule in the case of multi-receiver devices has the appropriate network's SIM inserted, the appropriate technology lock in place and that any further channel or band locks are set if required.
- Conduct the survey as a non-static location survey rather than a static spot survey if possible (and if safe or practical).
- Spend at least 5 min in the immediate vicinity of the surveyed address but extend the survey up to 50 or 100 m in all directions from the target address (if practical).
- Conduct the surveys in free running Idle Mode first and perform additional locked surveys or Connected Mode test calls (or test data sessions) only if the necessary information was not captured in Idle Mode.
- If possible, capture more than one simultaneous survey, using different devices or even different types of device to ensure a more broadly representative set of survey results – results from different devices can be combined into one overall set of results during post-processing.
- For 3G surveys make each test call last for an extended period – 1 min, for example – to ensure that a representative range of soft handover servers are captured.
- For 3G and 4G surveys, if the survey equipment does not automatically capture neighbour Cell IDs, consider recording a 'lock file' (if cell locks are supported by the survey equipment) to ensure that Cell IDs for neighbour cells are captured.
- Make sure that GPS fixes of the locations(s) of the survey are captured as evidence that the surveys were conducted in the reported location, possibly also take a photo of the survey address or location, if this can be done safely and without causing distress to victims or witnesses.

Spot/Location Survey Raw Data

The raw data for a spot/location survey will be a data table that shows, for each point in time at which radio measurements were captured; the GPS fix of the place at which the capture was taken; details of the serving cell at that time, including Cell ID, LAC/TAC, radio channel number, physical layer cell ID and signal strength measurement; the same details for each detected neighbour cell.

Further processing is usually employed to convert the raw results table into a survey summary table, which lists each detected cell in order of average signal strength and also indicates the cell or cells that were detected as serving cells during the survey. An example of the raw data table typically created from spot/location survey data is shown in Table 7.2 and further examples of both the raw and summary tables typically compiled from spot/location survey data are provided later in this chapter.

Table 7.2 Spot/location survey raw data table.

Time	Longitude	Latitude	Cell 0 (serving)				Cell 1 (first neighbour)			
			ARFCN	BSIC	Cell ID	RXLev	ARFCN	BSIC	Cell ID	RXLev
11:02:01	-1.462	52.521	89	13	27165	-89.4	64	15	6875	-90.1
11:02:02	-1.462	52.521	89	13	27165	-88.7	64	15	6875	-89.4
11:02:02	-1.462	52.521	64	15	6785	-88.9	89	13	27165	-90.2
11:02:03	-1.462	52.521	64	15	6785	-90.2	89	13	27165	-92.1
11:02:04	-1.462	52.521	64	15	6785	-86.3	89	13	27165	-88.4
11:02:05	-1.462	52.521	72	18	6784	-84.9	89	13	27165	-86.2
11:02:05	-1.462	52.521	72	18	6784	-85.6	64	15	6785	-85.9
11:02:06	-1.462	52.521	64	15	6785	-83.2	72	18	6784	-84.2
11:02:08	-1.462	52.521	72	18	6784	-81.6	651	16	33145	-84.1
11:02:09	-1.462	52.521	72	18	6784	-83.2	651	16	33145	-85.4
11:02:10	-1.462	52.521	72	18	6784	-85.8	651	16	33145	-85.9
11:02:11	-1.462	52.521	651	16	33145	-87.1	72	18	6784	-88.9
11:02:12	-1.462	52.521	651	16	33145	-88.0	72	18	6784	-90.4
11:02:12	-1.462	52.521	651	16	33145	-84.3	64	15	6785	-94.3

7.7.2   All-Network Profiles

An all-network profile is, in reality, a series of spot/location surveys captured at the same location for each network and each technology (and possibly each frequency band for some technologies).

An all-networks profile specification would include all of the survey types listed in the previous section in relation to spot/location surveys.

Multi-receiver devices, such as the CSurv MuNST, are far better suited to the requirements of capturing all-network, all-technology surveys as they can record multiple different captures simultaneously. Undertaking an all-networks profile with a single-receiver device such as a NEMO or TEMS handset, especially if only one device is used, can be enormously time-consuming, as the surveyor would have to perform all of the required captures consecutively rather than concurrently. An all-networks profile undertaken using a single device, in a country that had four or five operators all with 2G, 3G and 4G resources could easily take a full day to complete.

Given the complexities involved in undertaking all-network profiles, there can be a temptation to conduct them as static spot surveys, rather than mobile location surveys. The same concerns about the completeness of static survey results that were outlined in relation to spot surveys hold true for all-network profiles and surveyors are urged to consider the implications for the accuracy of their results if they decide to use the static procedure.

7.7.3   Cell Coverage Surveys

Cell coverage surveys are almost always conducted as drive surveys, so the survey safety recommendations outlined above should be taken into consideration.

Coverage Survey Actions

The usual methodology for a cell coverage drive survey is to pinpoint the location of the cell site and work out a route that travels out from the site and covers the area within which the target cell can be expected to provide serving or first neighbour coverage. The survey vehicle will typically drive the network of streets that surround the site, gradually working away from it in the direction indicated by its azimuth.

The edge of the cell's coverage area (or at least, of one contiguous section of the cell's coverage area) will be identified when the survey equipment reselects to a different serving cell. As cell coverage is sometimes non-contiguous (meaning that there can be small patches of serving coverage of other cells within the main coverage area of the target cell) the switch to an alternative serving cell may not always signal the end of coverage for the target cell. One technique that is often used is as follows: drive away from the target cell until the survey equipment reselects to another serving cell, continue to drive for another 500 m to see if the survey equipment reselects back to

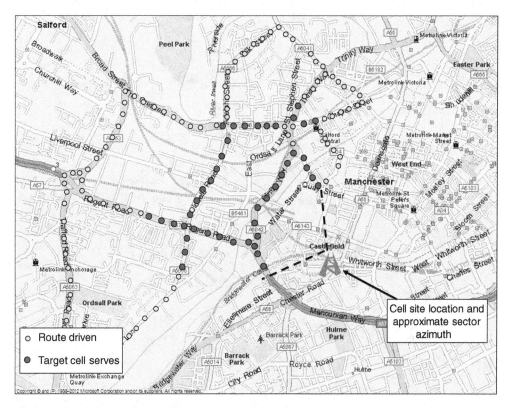

Figure 7.14 Cell coverage surveys: outline survey. Source: Microsoft AutoRoute map reproduced with permission from Microsoft

the target cell. If it does, continue driving; if it does not, then turn around and move the survey to another area of target cell coverage.

There are two commonly used variations on the cell coverage survey protocol: outline survey and full-detail survey.

The outline survey method attempts to plot just the basic outline and extent of serving coverage, to provide an indication of how far out from the site the cell serves (or, if requested, is also the first neighbour). This usually involves driving just the main roads in the area and misses out the detail of infill coverage along smaller roads. Outline surveys can be completed reasonably quickly in most cases and are sufficient if all that is required is a general idea of cell extent and likely coverage area. An example of the detail provided in an outline coverage map is shown in Figure 7.14.

A full detail survey (sometimes known as a 'fill in' survey) tries to map the extent and contiguity of the cell's coverage, looking especially for pockets or 'hot spots' of serving coverage from other cells within the area. Fully detailed surveys require the survey vehicle to drive all or most of the roads in the area and may take a considerable amount of time to complete. This type of survey is generally only required when there are questions related to the use of other cells at a time when a target phone was

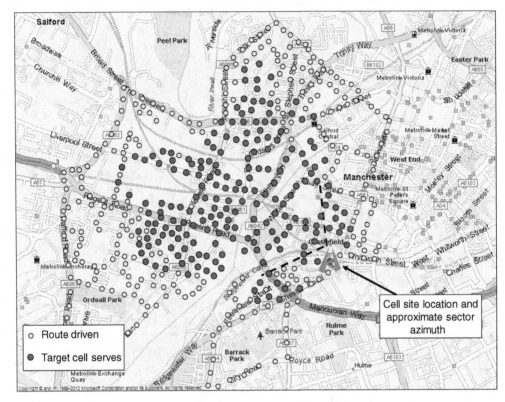

Figure 7.15 Cell coverage surveys: full detail survey. Source: Microsoft AutoRoute map reproduced with permission from Microsoft

assumed to be within the coverage of the target cell, or where it is necessary to determine target cell coverage with greater accuracy. Figure 7.15 provides an example of a full-detail survey, which should be compared with the level of detail provided in Figure 7.14 for outline surveys.

Coverage Survey Procedures

The recommended procedures to be taken during a coverage survey include:

- Ensure that each survey device or submodule in the case of multi-receiver devices has the appropriate network's SIM inserted and the appropriate technology, band or channel lock in place.
- Cell locks or Connected Mode surveying for cell coverage surveys are not recommended, as the purpose of the survey is to map the natural boundary of the cell's serving coverage. Setting a cell lock for the target cell would force the survey device to continue to treat it as the serving cell way beyond the point where a free running

device would have reselected to a neighbour cell. A cell lock can therefore distort the results of the survey.

- Ensure that the device's GPS receiver is online and that GPS fixes are being captured. Coverage survey data is valueless without accurate location information.
- Follow the drive survey safety recommendations; drive surveys should always be a two-person job.
- Work out the route to be driven in advance and record the route as it is being driven (some survey devices provide a map with the driven route overlaid on it to aid surveyors). This avoids unnecessary duplication and can shorten the survey time.
- If the survey is intended to capture details of only specific sectors of a site rather than all coverage, then it might be beneficial to confirm the azimuth of the required sectors first to avoid driving in unnecessary directions.
- Consider using multiple survey devices simultaneously, especially if the survey instructions require the surveyor to capture details of different cells on the same site that share the same sector azimuth, as would be the case in a 3G or 4G multi-layer single frequency network. In this case each survey device would need to have the appropriate band or channel lock in place to ensure that it captured details of the correct frequency layer.

In some cases the surveyor may be required to undertake surveys of cells belonging to different networks that happen to be broadcast from base stations that are sharing the same site. The site sharing agreements between operators in many countries mean that this scenario is becoming increasingly common. In these cases it is possible to capture details of multiple cells belonging to different networks during just one drive survey if multiple survey devices are used.

One final recommendation relates to the post-processing of survey data to create coverage maps. In a large majority of cases, if cell coverage surveys are required, the cell site analyst will request survey data for multiple cells that are located in the same general area. Surveyors would typically complete the whole survey set by conducting individual drive surveys for each cell or sector on their list, but some or all of these surveys may include overlapping areas.

Given the potential that exists for cells to provide non-contiguous patches of coverage beyond the main contiguous coverage area – for there to be small 'islands' of coverage separate from the main area of coverage – it is considered sensible to 'pool' the results of all of the surveys before extracting details of each individual cell to be mapped.

Figure 7.16 provides an example of this process: the right side of the map contains coverage details for a cell drawn only from that cell's specific coverage survey, the left side of the map shows some additional areas of non-contiguous coverage that were detected during surveys undertaken to map neighbouring cells. Details of these non-contiguous areas of coverage would have been missed if the surveyor had not pooled the data from all surveys before creating the coverage maps.

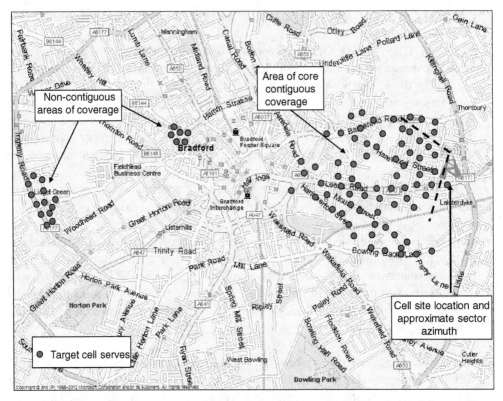

Figure 7.16 Non-contiguous cell coverage. Source: Microsoft AutoRoute map reproduced with permission from Microsoft

The example shown in Figure 7.16 may look unrealistic but is actually based on a real case that the author was involved in. The area depicted in the map, the city of Bradford in the United Kingdom, is built in a bowl-shaped valley and is surrounded by a ring of hills. The target cell site was on the eastern lip of the bowl and was transmitting its signal out over the top of the city centre, which is down in the bottom of the bowl. The area of non-contiguous coverage near to the city centre was surrounded on three sides by high-rise tower blocks, which appear to have blocked signals from more local base stations, and the area of coverage to the west of the map was up on the lip on the other side of the valley and was therefore in direct line of sight from the site antenna.

In this case the data from numerous drive surveys taken in and around Bradford were combined into one survey data file before the data for each target cell was mapped. Neither of these areas of non-contiguous coverage would have been detected if the map had been created using only the data from the specific drive survey of the target cell; the measurements for both of these areas were found in drive survey data for other cells in the same area. In this particular case, the areas of non-contiguous coverage to the west of Bradford proved useful as it helped to place the user of a target phone near to an offence location at a significant time. Without having details of the relevant non-contiguous coverage the investigators in the case would have been forced to accept the suspect's alibi statement, which placed him at an address within the main coverage area of the target cell.

Cell Coverage Survey Raw Data

The raw data for a cell coverage survey will be a set of GPS fixes (showing the latitude/longitude at which each set of measurements was captured) plus the Cell ID of the serving cell at each point. The raw data is then filtered to show only the points at which the target cell for the survey served (and optionally can also show points at which the target cell was first neighbour). A second table is usually created that shows the overall route that was driven during the survey.

It is a simple matter to then import these two sets of data into a mapping application to turn it into a visual representation of serving coverage along the surveyed route. The data points for the overall route are usually mapped first, with the points showing target cell coverage overlaid on top. This allows the serving coverage area to be shown in context to the wider surveyed area and can also help to indicate whether the drive survey was extensive enough to cover all of the areas in which the target cell might conceivably serve.

An example of the type of raw data table compiled from cell coverage survey data is shown in Table 7.3 and further examples of both the tables and the maps created from the table are provided later in this chapter.

Table 7.3 Cell coverage survey raw data table.

Longitude	Latitude	Cell ID	Status
−1.463342	52.521378	27165	0 – serving
−1.463342	52.521378	27165	0 – serving
−1.463257	52.521317	27165	0 – serving
−1.463257	52.521317	27165	0 – serving
−1.463257	52.521317	27165	0 – serving
−1.463257	52.521317	27165	0 – serving
−1.463257	52.521317	27165	0 – serving
−1.463170	52.521255	27165	0 – serving
−1.463170	52.521255	27165	0 – serving
−1.463170	52.521255	27165	1 – first neighbor
−1.463170	52.521255	27165	1 – first neighbor
−1.463170	52.521255	27165	1 – –first neighbor
−1.463085	52.521198	27165	1 – –first neighbor
−1.462998	52.521141	27165	0 – serving

7.7.4   Route Profile Surveys

Route profile surveys are arguably the simplest form of forensic radio survey to conduct as they only require the surveyor to follow a pre-determined route with the survey equipment set to record radio measurements

Route Profile Actions

The usual methodology for a route profile survey is really no more complicated than that outlined in the previous paragraph:

- Ensure that each survey device or submodule in the case of multi-receiver devices has the appropriate network's SIM inserted and the appropriate technology lock in place, and that any further channel or band locks are set if required.
- Ensure that the device's GPS receiver is online and that GPS fixes are being captured. Route profile survey data is valueless without accurate location information.
- Follow the drive survey safety recommendations; drive surveys should always be a two-person job.
- Work out the route to be driven in advance and stick to it; unscheduled deviations from the prescribed route could cause inaccurate results.
- Try to drive the route at the slowest speed that is practicable and safe, this should ensure that a greater amount of detail is captured at each point along the route. Survey accuracy can also be improved by stopping at regular intervals, for 1–2 min at a time, to allow the survey device to capture long-term variations in coverage.
- Alternatively, consider driving the route multiple times or use multiple survey devices simultaneously to broaden the depth of information captured.
- If target phones were connected to different networks or used different technologies, consider using multiple devices, each set to capture a different network/technology to minimise the number of repeat drives along the route that need to be taken.

Typically, when compiling route profile data, only details of the serving cell at each location are required; details of neighbour cells are usually only examined if the serving cell data fails to detect an expected cell or a cell that was used by a target phone for a significant call.

Route Profile Raw Data

The raw data for a route profile will be a set of GPS fixes (showing the latitude/longitude at which each set of measurements was captured) plus the Cell ID of the serving cell at each point. It is a simple matter to then import this data into a mapping application to turn it into a visual representation of serving coverage along the surveyed route.

Table 7.4   Route profile raw data table.

Longitude	Latitude	Cell ID	Status
−1.462905	52.521084	27165	0 – serving
−1.462897	52.521064	27165	0 – serving
−1.462894	52.521057	6785	0 – serving
−1.462871	52.521062	6785	0 – serving
−1.462862	52.521067	6785	0 – serving
−1.462859	52.521076	6784	0 – serving
−1.462848	52.521071	6784	0 – serving
−1.462834	52.521068	6785	0 – serving
−1.462832	52.521055	6784	0 – serving
−1.462832	52.521051	6784	0 – serving
−1.462825	52.521043	6784	0 – serving
−1.462817	52.521039	33145	0 – serving
−1.462811	52.521032	33145	0 – serving
−1.462802	52.521028	33145	0 – serving

An example of the type of raw data table compiled from route profile survey data is shown in Table 7.4 and further examples of the tables and the maps created from the table are provided later in this chapter.

7.8   Survey Results: Checking and Confirmation

7.8.1   Confirming the Expected Results

When a survey is underway, most types of survey equipment provide an on-screen list of the cells currently being detected and measured. The forensic survey engineer will use this output to check the progress of a survey against their target cell list.

If all of the cells listed in the target cell list are detected, the survey can be deemed a success.

7.8.2   Expected Results Not Found

If one or more of the cells on the target cell list fails to be detected, the forensic survey engineer has a number of options:

- If the survey device supports channel or cell locking, they could lock to the relevant channel or Cell ID and see if the 'missing' cell is detectable that way, although this would be done purely to determine whether the 'missing' cell was on air or not and would not be used as the basis for radio measurements of that cell – measurements taken with a cell lock in place cannot be regarded as being representative of a cell's normal operation.

- They could try making test calls to see if the 'missing' cell is used in Connected Mode – this is particularly relevant in 3G mode, where some cells may only show up during soft handover, and so it is important that test calls are considered.
- If the survey equipment supports 'band scan' function, which scans through all channels in a radio band and identifies the cells that are detected, the surveyor could run this test to see if the missing Cell ID appears.
- If the survey equipment provides only details of the cells that are currently being detected, it might be that the 'missing' cell is appearing in the data only intermittently and too quickly for the equipment to register onscreen. If the surveyor is able to review or process the captured survey data after the survey period has ended the 'missing' Cell ID might appear in a summary of the collated data.
- If cell address details have been provided the surveyor could walk or drive towards the cell site to see if a signal can be detected as they move closer (noting the location at which a signal is eventually detected) bearing in mind the drive survey safety recommendations mentioned above.
- They could visit the cell address to check that the site is still there and still 'on air'.
- They could perform an 'orbit' test of the site in an attempt to determine the current orientation (azimuth) of the cell sectors to see if they agree with the information provided (making a note of any revised azimuth estimations they calculate).
- If they are working for a law enforcement agency, they could ask their service provider liaison department to check with the network operator to determine why the cell is not being detected. This would include asking for details of if/when a site had been reoptimised, reorientated, relocated or retired.

In all cases, if an expected cell is not detected the surveyor should make a note of the steps taken to try to detect it and if the cell is still not discovered a note to this effect should be made in the post-survey report.

7.8.3 Surveying Near Location/Tracking Area Boundaries

The Idle Mode cell reselection procedures employed by 2G, 3G and 4G networks seek to minimise the amount of 'location update' traffic generated by each mobile device.

This is done partly to reduce the administrative load on the network and partly to maximise the mobile device's battery life. It is typically achieved by specifying that negative offsets should be applied to the reselection measurements of cells that belong to location/tracking areas other than the one the device is currently being served by.

A scenario that illustrates this process is illustrated in Figure 7.17.

A forensic survey device in Idle Mode initially elects to camp on Cell ID 13423, which is in LAC 212 and which was the strongest local cell when the survey device was first switched on.

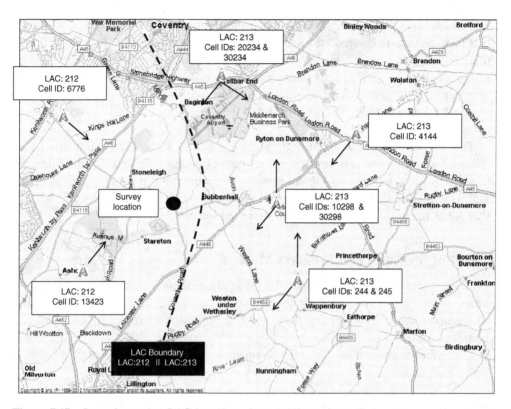

Figure 7.17 Surveying near a LAC boundary. Source: Microsoft AutoRoute map reproduced with permission from Microsoft

The device can detect a number of neighbour cells, some of which are also in LAC 212 and some of which are in LAC 213. The cells in LAC 213 indicate (on their BCCH) that a negative offset of –6 dB should be applied to measurements made of those cells while the mobile is camped on a cell belonging to a different LAC (which means that the phone reduces the measured signals of those cells to just 25% of their actual values).

The device takes a set of reselection measurements, which indicate that the strongest local cell is currently Cell ID 244, which is in LAC 213 (a different LAC to the one currently serving the device) and has a measured received signal strength of –75 dBm, but this is reduced to –81 dBm after applying the required negative offset. The second strongest cell is Cell ID 13423 (the current serving cell), which is in LAC 212 and has a measured received signal strength of –80 dBm.

In theory, the device should reselect to Cell ID 244 as it offers the strongest signal, but after applying the required negative offset, it decides to stay camped on Cell ID 13423.

When the results of the survey are processed they show that Cell ID 244 offered the strongest signal at the survey location but never served, whereas Cell ID 13423 offered a lower average signal strength but was selected to serve.

This apparent discrepancy is commonly highlighted by defence lawyers, who point to it in an attempt to discredit the accuracy of a set of survey results. In reality, though, there is a perfectly reasonable explanation for it. As the survey device initially elected to use a cell in LAC 213, cells in other LAC s would be required to be significantly stronger than the serving cell in order to be selected. Had the survey device initially selected a serving cell that was in LAC 213, cells in LAC 212 would be at a disadvantage.

This scenario is encountered quite regularly and is a common explanation for a 'non-serving strongest cell' result in survey data. A common form of words to describe the signals provided by a strong but non-serving neighbour in these scenarios is 'that while the measured cell was not selected as a serving cell its signal strength indicates that is has the potential to serve at the surveyed location'.

7.9 Survey Notes and Progress Maps

Forensic radio surveyors are under the same obligations to make contemporaneous notes during the course of their investigations as any other forensic investigator. These notes must be retained as they could be requested by the court when the case comes to trial.

Contemporaneous notes for cell site surveys usually take the form of a series of notes indicating:

- The location or address being surveyed, possibly including a GPS fix
- A note of the specific spot at which the survey was taken (e.g. balcony outside Flat 7) for a static spot survey or a basic description of the route followed (e.g. walked around perimeter of building) during a non-static location survey
- The time and date the survey started
- An indication of the weather conditions and any other significant factors
- The network and technology being surveyed
- Details of any channel or cell locks applied before or during the survey
- A list of each detected Cell ID, with an indication of whether it served, was a first neighbour or was simply detected
- The list of Cell IDs could be added to during the survey with details of the changes in signal strength
- An indication of whether additional survey types were captured (e.g. lock files) or whether multiple devices were used simultaneously
- If any specific conclusions are reached during the survey then these should be noted for future reference
- Details of any expected but missing cells should be noted, as should the steps taken by the surveyor to determine whether the missing cell was actually on air or not.

It can be argued that if the surveyor is using a forensic survey device that automatically captures and records survey details then it may not be necessary to manually record some of the detail listed above. Information such as a description of the survey location, the weather conditions and the types of survey undertaken should always be recorded in the surveyor's contemporaneous notes, as should details of any specific conclusions and actions taken to 'find' a 'missing' cell.

7.10   Survey Equipment Types

A variety of forensic radio survey devices are available ranging from very simple techniques (such as using an iPhone's 'engineering mode' to view serving cell details) through to the use of professional-grade RF analysers (such as those produced by Anritsu or Rhode and Swartz) which cost tens of thousands of dollars. In the middle ground between these extremes there are a number of commonly used and moderately priced survey tools, including the types outlined below (in a non-exhaustive list).

7.10.1   3G Forensics CSurv

CSurv has been available in two models; CSurv M-Tek (which was discontinued in 2012) and CSurv MuNST and is one of the few survey devices specifically designed for forensic survey purposes. It is manufactured by 3G Forensics (www.3gforensics. co.uk) and is widely used by law enforcement agencies in the United Kingdom and elsewhere [2].

CSurv M-Tek was able to survey 2G and 3G networks in Idle or Connected modes. Each MTek unit was able to connect to one 2G network and one 3G network at a time (which did not need to belong to the same provider).

CSurv MuNST (Multi Network Survey Tool) is the currently available device from 3G Forensics and is able to survey cellular coverage (the original model can survey 2G GSM and 3G UMTS networks, while the most recent version can also handle 4G LTE) and Wi-Fi hotspots. CSurv does not survey CDMA2000 networks.

2G, 3G and 4G surveys can be conducted in either Idle or Connected Mode. If Connected Mode surveying is requested then test calls or test data connections are made approximately every 20s.

The original version of the MuNST unit contained 10 separate 2G/3G radio modules and was able to survey up to 10 networks/technologies at the same time – this meant that it could undertake 10 simultaneous 2G surveys, 10 3G surveys or 10 surveys connected to a mix of 2G and 3G resources. The latest version of MuNST supports 20 2G/3G/4G-capable radio modules that cover all GSM and UMTS bands plus a wide selection of commonly used LTE bands.

If no 3G service is detected by a radio module during a 3G survey, the MuNST records details of any 2G service that is detected for the same operator. If no 4G service is detected the MuNST records same network 3G or 2G coverage details.

All CSurv versions support technology, band, channel and cell lock options and are able to record survey results for later processing and review. CSurv typically captures all relevant identification information, including Cell ID, about serving cells and neighbours which means that surveyors can usually avoid the necessity to 'hunt' for neighbour Cell IDs (using techniques such as cell lock) as they sometimes need to when using other types of survey device.

CSurv MuNST also supports a 'spectrum scan' function, which passively scans through all 2G, 3G or 4G channels and compiles details of the channels that are detectable at a location. In 2G mode the scan reports the set of channels across all networks that were detectable at a survey location and indicates whether each detected channel is a BCCH (Broadcast Control Channel) cell or a TCH-only cell, together with their network operator, LAC, Cell ID, power level and barring status details. 3G and 4G scans also show channel usage across all networks and indicate the RNC-ID, PSC for 3G cells and eNB ID, PCI for 4G cells among other parameters.

In respect of cell identification information, CSurv captures the following data for the following technologies during both normal operation and in spectrum scan mode:

- For GSM surveys: ARFCN, BSIC, LAC and Cell ID
- For UMTS surveys: UARFCN, PSC, LAC and Cell ID
- For LTE surveys: EARFCN, PCI, TAC and Cell ID
- For CDMA2000 surveys: not supported.

CSurv MuNST includes built-in GPS and GLONASS satellite receivers so that accurate location information can be added to each captured set of measurements.

The CSurv software is able to store survey results in a historical database on the laptop supplied with each unit that can be queried by location/date range/network to retrieve 'old' survey results for later review.

Sample rates using CSurv are determined by the operating mode. In Idle Mode, samples are captured approximately every 2.5s and in the intervening period the unit will check and confirm its attachment to the serving network. In Connected Mode the sample rate differs slightly for each technology and is determined by the sample method and the network response time. The minimum Connected Mode sampling interval is around 8s, with the standard sample interval being around 16 s.

The CSurv range was designed in cooperation with United Kingdom law enforcement agencies and is highly optimised for United Kingdom requirements, although the range of technologies and frequency bands covered makes it applicable for most regions around the world.

7.10.2   Anite NEMO Handy

The Anite NEMO Handy is part of the wider range of Anite radio survey devices (see www.anite.com) and is supplied as NEMO software running on an (essentially) standard mobile phone – the current version of NEMO runs on a Samsung Galaxy S4, for example. The NEMO Handy and other Anite survey devices are primarily designed to be used by cellular network operators and are not specifically designed to serve the needs of forensic surveyors; they therefore support a much wider range of functions than are necessarily of interest to forensic users [3].

The most recent versions of NEMO are available on a range of handset types that, between them, are able to survey GSM/UMTS/LTE, CDMA2000, WIMAX and other network types in both Idle Mode and Connected Mode (and most also survey Wi-Fi) and are able to record survey results for later processing and review.

In respect of cell identification information, NEMO captures the following data for the following technologies:

- For GSM surveys: ARFCN, BSIC, LAC and Cell ID
- For UMTS surveys: UARFCN, PSC, LAC and Cell ID
- For LTE surveys: EARFCN, PCI, TAC and Cell ID
- For CDMA2000 surveys: Band, Channel Number, SID and NID. Whether Base ID is captured or not is not clear from the documentation.

NEMO output data, which is initially presented in the proprietary .nbl format, must be converted using the (free) NEMO File Manager application, which makes it available to view in a .txt-based file format (known as .nmf).

NEMO Handy devices support technology, radio band and channel lock options. 2G/3G Cell lock is supported on older (Nokia/Symbian-based) handset models but only 2G cell lock (and not 3G/4G cell lock) is supported on the newer (Android-based) models. The cell lock feature allows the surveyor to instruct the NEMO to select a specific cell, even if that cell is not a normal serving cell.

The current ranges of the NEMO handset do not capture full details of 3G and 4G neighbour cells; they capture physical layer 3G PSC/UARFCN or 4G PCI/EARFCN details but do not capture the full Cell ID, which makes it difficult to tie a particular measured cell with a target phone's call records (a call records usually only show the logical Cell ID and not the physical layer attributes of each cell). Lack of a cell lock function makes it more difficult to get full details of 3G/4G neighbour cells.

Each NEMO Handy, as it is really just a standard mobile phone, can survey only one network at a time but Anite also offers the NEMO Outdoor product, which can survey using up to six mobile devices simultaneously.

NEMO Handy devices have been observed to have a typical measurement sample reporting rate of 0.5–1.0s.

7.10.3   Ascom TEMS

TEMS (which was formerly part of Ericsson) is now produced by Ascom (see www. ascom.com) and, like Anite NEMO, offers a range of survey solutions. The one most often employed for forensic radio survey work is a handset-based solution known as TEMS Investigation [4].

Due to the historic Ericsson link, TEMS devices tend to be based on Sony (formerly Sony–Ericsson) handsets but they also use other makes, such as Samsung.

The most recent versions of TEMS have handset options that, between them, are able to survey GSM/UMTS/LTE, CDMA2000, WIMAX and other network types in both Idle Mode and Connected Mode (and most handset types also survey Wi-Fi) and are able to record survey results for later processing and review, although this is only possible using the TEMS Analysis package which must be purchased additionally to the handset.

In respect of cell identification information, TEMS captures the following data for the following technologies:

- For GSM surveys: ARFCN, BSIC, LAC and Cell ID
- For UMTS surveys: UARFCN, PSC, LAC and Cell ID
- For LTE surveys: EARFCN, PCI, TAC and Cell ID
- For CDMA2000 surveys: Band, Channel Number, SID, NID and Base ID.

TEMS supports technology, radio band and channel lock options. 2G/3G cell lock is supported on older (Sony–Ericsson/Symbian-based) handset models but only 2G/3G cell lock (and not 4G cell lock) is supported on the newer (Sony and Samsung Android-based) models.

TEMS, like NEMO, currently does not capture full details (e.g. logical Cell ID) of 4G neighbours and the lack of a cell lock feature makes it difficult to get full details of non-serving 4G cells.

Each TEMS handset can survey only one network at a time but Ascom also offer multi-phone survey solutions.

TEMS devices have a measurement sampling rate of around 1s.

7.10.4   Forensic Mobile Services CSU-4 L

The CSU-4 L is a small scale, multi-radio device that can survey up to four networks/technologies simultaneously. It is produced by FMS (Forensic Mobile Services Ltd) a United Kingdom-based company that offers cell site analysis, expert witness services and forensic radio survey equipment. The CSU-4 L is another example of a radio survey device that was designed specifically for the needs of forensic cell site analysis [5].

The CSU-4 L is able to survey GSM, UMTS and LTE cellular networks plus Wi-Fi and also has an in-built GPS receiver.

In respect of cell identification information, CSU-4 L captures the following data for the following technologies:

- For GSM surveys: ARFCN, BSIC, LAC and Cell ID
- For UMTS surveys: UARFCN, PSC, LAC and Cell ID

- For LTE surveys: EARFCN, PCI, TAC and Cell ID
- For CDMA2000 surveys: not supported.

It stores survey data in an accessible .csv format on an SD card and in-progress surveys can be monitored from the browser of any PC, phone or tablet that is in range of the Wi-Fi signal from the unit.

The advantages of the CSU-4 L include its small size and the external battery pack, which allow it to be used outside of a vehicle for spot/location, indoor or walk surveys.

Like some other devices, the CSU-4 L does not capture neighbour Cell IDs in 3G or 4G modes and does not support a cell lock feature, meaning that other methods (such as cell site visits or wider area drive surveys) might need to be employed to ensure that Cell IDs for all of the significant non-serving neighbours at a location can be captured.

Although it does not support a channel lock command, the CSU-4 L does come with an optional external 3G bandpass unit that will allow signals for one UMTS channel at a time to be passed through to the receivers. Due to the single frequency network configuration of UMTS networks, where all cells on a frequency layer share the same channel, a bandpass device like this will allow the receiver to concentrate on a single carrier at a time but will not allow it to isolate the signals for an individual cell if more than one cell is detectable.

7.11   Raw Survey Results

The exact layout of raw survey results data will depend upon the type of survey device being used.

Examples of raw data from various device types and an interpretation of them are provided below. Only examples of 3GPP networks types – 2G GSM, 3G UMTS and 4G LTE – have been provided. One example file for each technology has been annotated to show where each presented parameter is discussed in this book – the annotations provide the section reference for the relevant paragraphs.

7.11.1   CSurv Data (2G)

An example of the CSurv 2G data output format is shown in Figure 7.18.

7.11.2   CSurv Data (3G)

An example of the CSuv 3G data output format is shown in Figure 7.19.

CSurv output is provided in a tabulated format, as shown in Figures 7.18 and 7.19, which is simple and readily understandable, especially when compared with the more

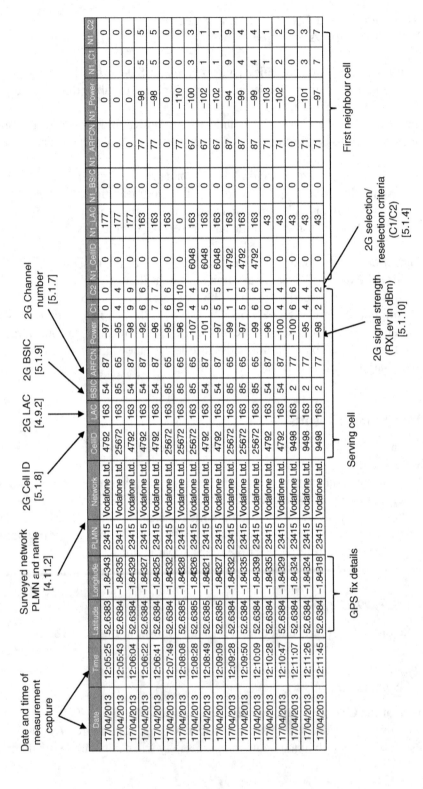

Figure 7.18 Annotated CSurv 2G data. Source: CSurv data format, reproduced with permission from 3G Forensics Ltd

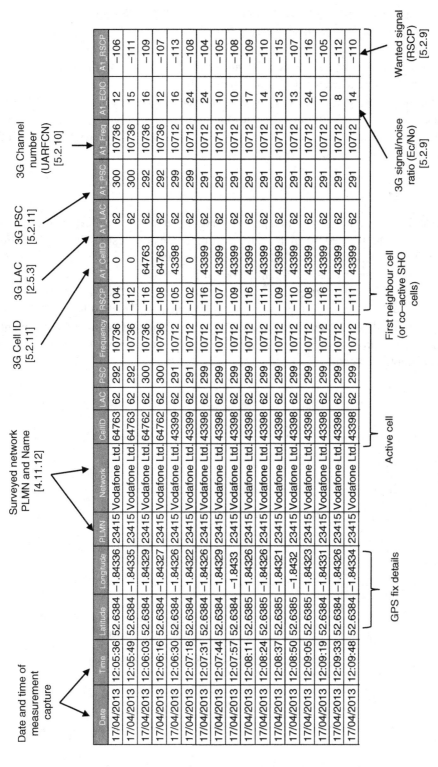

Figure 7.19 Annotated CSurv 3G data. Source: CSurv 3G data format, reproduced with permission from 3G Forensics Ltd

complex 'raw' formats employed by NEMO and TEMS. The layout of the output table is open to a degree of customisation by users, who are able to choose which of a variety of data columns are included in their preferred output format.

The tabulated form can be difficult to understand, however, and is usually subject to further processing to create usable summary tables.

The Forensic Analytics CSAS (Cell Site Analysis Suite) tool processes the CSurv tabulated output into a Summary format that shows the average signal strength and other details of all cells detected. Other methods of creating summary tables, including manual methods, are also available.

7.11.3 NEMO Data

The NEMO Handy outputs data in a proprietary format (with a .nbl extension); this is converted to a text-based format (with a .nmf extension) using the NEMO File Manager application that is bundled with the Handy device.

An example of the Anite NEMO .nmf format is shown in Figure 7.20.

The data consists of a series of capture events, each of which is encoded into a separate row of CSV (comma separated value) information. The first few rows in each file provide details of the equipment used and the version of NEMO File Manager software used to process the original .nbl file. The remaining rows provide details of individual capture events.

```
#PRODUCT,,,"NEMO HANDY","3.40.171"
#FF,,,"2.09"
#ID,,,"{1F8E1A36-428A-ED43-788E-221218A2E18A}"
#HV,,,"5.81.9"
#DS,,,2,1,5
#DT,,,1
#SW,,,"V ICPR82_10w34.2, 28-01-11, RM-645, NTM 69 v2.0"
#EI,,,"357409043034213"
#SI,,,"234330051298765"
#DN,,,"Nokia 6300"
#START,12:47:07.066,,"10.06.2013"
SEI,12:47:07.132,,1,1,891,234,30,,,,,
CHI,12:47:07.132,,1,11800,1,664,58082,891,,,,,664
NLIST,12:47:07.132,,0,0
CELLMEAS,12:47:07.132,,1,0,16,16,1,11800,664,8,-67.0,-67.0,39.0,39.0,,,,,58082,891,1,,0,11800,668,32,-86.0,,18.0,18.0
ROAM,12:47:07.132,,,1,2
DSC,12:47:07.132,,1,,
CI,12:47:07.132,,1,17.5,0,1,3,664,17.5,
ADJMEAS,12:47:07.132,,1,0,1,11,664,20.0,-67.0,20.0,-87.0,20.0,-87.0,38.0,-105.0,40.0,-107.0
PCHI,12:47:07.132,,1,1,1,1,,,,,,0,0,,,0,,,
TBFI,12:47:07.132,,1,1,,2,0,0
TAD,12:47:07.132,,1,
LOCK,12:47:07.187,,2,3,1,1,5,1,2
GPS,12:47:07.863,,-1.462270,52.520695,84,0,1,9,0
CELLMEAS,12:47:08.526,,1,0,16,16,1,11800,664,8,-68.0,-68.0,38.0,38.0,,,,,58082,891,1,,0,11800,668,32,-86.0,,18.0,18.0
CI,12:47:08.526,,1,17.9,0,1,3,664,17.9,
ADJMEAS,12:47:08.526,,1,0,1,11,664,19.0,-68.0,19.0,-87.0,19.0,-87.0,38.0,-106.0,39.0,-107.0
GPS,12:47:08.863,,-1.462272,52.520695,84,0,1,9,0
GPS,12:47:09.863,,-1.462275,52.520695,85,0,1,9,0
CELLMEAS,12:47:09.939,,1,0,16,16,1,11800,664,8,-66.0,-66.0,40.0,40.0,,,,,58082,891,1,,0,11800,668,32,-85.0,,21.0,21.0
CI,12:47:09.939,,1,24.1,0,1,3,664,24.1,
ADJMEAS,12:47:09.939,,1,0,1,11,664,18.0,-66.0,18.0,-84.0,18.0,-84.0,40.0,-106.0,38.0,-104.0
GPS,12:47:10.863,,-1.462277,52.520695,85,0,1,9,0
CELLMEAS,12:47:11.351,,1,0,16,16,1,11800,664,8,-64.0,-64.0,42.0,42.0,,,,,58082,891,1,,0,11800,668,32,-85.0,,21.0,21.0
CI,12:47:11.351,,1,23.5,0,1,3,664,23.5,
ADJMEAS,12:47:11.351,,1,0,1,11,664,20.0,-64.0,20.0,-84.0,20.0,-84.0,42.0,-106.0,40.0,-104.0
GPS,12:47:11.850,,-1.462280,52.520695,85,0,1,9,0
CELLMEAS,12:47:12.762,,1,0,16,16,1,11800,664,8,-65.0,-65.0,41.0,41.0,,,,,58082,891,1,,0,11800,668,32,-84.0,,21.0,21.0
CI,12:47:12.762,,1,28.9,0,1,3,664,28.9,
ADJMEAS,12:47:12.762,,1,0,1,11,664,19.0,-65.0,19.0,-84.0,19.0,-84.0,40.0,-105.0,39.0,-104.0
GPS,12:47:12.851,,-1.462283,52.520695,86,0,1,9,0
GPS,12:47:13.866,,-1.462285,52.520695,86,0,1,9,0
CELLMEAS,12:47:14.174,,1,0,16,16,1,11800,664,8,-64.0,-64.0,42.0,42.0,,,,,58082,891,1,,0,11800,585,8,-84.0,,13.0,13.0,
CI,12:47:14.174,,1,23.2,0,1,3,664,23.2,
ADJMEAS,12:47:14.174,,1,0,1,11,664,19.0,-64.0,20.0,-84.0,19.0,-83.0,41.0,-105.0,40.0,-104.0
GPS,12:47:14.851,,-1.462290,52.520695,87,0,1,9,0
```

Figure 7.20 Example of the NEMO Handy .nmf format. Source: NEMO .nmf data format, reproduced with permission from Anite Ltd

There are dozens of different capture event types, but only a limited number of them provide forensically useful information, including:

- **CELLMEAS** – Cell measurement event, this provides details of the current serving cell(s) and any neighbours that are detected. For each listed cell, the captured data indicates received signal strength, channel number and physical layer ID, and may, in the case of 2G readings, also capture the serving and neighbour cells' Cell ID. 3G and 4G readings do not capture the Cell ID in the CELLMEAS event itself, they are captured in a separate CHI (Channel Information) event, which can lead to uncertainty as to which cell each CELLMEAS event refers to.
- **GPS** – This captures GPS data from an internal or external receiver and records latitude, longitude, geodetic height and other details.
- **CHI** – Channel information, captures details of the serving Cell ID in 3G and 4G readings and is captured following a change of serving cell.
- **SEI** – Provides details of surveyed network MCC/MNC.
- **CAA, CAC and CAD** – Capture details of the progress of voice call events.

The format of the .nmf data can vary slightly depending upon the version of NEMO File Manager that is used to process the source data, with each successive version potentially using a slightly different 'file format'. Different file formats can support different numbers and arrangements of data fields in each event type. Anite publish file format definition documents for each version of file manager software.

The 'raw' .nmf format is of limited forensic usefulness in its basic state and is usually subjected to further processing to create usable summary tables. The Forensic Analytics CSAS tool processes NEMO .nmf data into a tabulated output format and also into a summary format that shows the average signal strength and other details of all cells detected. Other methods of creating summary tables, including manual methods, are also available.

7.11.4 TEMS Data

TEMS devices capture raw survey data in an encrypted xml-based format, which is inaccessible without purchasing the TEMS analysis pack software. The analysis pack provides multiple analysis and 'playback' features and also supports the output of survey data in a variety of formats, including an .xls format. TEMS allows the user to select the types of data columns and their arrangement when exporting data, so there is not really a 'standard' output format.

Raw TEMS data is similar to raw NEMO data in the sense that it consists of a series of capture events; unlike NEMO data, however (which only captures cell measurement data in the CELLMEAS event type), TEMS captures cell-related data in most capture event types.

The 'raw' TEMS output format is of limited forensic usefulness in its basic state and is usually subjected to further processing to create usable summary tables. The

Forensic Analytics CSAS tool processes TEMS data into a tabulated output format and also into a summary format that shows the average signal strength and other details of all cells detected. Other methods of creating summary tables, including manual methods, are also available.

7.11.5   FMS CSU-4 L Data

The FMS CSU-4 L outputs data in a tabulated .csv format that shows the time and date of each measurement capture, the GPS fix of the location at which it was captured, the state of the survey device at the time (idle or connected) plus details of the set of detected cells.

An example of the CSU-4 L 4G data output format is shown in Figure 7.21 and the 2G and 3G output tables are similar.

The tabulated CSU-4 L output format is of reasonable forensic usefulness in its basic state but is usually subjected to further processing to create more accessible summary tables. The Forensic Analytics CSAS tool processes CSU-4 L data into a summary format that shows the average signal strength and other details of all cells detected. Other methods of creating summary tables, including manual methods, are also available.

7.12   Processing Survey Results

In most cases, raw survey data will not be in a state that makes it easily understood by analysts and some form of post-processing will be required.

Some survey devices go some way towards making their output understandable; CSurv, TEMS and CSU-4, for example, collate the captured data and present it in a tabulated format. Other devices, such as NEMO, provide data in a reasonably complicated raw format.

Most post-processing is performed manually or using some form of macro, but there are commercial products available (such as Forensic Analytics' CSAS software) that offer this functionality [6].

Generic instructions on how to convert a tabulated set of raw survey results into a summary table are provided in Chapter 9.

The following sections provide examples of the raw tabulated data formats and, where applicable, the processed summary tables for the three basic types of forensic radio survey plus the all-network profile option.

7.12.1   Spot/Location Raw Survey Results

An example of a raw radio survey results table was previously presented but is shown again in Table 7.5 and is explained in more detail below.

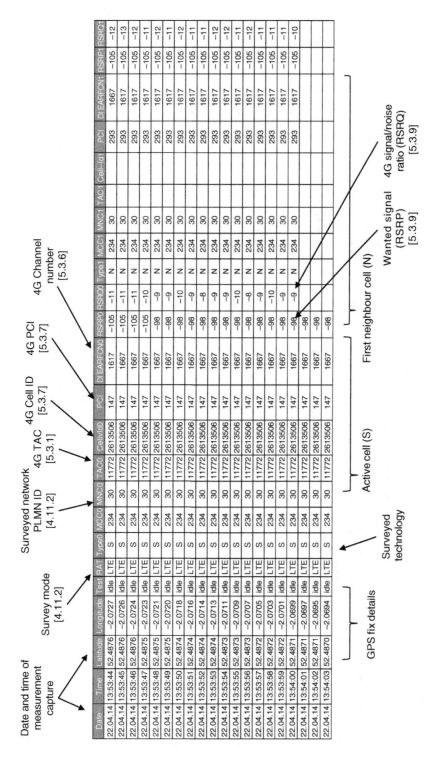

Figure 7.21 Example of the CSU-4 L 4G output format. Source: CSU-4 L data format, reproduced with permission from Forensic Mobile Services Ltd

Table 7.5 Spot/location survey raw data table.

Time	Longitude	Latitude	Cell 0 (serving)				Cell 1 (first neighbour)			
			ARFCN	BSIC	Cell ID	RXLev	ARFCN	BSIC	Cell ID	RXLev
11:02:01	-1.462	52.521	89	13	27165	-89.4	64	15	6875	-90.1
11:02:02	-1.462	52.521	89	13	27165	-88.7	64	15	6875	-89.4
11:02:02	-1.462	52.521	64	15	6785	-88.9	89	13	27165	-90.2
11:02:03	-1.462	52.521	64	15	6785	-90.2	89	13	27165	-92.1
11:02:04	-1.462	52.521	64	15	6785	-86.3	89	13	27165	-88.4
11:02:05	-1.462	52.521	72	18	6784	-84.9	89	13	27165	-86.2
11:02:05	-1.462	52.521	72	18	6785	-85.6	64	15	6785	-85.9
11:02:06	-1.462	52.521	64	15	6784	-83.2	72	18	6784	-84.2
11:02:08	-1.462	52.521	72	18	6784	-81.6	651	16	33145	-84.1
11:02:09	-1.462	52.521	72	18	6784	-83.2	651	16	33145	-85.4
11:02:10	-1.462	52.521	72	18	6784	-85.8	651	16	33145	-85.9
11:02:11	-1.462	52.521	651	16	33145	-87.1	72	18	6784	-88.9
11:02:12	-1.462	52.521	651	16	33145	-88.0	72	18	6784	-90.4
11:02:12	-1.462	52.521	651	16	33145	-84.3	64	15	6785	-94.3

A generic set of raw survey results will consist of a table showing the following types of information (which is more detailed than the abbreviated format shown in Table 7.5):

- Date and time of the measurement capture event.
- GPS fix (at least Latitude/Longitude) of the location at which the capture took place.
- Some formats also provide an indication of the network being surveyed (MCC and MNC) and the state of the survey device (Idle or Connected).
- There are then separate 'clusters' of radio measurements for the detected set of serving and neighbour cells (remembering that due to soft handover in Connected Mode, 3G devices might record up to three concurrent serving cells). Each cluster would typically list these features of associated cells:
  - Status – serving or neighbour
  - Channel number and radio band (or Band Class for CDMA2000)
  - Physical layer ID – BSIC, PSC, PCI or PN Offset
  - LAC (GSM and UMTS), TAC (LTE) or SID/NID (CDMA2000)
  - Cell ID (or Base ID for CDMA2000)
  - RF signal measurement – RXLev (GSM), Ec/No and RSCP (UMTS), RSRQ and RSRP (LTE), Ec/Io and RSCP (CDMA2000).

This type of tabulated results format offers the opportunity to view the changing relationship between the detected cells during the survey period – in areas of non-dominance, it shows the reselection process taking place, as the signals from some cells become stronger enabling them to move up the list, while other cells drop down the list.

7.12.2   Spot/Location Survey Summarised Results

Although the tabulated form of results data has its uses, it does not help to provide an immediate understanding of which cells, in areas of non-dominance, serve most often or offer the most consistently strong signals. A summarised table is much more useful for this kind of understanding and analysis.

An example of a set of processed and summarised spot/location survey results is shown in Figure 7.22.

A typical summarised survey results table will provide some or all of the following types of information:

- Survey location, start date/time of survey and surveyor name
- Duration of survey
- GPS fix (or average GPS fix) of survey location or locality
- Network and technology surveyed

CSAS   <Case name>   Spot/Location Survey Summary

Location	A - 14 High St		
Ave GPS	Lat: 52.123	Long: -1.2345	Height: 110m

Network	Technology	# of Test Calls
Vodafone UK	2G	3

Survey Results

Cell ID	LAC/TAC	Channel/PCI	Ave. Strength	Description	Detected (Serving)	Serving Percentage	Azimuth	Site Name	Served in Call
12345	345	654 12	-84.12	Strong	123 (123)	100%	120	Town Hall	Yes
2468	345	101 14	-94.32	Moderate	123 (0)	0%	240	Town Hall	No

source filename - C:/surveys/operation cassidy/feb 2014/asdfasdfasdf.csv

Surveyor	Jim Smith	Organisation	RF Group Ltd	Comments: survey conducted 15m from front door,
Survey Date	01/02/2014 13:45-13:55	Equipment	Csurv MuNST	with walk survey around perimeter of building
Processed By	Jill Jones	Date	02/02/2014 15:12	after 5 minutes

Figure 7.22 Spot/location survey results. Source: CSAS data format, reproduced with permission from Forensic Analytics Ltd

- List of detected cells, ranked in order of signal strength, showing:
  - Cell ID
  - Possibly, cell site name details (if known)
  - Channel number and physical layer cell identifier (BSIC/PSC/PCI/PN Offset)
  - Average signal strength (RXLev, Ec/No, RSRQ and Ec/Io)
  - Number of times the cell was detected in total and the number of times it was selected as serving
  - Indication of whether the cell served in Idle Mode
  - Indication of whether the cell served during a test call
- Overall number of test calls made
- Surveyors comments.

This type of summarised table is often the forensic radio survey 'product' that is passed to the case investigators or cell site analysts and is therefore used to enable them to form conclusions on the potential locations of target phones during significant calls.

7.12.3   All-Network Profile Results

An all-network profile is essentially a 'summary of summaries' for each surveyed spot or location and is designed to draw together the summary data for each separate network/technology survey into one overall coverage summary table.

An example of an all-network profile report – taken in a hypothetical country that has three network operators who offer 2G, 3G and 4G services – is shown in Table 7.6:

A typical all-network profile presents an abbreviated set of details for each network and technology that was surveyed and will typically show, for each network and each technology, the set of detected cells ranked in order of average signal strength along with these features of each cell:

- Channel number and physical layer ID (BSIC, PSC, PCI or PN Offset)
- Cell ID (or Base ID) and status (serving or neighbour)
- Average signal strength (RXLev, Ec/No, RSR or Ec/Io).

All-network profile reports are often undertaken to allow investigators to capture a 'snapshot' of network coverage at a crime scene or other significant location in the immediate aftermath of an event. This allows them to preserve evidence of the cellular coverage at the site as it was at around the time of the event and guards against changes in network configuration affecting the accuracy of any cell site reports or conclusions that are reached later in the investigation.

7.12.4   Coverage Survey Results

The output of a coverage survey will usually be a map showing the route driven and indicating measurement locations that selected the target cell as serving (and optionally another set of markers showing where the target cell was first neighbour).

Table 7.6 Example of all-network profile report.

	Network 1				Network 2				Network 3		
	Channel/PCI	Cell ID (status)	Average signal		Channel/PCI	Cell ID (status)	Average signal		Channel/PCI	Cell ID (status)	Average signal
2G	45 27	20456 (s)	−82.34	2G	23 12	234 (s)	−75.12	2G	787 34	45398 (s)	−82.04
	47 29	30456 (s)	−87.23		67 13	345 (n)	−98.12		821 42	2123 (n)	−87.38
	105 23	10456 (n)	−93.87		97 16	12765 (n)	−100.67		698 31	54901 (s)	−88.13
	51 23	38765 (n)	−99.56		—	—	—		670 32	19801 (n)	−90.87
	—	—	—		—	—	—		701 36	20700 (n)	−91.45
3G	10637 145	10987 (s)	−4.56	3G	10712 87	32154 (s)	−7.38	3G	10836 198	56901 (n)	−5.78
	10637 146	20987 (n)	−14.56		—	—	—		10811 198	55901 (s)	−6.87
	—	—	—		—	—	—		10811 199	55902 (s)	−9.21
	—	—	—		—	—	—		10836 198	56903 (n)	−5.78
	—	—	—		—	—	—		10836 199	56902 (n)	−12.72
	—	—	—		—	—	—		1076 114	8790 (n)	−16.87
4G	640031	10976542 (s)	−8.32	4G	6300121	435213 (s)	−7.02	4G	1617341	62541 (s)	−8.3
	640032	10976543 (n)	−14.34		6300122	435212 (s)	−11.59		—	—	—
	—	—	—		285021	143256764 (n)	−18.34		—	—	—
	—	—	—		—	—	—		—	—	—
	—	—	—		—	—	—		—	—	—

Table 7.7 Example of coverage survey results raw data, showing details of only the 'target' cell.

Longitude	Latitude	Cell ID	Status
–1.463342	52.521378	27165	0 – serving
–1.463342	52.521378	27165	0 – serving
–1.463257	52.521317	27165	0 – serving
–1.463257	52.521317	27165	0 – serving
–1.463257	52.521317	27165	0 – serving
–1.463257	52.521317	27165	0 – serving
–1.463257	52.521317	27165	0 – serving
–1.463170	52.521255	27165	0 – serving
–1.463170	52.521255	27165	0 – serving
–1.463170	52.521255	27165	1 – first neighbourneighbor
–1.463170	52.521255	27165	1 – first neighbour
–1.463170	52.521255	27165	1 – first neighbourneighbor
–1.463085	52.521198	27165	1 – first neighbourneighbor
–1.462998	52.521141	27165	0 – serving

The raw material for such a map is a set of GPS coordinates followed by details of the serving cell (and optionally the first neighbour cell) from the measurement events captured at each surveyed point along the driven route.

An example of the raw results data table for a coverage survey is shown in Table 7.7.

Such a data set could then be manually filtered, so that only events that showed the target cell as serving or first neighbour are retained, and the filtered results could be imported into a mapping tool such as Microsoft AutoRoute.

Depending upon the level of detail required in the coverage map, it may be necessary to import three separate data sets into each cell coverage map:

- The first data set shows the entire route driven during the survey and consists of just the GPS latitude/longitude data for the survey. The route would usually be represented on the map using small pushpins in a neutral colour.
- The second data set shows the locations of survey measurements where the target cell was detected as the first neighbour cell. These locations would usually be shown on the map using a larger pushpin and a different colour to the route pins.
- The last data set shows the locations of survey measurements where the target cell was detected as the serving cell. These locations would use the same size pushpins as for first neighbour locations but would be in a third colour.
- The map should contain a legend indicating the significance of each type and colour of pushpin shown.

An example of a completed cell coverage survey map, created using Microsoft AutoRoute and where all three levels of data have been imported, is shown in Figure 7.23.

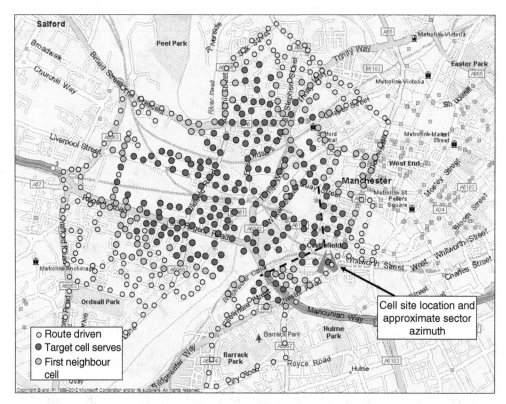

Figure 7.23 Cell coverage survey map. Source: Microsoft AutoRoute map reproduced with permission from Microsoft

This type of map is usually the form in which the forensic radio survey product for cell coverage surveys is passed to the investigators or cell site analysts in a case and allows them to draw conclusions as to the area within which a target phone could have been located when making significant calls via that cell.

7.12.5   Route Profile Results

Like coverage survey results, the processed output of a route survey is usually a data set containing GPS fixes and serving cell details, which can then be turned into a route map that indicates the progression of serving cells detected along a route.

Unlike coverage survey data, which details the locations at which a specific target cell was selected as serving, route profile data usually lists numerous serving cells, depending upon the length of the surveyed route and the progression of coverage encountered along that route. Table 7.8 provides a generic example of a route profile raw data table.

The raw data contained in such a table can then be converted into a more visual representation of serving coverage along a route by importing it into a mapping application such as Microsoft AutoRoute.

Table 7.8 Example of route survey results raw data showing a succession of serving cells along a surveyed route.

Longitude	Latitude	Cell ID	Status
−1.462905	52.521084	27165	0 – serving
−1.462897	52.521064	27165	0 – serving
−1.462894	52.521057	6785	0 – serving
−1.462871	52.521062	6785	0 – serving
−1.462862	52.521067	6785	0 – serving
−1.462859	52.521076	6784	0 – serving
−1.462848	52.521071	6784	0 – serving
−1.462834	52.521068	6785	0 – serving
−1.462832	52.521055	6784	0 – serving
−1.462832	52.521051	6784	0 – serving
−1.462825	52.521043	6784	0 – serving
−1.462817	52.521039	33145	0 – serving
−1.462811	52.521032	33145	0 – serving
−1.462802	52.521028	33145	0 – serving

Depending upon the level of detail required in the route profile map, it may be necessary to import two separate data sets into each map:

- The first data set shows the entire route driven during the survey and consists of just the GPS latitude/longitude data for the survey. The route would usually be represented on the map using small pushpins in a neutral colour.
- The second data set shows the locations of survey measurements where cellular coverage was detected and should indicate the details of serving cell at each point. These locations would usually be shown on the map using a larger pushpin and a different colour for each serving cell.

An example of a completed route profile survey map, created using Microsoft AutoRoute and where both levels of data have been imported, is shown in Figure 7.24.

This type of map is usually the form in which the forensic radio survey product for route profile survey is passed to the investigators or cell site analysts in a case and allows them to draw conclusions as to the parts of a route within which a target phone could have been located when making particular significant calls.

7.13 Understanding Survey Results

Forensic radio surveyors need to acquire the skills necessary to capture accurate radio survey data if they are going to be able to aid the investigations to which they are assigned. Being able to capture survey data is only part of the process, however, as surveyors also need to be able to correctly interpret and understand the results of those surveys. The main reason for this is that surveyors need to be able to determine

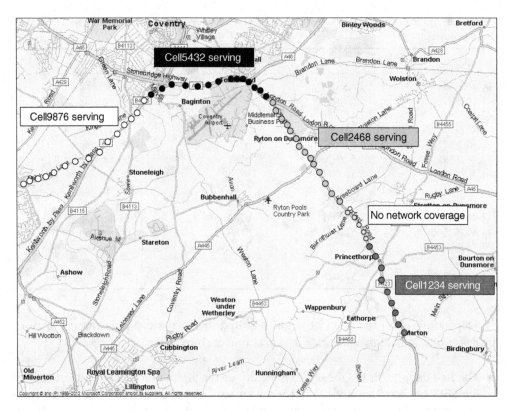

Figure 7.24   Route profile survey map. Source: Microsoft AutoRoute map reproduced with permission from Microsoft

when surveys have been completed correctly and they need to be able to quickly spot when surveys have encountered problems and understand how to deal with those problems.

It is therefore necessary for forensic surveyors to be able to understand the nature of the information presented in their 'product'.

7.13.1   Understanding Coverage and Route Survey Results

Coverage and route profile survey results are generally easy to understand as their data is essentially self-explanatory. In both cases the survey data and any resulting maps simply identify the points at which measurements were captured and indicate details of points at which the particular cells were measured as being the serving cell.

The only things that generally go wrong with coverage or route surveys are related to a lack of GPS data – if the GPS receiver associated with the survey device was switched off or not working – or are related to inappropriately applied technology, channel or cell locks.

Coverage and route profile data is useless if it is not accompanied by accurate GPS data showing where each set of measurements was captured. If the raw data for a survey does not contain any GPS data it is generally not worth processing the data into a table or a map and the survey must be completed again.

Coverage and route profile surveys are usually conducted in Idle Mode with a technology lock in place. In some circumstances, especially for surveys of 3G and 4G cells, it may be necessary to apply a channel lock to ensure that the survey device stays on the correct frequency layer throughout the survey. It is not recommended that a cell lock be applied as this would affect the accuracy of the survey results – a coverage survey is intended to map the natural serving footprint of a cell and a route profile is intended to map the progression of surveying cells encountered along a route, neither of which objectives can be achieved with a cell lock in place.

7.13.2   Understanding Spot/Location Survey Results

Spot/location survey results often require a degree of interpretation in order to fully understand the information that is being presented. The most common issues with survey results relate to non-serving strong or strongest cells.

Spot/location survey summaries usually list the strongest cell detected in the area as a serving cell.

In areas of dominance of one cell, this type of result is intuitive – the strongest cell in an area will be selected as the serving cell – and is generally to be expected as the Idle Mode cell selection/reselection algorithms essentially instruct a mobile device to camp on the strongest local cell.

Although this type of result is intuitive, it is not always what actually happens and spot/location survey results often include a strongest cell that is not shown as serving, or a set of strong cells within the top five or 10 cells that do not serve.

An example of a spot/location summary table that shows non-serving strong cells is shown in Figure 7.25. The shading on the Cell ID column indicates a serving cell.

CSAS			R v Moriarty					Spot/Location Survey Summary	
	Location	B - Four Shires Bank,							
	Ave GPS	Lat: 52.123		Long: -1.2345		Height: 110m			
	Network	O2 UK		Technology	2G	# of Test Calls	3		
Survey Results									
Cell ID	LAC/TAC	Channel/PCI	Ave. Strength	Description	Detected (Serving)	Serving Percentage	Azimuth	Site Name	Served in Call
61897	2987	1013 23	-80.91	Very Strong	178 (0)	0%	0	Middle St	No
21897	2987	101 14	-86.87	Strong	178 (154)	86%	220	Middle St	Yes
40456	2988	48 22	-90.51	Moderate	178 (24)	14%	270	Green Hill	Yes
12040	369	33 24	-100.99	Poor	134 (0)	0%	120	Howden Rd	No

Figure 7.25   Survey results with non-serving strong cells. Source: CSAS data format, reproduced with permission from Forensic Analytics Ltd

Defence reports sometimes highlight such instances and attempt to use them to imply that the survey results are flawed or inaccurate, but there are a number of perfectly valid reasons for having non-serving strongest cells:

- A typical cause is that the survey was taken near a LAC (2G/3G) or TAC (4G) boundary and that the 'non-serving' strong cells were on the other side of the boundary – see Section 7.8 above for a more detailed explanation of this topic.
- One or more of the 'non-serving' cells had been deliberately optimised to be non-serving; examples of where this might happen include 3G HSPA carriers, EGSM900 cells and GSM1800 cells, all of which are commonly optimised to be 'unattractive' to reduce the likelihood that devices will choose to camp on them.
- 3G and 4G carriers can be optimised to be unattractive by specifying permanent negative offsets in their BCCH neighbour cell measurement parameters, which has the effect of artificially reducing the value of the received neighbour signal strength measured by a mobile device when surveying those cells.
- 2G carriers can also be optimised to be unattractive by specifying a negative offset that is applied to the C2 value calculated for those neighbour cells, again making them artificially less attractive.
- Networks are also able to set 'Idle Mode behaviour' cell and carrier priority settings, which ensure that some carriers are artificially promoted in the neighbour cell list when considering reselection candidates.
- Detected cells in a survey summary might be assigned to a CSG of which the survey device is not a member.

Optimisation parameters are often set on cells that are configured to be used as 'expansion carriers' or that are set aside for specific functions, such as being data carriers. In these cases network operators attempt to ensure that Idle Mode devices do not camp on to these cells, which is why survey devices might not be able to select them as serving cells even if they offer the strongest signal.

In general, the carrier types that most commonly have these types of parameter set include:

- EGSM900 carriers in 2G networks:
  - ARFCNs – 0, 975–1023
- GSM1800 carriers in 2G networks:
  - ARFCNs – 512–885
- Designated 3G HSPA carriers, which will differ for each network.

In 4G LTE networks, at least while CSFB (Circuit Switched Fallback) is still in use, it may be undesirable for CSFB-enabled mobile devices to camp on LTE800 band cells, meaning that operators are likely to set Idle Mode behaviour parameters that discourage camping on 800 band cells. This is more likely to be the case with operators who use GSM1800 for 2G services and less likely with those who use GSM900.

The reason is that 800 band cells are likely to have different (e.g. larger) serving footprint sizes than GSM1800 cells, making it difficult to guarantee that a mobile device that needs a fallback from an 800 band cell is still within the coverage area of the corresponding 1800 band cell at the time. That would cause the fallback to either fail or be delayed while the call is rerouted via a different 2G base station. 4G surveys (captured using a survey device that supports CSFB rather than VoLTE) may therefore detect strong but non-serving cells using LTE800 band EARFCNs 6150–6449.

Other indicators to look for regarding strong but non-serving cells include the number of times that the cell was detected – if it has a very low detection count (just 1 or 2 measurements, for example) it may be the case that, although it was very strong, it was not detected for long enough to be considered for a reselection.

Another possibility is that a strong non-serving cell, while it was stronger than cells that were selected as serving, may not have been stronger by a sufficiently large margin. Reselections and handovers are usually triggered if a neighbour is stronger than the current server plus a margin of 3 or 6 dB, so a non-serving cell that was only 1 or 2 dB stronger may not have been strong enough to be selected.

7.13.3   Finding 'Missing' Cell IDs

It has been noted elsewhere in this text that some survey devices have difficulty in capturing neighbour Cell IDs for 3G and 4G cells. This is mainly due to the single frequency nature of 3G and 4G deployments, in which the signals from a dominant nearby cell can overwhelm and 'drown out' signals from more distant neighbours. There will also be situations in which any survey device struggles to capture the Cell IDs of weak neighbour cells or of those that are suffering interference on any technology.

All of these scenarios could result in a survey measurements table that contains one or more cells with unknown or 'missing' Cell IDs.

A survey measurement without a Cell ID would have usually correctly captured the channel number and physical layer cell ID used by the cell – for GSM cells this would be ARFCN + BSIC; for UMTS cells UARFCN + PSC; for LTE cells EARFCN + PCI; and for CDMA2000 cells Band Class + Channel Number + PN Offset.

Missing Cell IDs in a set of survey results, like the example shown in Figure 7.26, may be acceptable if the cells concerned were near the bottom of the list (in order of signal strength) and were not serving cells. Having serving or very strong neighbour cells listed (serving cells are indicated by the shading of the Cell ID column in Figure 7.26) without a Cell ID could mean that the survey is missing details of an important cell; this possibility is even more important to consider if the survey failed to detect one or more of the 'target' cells that were used by significant phones.

There are a number of ways in which the missing Cell IDs can be obtained.

First, the simplest (but most expensive) method is to revisit the site of the survey and attempt to capture the missing Cell IDs by re-running the survey. This option

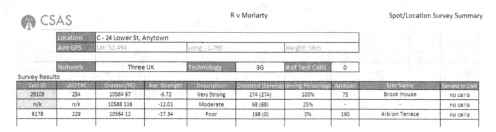

Cell ID	LAC/TAC	Channel/PCI	Ave. Strength	Description	Detected (Serving)	Serving Percentage	Azimuth	Site Name	Served in Call
29109	254	10564 97	-6.72	Very Strong	274 (274)	100%	75	Brook House	no calls
n/k	n/k	10588 116	-12.01	Moderate	68 (68)	25%	-	-	no calls
6178	229	10564 12	-17.34	Poor	198 (0)	0%	190	Albion Terrace	no calls

Figure 7.26 Example of survey with 'Missing' Cell IDs. Source: CSAS data format, reproduced with permission from Forensic Analytics Ltd

might be best kept in reserve as a last resort, however, as other options may be available.

If a lock file was taken as part of the survey, the missing Cell ID might have been captured as part of that. The way to check is to see if the channel number and physical layer cell ID of the 'missing' cell in the survey results was also captured in the lock file. As no two cells in the same area *should* have the same channel number and physical layer ID as each other it can reasonably be assumed that the two instances of channel/PCI relate to the same cell and so the Cell ID captured in the lock file can be copied into the survey results. It is considered good practice to make a note of this substitution in the 'comments' section of the amended summary table if this occurs.

If a lock file was not taken at that location, or if one was taken but the missing cell was not captured in the lock file either, then it may be possible to find the Cell ID in the data captured in surveys at other nearby locations.

If other surveys were taken in reasonably close proximity to the location with the missing Cell ID, the same rules about the reuse of a channel number/physical layer ID pair should hold true. If the missing cell's details were captured during a nearby survey then, again, they may be copied across to the incomplete survey data (and a note made in the summary comments field).

The interpretation of 'reasonably close proximity' in the above paragraph is key to the success of this procedure. Frequency and physical layer ID reuse distance are determined in part by cell radius values; if the two surveys were performed in a rural or suburban area, where cell radius can be assumed to be relatively large, an acceptable level of proximity for survey locations might be as high as 5–10 km (3–6 miles); in a dense urban area, where cell radius may be as little as few hundred metres, an acceptable level of proximity might be as low as 500–1000 m.

The processes outlined above can be effective and can improve the accuracy of survey results, but their use is not without attendant risks. If the channel/physical layer ID of the missing cell was reused within less than the anticipated reuse distance, the 'copying' process could result in an incorrect attribution of Cell ID to measurement result, which in turn could lead to incorrect cell site conclusions being drawn.

If the 'finding and copying' technique is used, it should be recorded in the surveyor's (or data processor's) contemporaneous notes to ensure full disclosure of any potential risks. Ultimately, if there is any doubt about the veracity of the Cell ID information captured in nearby surveys and if it is deemed necessary to obtain the missing Cell ID, then the only option is to return to the location and attempt to capture the Cell ID that matches the incomplete channel number/physical layer ID.

7.14   Storage of Survey Data

Forensic radio surveys are typically undertaken on a case by case basis and are used to test the allegations being made in each case. The survey data collected during each case has the same significance as any other type of evidence and should be stored in a way that allows it to be retrieved and presented in its original format if the court requests it.

The data collected during a survey has a wider usefulness, beyond the scope of an individual case, however, and organisations that make use of forensic survey data should take this into consideration.

The storage of survey data therefore has two main aspects: (i) storage in relation to continuity of evidence in a case; (ii) long-term storage of survey data for historical purposes.

7.14.1   Continuity of Evidence

If forensic radio survey data is captured for use in an investigation or a court case, it comes under the same rules of retention and disclosure/discovery as any other form of evidence.

Most forensic survey devices allow the user to store survey results on a removable data card (an SD card or similar) or at the very least allow the data collected during a survey to be backed up to a computer or uploaded to a network or database.

In whichever format the original 'raw' data from each survey is presented, most legal jurisdictions will require the surveyor to keep a copy of it in some form of non-volatile, long-term storage. This makes the original data available for disclosure to defence lawyers or experts, if required, and means that the accuracy of any summaries, reports or conclusions that were based on the data can be checked against the source information at a later date.

Some forensic surveyors physically remove the data card used to store data for each survey and store it with the case files; others take the data card and capture an 'image' of the data held on the card (using an application such as EnCase) which is then stored with the case files; others simply back up the data files from the data card to a computer file.

The exact methods used are determined by the rules governing storage and disclosure of digital evidence that are in force in each jurisdiction, but the very least that can be said is that it is good practice to back up the original raw data using one of the methods listed above.

7.14.2   Historical Survey Data

As stated above, forensic radio surveys are usually undertaken in support of a specific investigation or court case. In addition to providing specific information in relation to each case, forensic radio surveys also have a wider significance in that they provide a snapshot of cellular coverage at the times and in the places that the surveys were undertaken.

One objection to the use of cell site evidence, especially to the forensic radio survey element in cell site evidence, is that each survey provides an indication of cellular coverage *at the time that the survey was conducted* and does not necessarily provide details of the state of cellular coverage *at the time that the incidents being investigated occurred.*

For example, if an investigation is focused on the events surrounding a bank robbery that occurred in March, but the case-specific forensic radio surveys were not conducted until November, there is ample opportunity for the coverage provided by the networks and cells used by the suspects to have changed in the intervening period.

If forensic radio surveys are conducted on a case by case basis and the results of each survey are examined in isolation, then there is a possibility that the results may not accurately reflect the cellular coverage at the time of the significant calls.

One way of guarding against this type of inaccuracy, or at least, one method of determining whether there is a risk that survey results might be inaccurate, is for law enforcement agencies and other organisations that undertake radio survey work to maintain a historical database of previous survey results. Such a database would contain the details of all surveys that had been undertaken and would list the networks and cells that were surveyed for cell coverage, the routes driven for route profiles and the addresses surveyed for spot/location surveys. The database could also link to the stored copies of the raw data, summary reports and maps created for each survey.

The benefit of such a system would be that it gave investigators an opportunity to check to see if other surveys were conducted of cells or areas closer in time to the events that they were looking into.

To return to the example of the bank robbery put forward above: if the bank raid happened in March but forensic radio surveys were only requested in November, the surveyors could check their historical survey database to see if any surveys were conducted in that same area in March or at any other time closer than November. If historical data is available it can be compared against the results obtained in the case-specific surveys taken in November to see if they highlight any significant changes in cell coverage in the intervening period.

The author has had indirect experience of the benefits of such a historical database: a colleague was involved in a case in which the significant events happened some two years prior to the investigation commencing. When the suspect's call records were examined it became apparent that, due to changes and upgrades in the cellular network used by the suspect's phone, many of the cells that had been used by the suspect had since been decommissioned and removed. This meant that it was not possible to take a comprehensive set of spot/location or cell coverage surveys for the case as many of the cells simply were not there anymore.

Luckily, the agency prosecuting the case had the foresight to create a historical database of old survey results and there were details of several surveys that had been captured in roughly the same areas as were relevant to the new case. It was therefore possible to supplement the results of new radio surveys with details from the historical surveys and a successful case was built from the data.

7.14.3   Proactive Surveys

Traditionally, forensic radio surveying for cell site analysis has been a 'reactive' tool – an incident occurs, an investigation begins, cell site analysis is performed and forensic radio surveys are requested.

There are some within the law enforcement community who advocate that a more 'proactive' approach should be taken to survey data gathering. The argument put forward is that, as law enforcement agencies have access to forensic radio survey tools and given the proven benefits of having a historical database of survey results, it would be advisable for agencies to send their surveyors out to capture regular snapshots of cellular coverage that can then be added to the database.

These regular, non-case-specific surveys would be undertaken of a number of targets:

- The surveyors may be tasked with conducting an 'outline' coverage drive survey of the major roads in an region or a city to capture details of the general cellular coverage of each network and technology.
- Surveyors may be tasked with conducting regular drive or walk surveys in and around key strategic buildings and locations – airports, major train stations, government buildings or diplomatic districts – to capture cell coverage details.
- Surveyors might conduct regular drive surveys in and around known crime hotspots or areas in which investigations are likely to be conducted.

The main purpose of such proactive surveys would be to give investigators immediate access to recent, general cellular coverage information for key areas, which can be used in the initial phase of an investigation while they are waiting for specific, reactive forensic radio survey data to be captured.

There are arguments to be made regarding the cost and time associated with performing regular surveys of this type, but the proponents of the idea suggest that these surveys can be performed in the 'down time' experienced by forensic surveyors when there are no more pressing, reactive surveys to be undertaken. Cost savings may also be made by eliminating the need for case-specific All-Network profiles to be undertaken, if general network coverage details for the area in question had be recently obtained from a proactive survey.

7.15 Quality and Best Practice

Cell site analysis and forensic radio surveys may be new and untested disciplines in some parts of the world, but in other countries, such as the United Kingdom, these techniques have been in development and use for over 15 years.

Ultimately, it is for each surveyor or organisation to decide on the standards and practices that are appropriate for and applicable to the country in which they operate but the following suggestions for survey best practice might provide some indications of how to ensure that surveys are successfully completed.

It should be noted that there is no commonly accepted standard for cell site or forensic radio survey best practice. Indeed, it is commonly said that if four cell site experts were in the same room you would get a least five opinions about how best to undertake investigations. The suggestions outlined below are a synthesis of the author's own experience and preference (and must therefore be regarded as at least partly subjective) but are also based on consultation with other forensic survey and cell site experts. Some or all of the points listed below will undoubted be disputed by some sections of the cell site community.

The best practice suggestions cover a number of areas:

- Survey preparation
- The choice of Idle Mode or Connected Mode for each survey type
- Specific suggestions for each main survey type
- Contemporaneous note taking
- Post-survey reporting.

7.15.1   Survey Preparation

It is suggested good practice for surveyors to prepare or to be provided with a target cell list, cell site locations map and survey instructions for each survey undertaken.

This should indicate the type of survey being requested, list the 'target' cells that the survey is expected to find serving and should also provide details of the site locations and cell azimuths to allow further investigations to take place if the expected cells cannot be detected.

7.15.2   Idle Mode versus Connected Mode

It is suggested that Idle Mode is employed for the majority of survey types and in the majority of circumstances, supplemented when necessary with intermittent Connected Mode test calls.

7.15.3   Best Practice for Survey Types

Spot/Location Surveys

Idle Mode is recommended for the main capture of all spot/location surveys, unless a survey is commissioned to prove a specific point that could only be determined using constant connection Connected Mode techniques. An example of such a requirement could be when a survey is required to indicate whether a specific cell could be used as an 'end' cell or when details relating to the selection of or handover to 'capacity expansion' cells (e.g. EGSM900 or GSM1800 cells) are needed.

Industry opinion tends to agree that an Idle Mode survey has the best chance of capturing a full and representative picture of coverage at a location if the survey is undertaken over an extended period – 5–10 min is a generally accepted duration.

It is also recommended that 'static' spot location surveys should be avoided if possible, with a limited local coverage survey being preferable (if it is safe and practical to do so). Ways of undertaking a less static survey include walking around the perimeter of the building/site being surveyed during the survey or, if drive survey techniques are being employed, parking outside the location for a period of a few minutes and then driving round the block and up and down nearby streets during the survey. It is recommended that the surveyor stays within 50–100 m of the location being surveyed during the survey period if possible.

Additionally, if the surveyor has access to multiple survey devices – more than one NEMO/TEMS handset or is using a multi-radio device such as CSurv – then, if possible, they should capture multiple readings for the same network/technology. This can allow a greater depth of information relating to the changing cellular environment to be captured when the separate results are combined.

If an initial Idle Mode survey captures details of all of the cells listed in the target cell list, the surveyor might elect not to make test calls. However, it may be prudent to incorporate a minimum number of test calls into each Idle Mode survey just for the sake of completeness.

If test calls are being made during a 3G survey, it is suggested that each test call should last for a reasonably long time – 1 min, for example. This should allow a representative variety of 'soft handover' neighbours to be added to/dropped from the phone's active set and may capture details of active cells that were not selected (or even detected) during the Idle Mode survey.

If expected cells are still not detected following the use of both Idle and Connected Mode techniques, it is recommended that the surveyor visits each expected but non-detected cell site to confirm that the cell is on air and that its azimuth is as expected.

Depending upon the survey device being used, one further recommendation, especially for 3G and 4G surveys, is that the surveyor captures 'lock on' details of neighbour cells that have not been selected as serving/active cells (if the survey equipment being used does not capture neighbour Cell IDs automatically and if it supports the cell lock function). Cell lock files are taken with the survey device in 'recording' mode and the surveyor employs the cell lock function for each detected (or expected) neighbour cell in turn, leaving the lock in place for 10–20s each time. This forces the survey device to treat each of the selected cells as a serving cell and should enable it to capture details such as the Cell ID, which are not captured for neighbour cells by all survey devices. Cell lock files need to be clearly labelled as such as they need to be interpreted differently to 'normal' capture files.

If a cell lock feature is not available, the surveyor may have to consider driving to (or at least towards) each expected but non-detected or detected but non-serving cell site and attempting to capture a Cell ID for each cell that way.

Cell Coverage Surveys

It is suggested that cell coverage surveys should only be undertaken in Idle Mode.

Route Profile Surveys

It is suggested that route profile surveys are usually undertaken in Idle Mode, supplemented with Connected Mode surveying if the survey is required to prove that certain cells could be used as 'end' cells.

7.15.4   Contemporaneous Note Taking

It is strongly recommended that surveyors take contemporaneous notes during each survey and retain those notes for disclosure should they be required to do so by a court.

Contemporaneous notes taken could include:

- Details of the address (possibly including a photo) and duration of each survey
- Specific description of the survey location (e.g. balcony outside Flat 7) for a spot survey or the route taken (e.g. circled perimeter of builder's yard) for a location survey
- Number and duration of test calls
- Details of standard and lock-on files captured and their filenames
- List of non-detected target cells and the actions taken to ascertain the reason for the non-detection
- Weather conditions and general comments
- Details of any events or observations that lead to specific conclusions being drawn.

7.15.5   Post-Survey Reports

It is recommended good practice that the surveyor captures contemporaneous notes about the progress, results and exceptions encountered during surveys and that these are written up into a post-survey report along with the survey 'product' with comments that can be provided to the investigator, cell site analyst or expert for whom the readings were undertaken.

The survey product for spot/location surveys is typically a summary results table for each location.

The survey product for cell coverage surveys is typically a coverage map for each target cell, which shows the route travelled, locations at which the target cell was detected as serving and (optionally) locations at which the target cell was detected as first neighbour.

The survey product for route profile surveys is typically a map showing the surveyed route and the progression of serving cells detected along that route.

7.15.6   Summary of Survey Best Practice

A summary of the suggested best practice for forensic radio surveys is as follows:

- Survey safety recommendations should be followed at all times.
- Detailed survey preparation information should be available to the surveyor, including target cell list, cell addresses and azimuths and a cell locations map.
- Idle Mode should be employed for spot/location surveys, supplemented by Connected Mode test calls if the expected serving cells are not detected in Idle Mode.
- Connected Mode test calls can also be made at locations to prove specific points, such as the use of cells as 'end cells' and the likelihood of them being used for handovers.
- 3G test calls should last for up to 1 min (if call length is controllable) to provide a decent opportunity for any 'soft handover only' cells to be selected.
- 'Static' location surveys are likely to be less representative than non-static 'location coverage' surveys.
- Surveys should capture measurements of the same network/technology using multiple devices, if possible and then combine the results into one overall set of measurements.
- Location surveys will be more representative if captured over a relatively long duration of 5–10 min.
- If spot/location surveys are conducted using a drive survey, the vehicle should spend a reasonable proportion of the survey duration in the immediate vicinity of the address being surveyed.
- Expected but non-detected cells should be investigated, by visiting the cell site if necessary.
- Lock-on surveys could be undertaken if required (and if supported by the survey device) to capture Cell ID details of 3G/4G neighbours.
- Cell coverage surveys should be undertaken in Idle Mode only.

- Route Profile surveys should be undertaken in Idle Mode, supplemented by Connected Mode test calls in specific circumstances, such as to prove the use of 'end cells'.
- Contemporaneous notes should be taken and post-survey reports compiled.

7.16   Summary of Typical Survey Results

The different types of data provided from surveys of different network types can be summarised as follows.

7.16.1   GSM Measurements

Measurements of a 2G GSM cell will typically provide the following data:

- **RXLev** (Received signal level) in dBm. RXLev is the received signal strength on one ARFCN; it is measured in dBm and has a typical range of:
   ○ greater than −84 dBm (very strong) to
   ○ less than −100 dBm (very weak).

RXLev is the most useful measurement for GSM and is also known as RSSI.
   Cell differentiation for GSM is provided at the physical layer by ARFCN + BSIC and at the logical level by MCC-MNC-LAC-Cell ID.

7.16.2   UMTS Measurements

Measurements of a 3G UMTS cell will typically provide the following data:

- **RSCP** (Received Signal Code Power) in dBm (wanted signal)
- **RSSI** (Received Signal Strength Indicator) in dBm (channel noise)
- Typical ranges for RSCP and RSSI are from around:
   ○ −90 dBm (very strong) to
   ○ −120 dBm (very weak).
- **Ec/No** (RSCP/RSSI) in dB (signal to noise ratio). Typical ranges for Ec/No are from:
   ○ greater than −6 dB (very strong) to
   ○ less than −20 dB (very weak).

Ec/No is the most useful measurement for UMTS. Knowing RSCP only is not much use as, without knowing the RSSI noise level, it is impossible to tell whether the signal to noise ratio makes the cell usable or not.
   Cell differentiation for UMTS is provided at the physical layer by UARFCN + PSC and at the logical level by MCC-MNC-RNC ID-Cell ID.

7.16.3   LTE Measurements

Measurements of a 4G LTE cell will typically provide the following data:

- **RSRP** (Received Signal Reference Power) in dBm (wanted signal). Typical ranges for RSRP are from around:
  o –40 dBm (very strong) to
  o –90 dBm (very weak).
- **RSSI** (Received Signal Strength Indicator) in dBm (channel noise). Typical ranges for RSSI are from around:
  o –70 dBm (very strong) to
  o –120 dBm (very weak).
- **RSRQ** (Received Signal Reference Quality – RSRP/RSSI) in dB (signal to noise ratio). Typical ranges for RSRQ are from:
  o greater than –7 dB (very strong) to
  o less than –25 dB (very weak).

RSRQ is the most useful measurement for LTE, as it offers a measurement of signal to noise ratio.

RSRP is also quite commonly used (partly because there is some ambiguity among handset vendors about the correct calculation of RSRQ) but can be of less use as, without knowing the RSSI noise level it is impossible to tell whether the signal to noise ratio makes the cell usable or not.

Cell differentiation for LTE is provided at the physical layer by EARFCN + PCI and at the logical level by MCC-MNC-eNB ID-Cell ID.

7.16.4   cdmaOne and CDMA2000 Measurements

Measurements of a 2G cdmaOne or a 3G CDMA2000 cell will typically provide the following data:

- **RSCP** (Received Signal Code Power) in dBm (wanted signal). Typical ranges for RSCP are from around:
  o –45 dBm (very strong) to
  o –120 dBm (very weak).
- **RSSI** (Received Signal Strength Indicator) in dBm (channel noise). Typical ranges for RSSI are from around:
  o –40 dBm (very strong) to
  o –100 dBm (very weak).
- **Ec/Io** (Energy per chip over Interference) in dB (signal to noise ratio). Typical ranges for Ec/Io are from:
  o greater than–6 dB (very strong) to
  o less than –28 dB (very weak).

Ec/Io is the most useful measurement for CDMA-based systems, as it offers a measurement of signal to noise ratio.

Cell differentiation for cdmaOne/CDMA2000 is provided at the physical layer by Band Class + Channel Number + PN Offset and at the logical level by MCC-MNC-SID/NID/Base ID.

References

[1] 3GPP Technical Specification (2013) *Non-Access Stratum (NAS) Functions Related to Mobile Station (MS) in Idle Mode*, TS 23.122 v11.4.0 Section 2, www.3gpp.org (accessed 29 July 2014).

[2] 3G Forensics Ltd. (2014) *Home Page*, http://www.3gforensics.co.uk (accessed 2 June 2014) .

[3] Anite plc (2014) *Network Testing Solutions*, http://www.anite.com/businesses/network-testing/ (accessed 2 June 2014).

[4] Ascom Network Testing (2014) *Network Testing Products*, http://www.ascom.com/nt/en/ index-nt/tems-products-3/tems-investigation-5.htm (accessed 2 June 2014).

[5] Forensic Mobile Services Ltd. (2014) *CSU-4 L Survey Tool*, http://www.fmsgroup.co.uk/ csu-4lsurveytool.html (accessed 2 June 2014).

[6] Forensic Analytics Ltd. (2014) *Products – CSAS*, http://www.forensicanalytics.co.uk/products/ CSAS (accessed 2 June 2014).

8

Cell Site Analysis

Cell Site Analysis (or cell tower tracking as it is sometimes known in the United States) attempts to provide evidence of where a mobile phone may have been located when certain calls were made. It is a useful tool for investigators and is based on a combination of network-provided CDR (Call Detail Record) data and forensic radio survey results.

CDR data provides details of the calls made and the cells used by a target phone during a significant period. Typically the CDR data provides details of the cells used for each call/text/data session, including the cell site's name, its location and the orientation (azimuth) of the specific sector used.

By determining the location of the serving cell and its orientation, some bounds on the potential location of the subscriber can be set. This level of location determination is often all that is required, especially if the cell site evidence is simply being used to prove or disprove a suspect's alibi.

Forensic radio surveys can be commissioned to provide more detailed information about the coverage provided by individual cells or at specific locations.

Most cell site investigations combine an analysis of an individual's call records with details of the location and coverage of the cells they used.

Forensic Radio Survey Techniques for Cell Site Analysis, First Edition. Joseph Hoy.
© 2015 John Wiley & Sons, Ltd. Published 2015 by John Wiley & Sons, Ltd.

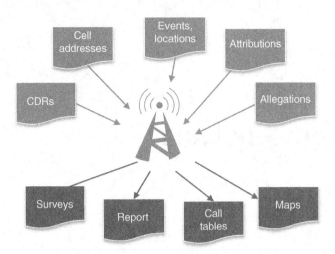

Figure 8.1   Cell site analysis

8.1   Cell Site Concepts

The elements that combine as 'inputs' to cell site analysis are shown in Figure 8.1 and include:

- The CDR billing data for each 'target' phone;
- Details of the locations of the cells used by the phone;
- Details of the events, times and locations significant to the case;
- Solid attribution of the target phone(s) to the suspected individual(s);
- Ideally, a description of the specific allegations or scenarios the investigators would like tested against the cell site data.

The elements that form the output of cell site analysis include:

- Forensic radio survey results;
- Cell Site Analysis report, providing conclusions related to calls or groups of calls and indicating if the evidence supports the allegations made or not;
- Call schedules, which list the call data provided for the target phone(s) in a common format and usually cut down to only the 'periods of interest' in the case;
- Cell site maps, which provide a graphical view of the relationship between significant locations/events and the locations of the cells used by the target phone(s).

Collectively, the results of a cell site analysis investigation can be used to prove (or disprove) the specific allegations made in a case. It is important to remember that cell site analysis can be just as useful to the Defence in a case as it is to the Prosecution.

In the event that an investigation goes to court, the cell site analysis report is often put forward as evidence, requiring the cell site analyst or expert to attend court to present their findings. It is common in these scenarios for the defence team(s) to engage cell site experts of their own to examine and comment upon the prosecution's cell site evidence.

8.2   Uses and Limitations of Cell Site Evidence

8.2.1   *Limitations*

Cell site evidence works best as supporting evidence.

On its own, cell site evidence is generally considered to be too open to interpretation to be used as the sole or the primary evidence in a case. There have been cases where the cell site evidence was so strong that it could be used as the primary evidence, but there are dangers inherent in using cell site evidence in this way.

The simplest thing that cell site evidence can prove, as illustrated in Figure 8.2, is that a target phone used a specific cell to make a call that started at a certain time. The target phone *must* therefore have been somewhere within the coverage footprint of that cell when the call started. If details of the 'end cell' for each call are listed then it may also be possible to state that the target must have been within the coverage of the end cell when the call ended. As only start and end cell details are usually included in call records, it is not usually possible to infer a location for target phones in between the start and end times of calls, unless the calls are of very short durations.

If the coverage area of a cell can be measured (e.g. by undertaking a cell coverage survey), then a reasonably representative impression can be determined of the general area in which the phone must have been located.

If there are other types of evidence available – eyewitness statements, CCTV images, ANPR (Automatic Number Plate Recognition) camera activations, credit card usage records – that help to place the alleged user of the target phone at the

Figure 8.2   Simple cell site conclusion

significant location, the cell site evidence can be used to provide compelling reinforcement of that evidence.

If the cell site evidence is all that the investigator has to tie the suspect to a location, then a level of uncertainty must be accepted.

In general, except in very specific and unusual circumstances, cell site evidence cannot be used to prove that the user of a phone was *definitely* at a particular location and nowhere else. At best, cell site evidence can be used to show only that it is *possible* for the user of the phone to have been at a particular location when significant calls were made.

8.2.2   Cell Site Uses

There are several typical uses for cell site evidence, each of which has qualities that can be of benefit to the prosecution and/or the defence in a case.

Cell site evidence is commonly used to help test a suspect's alibi, for example, and is also used to test investigator's allegations relating to whether a suspect could have been at or near the scene of an incident.

Looking at the alibi aspect first; if a suspect provides an alibi stating that they were in Paris at the time of an offence but their call records indicate that their phone (or more accurately, that a phone that has been attributed to them) was using cells in Berlin at the time, the alibi can be shown to be potentially false.

For alibis that place the suspect closer to the site of the incident, where they state that they were in a different part of the same city at the time of the incident, for example, cell site evidence can be used to indicate whether the cells used by the suspect's phone are likely to provide coverage at the alibi location. This can be determined in a simple 'high-level' fashion by comparing the alibi location to the used cell's location and azimuth. If the alibi location is several kilometres away from the used cell site, as in the example in Figure 8.3, and is in a different direction than that to which the azimuth is pointing then, again, the alibi can be shown to be potentially false.

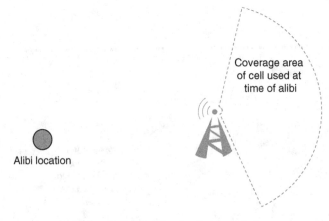

Coverage area
of cell used at
time of alibi

Alibi location

Figure 8.3   High-level cell site analysis

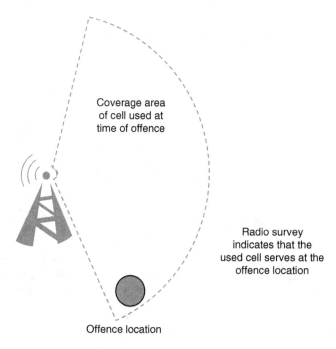

Coverage area
of cell used at
time of offence

Radio survey
indicates that the
used cell serves at the
offence location

Offence location

Figure 8.4   Low-level cell site analysis

The use of cell site evidence to test alibis can be used in an attempt to support the assertions made by a suspect or defendant and would therefore be a technique used by the defence in a case, or it could be used by the prosecution in an attempt to disprove an alibi.

The prosecution and investigators in cases also use cell site evidence in an attempt to provide support for the assertions and allegations that they are making against individuals. In this sense, cell site evidence is often employed to show that the use of certain cells by a suspect's phone is consistent with the allegation that they could have been located at or near the scene of an incident when those calls were made.

In circumstances in which a high-level analysis of a suspect's call records shows that their phone could potentially have been in use close to an incident location, forensic radio spot/location surveys can be undertaken, which will provide details of the 'serving' cells at that specific address or location. If one or more of the cells used by the target phone at or around the time of the significant event is seen to serve at that location then it is possible to conclude that the user of the target *could have been* at that location at that time. This type of 'low level' cell site analysis is illustrated in Figure 8.4.

Note the use of the words 'could have been' in the previous paragraph.

Unless a cell can be shown to provide coverage *only* to the surveyed address and nowhere else (which might be the case with, e.g., a femtocell), then the fact that a cell used by the target phone serves at that location does not necessarily make it any more

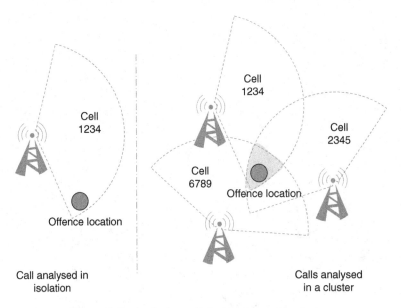

Figure 8.5   Analysing calls in isloation and in clusters

likely that the phone was actually there. The target phone may alternatively have been anywhere else in the cell's coverage area at that time. This point is made time and again by defence lawyers when cross-examining cell site expert witnesses and it is perfectly true, if calls are examined as individual events.

This level of uncertainty is always present if calls are examined in isolation – the evidence provided by one call can only place the target phone somewhere within the coverage of the used cell. More certainty can sometimes be gained by examining clusters of calls, especially if multiple cells were used in quick succession during the period covered by the call cluster.

A comparison of single call and call cluster analysis techniques is shown in Figure 8.5.

As illustrated in the left hand portion of Figure 8.5, the analysis of a single call shows that the target phone used Cell ID 1234 at the time of an incident and that Cell ID 1234 was detected during a spot/location survey as serving at the incident location. The only concrete conclusions that can be drawn from that one call is that the target phone was somewhere within the coverage area of Cell ID 1234 at the time of the call and could have been at or near the incident location but equally could have been anywhere else in the cell coverage area.

An analysis of the cluster of calls made at around the same time, however, shows that that target phone also used cells 2345 and 6789 in addition to cell 1234, all in a short period of time, as shown in the right hand portion of Figure 8.5. Cell coverage maps taken of cells 1234, 2345 and 6789, which are all transmitted by different base station sites, indicate that they have different coverage areas and that the only area in which the serving coverage of all three cells intersects is in the vicinity of the incident location.

This analysis of a call cluster therefore allows investigators to further constrain the size of the area in which a target phone 'could have been'. It should be noted that this 'further constraining' still does not usually reduce the area of potential location down to just the incident location, so a degree of uncertainty will still exist.

8.2.3   Historical versus Live Cell Site

Law enforcement agencies and investigators typically employ cell site analysis techniques to support two very different types of enquiry: historical investigations, in which cell site analysis is used in an attempt to show where a target phone *may have been* when calls were made; live investigations, in which cell site analysis techniques are used to try and determine where a target phone *is currently located* at that moment in time.

Historical cell site reports are used as part of the process of investigating incidents that have already occurred and are often commissioned as part of the process of preparing a case for court.

Live cell site analysis is typically used as part of an on-going incident, such as a missing person, a kidnap, or to track the perpetrators of an event such as a bank robbery.

8.2.4   Combining Cell Site Analysis with Other Evidence

As has been previously stated, cell site evidence on its own is generally considered to be too open to alternative interpretations to be used as the sole or primary evidence in a case. Cell site evidence often works best when being used to support other forms of evidence.

A word of caution when considering the combined use of multiple forms of electronic evidential data would be to ensure that there are common time and date benchmarks for the data to be analysed. For example, it is very common for CCTV timing to be different from a cellular network's timing. This is because cellular networks are generally timed from an accurate 'official' source, whereas CCTV systems (especially small, standalone systems) will typically have had their timing set from a local source and may not have had it updated since it was installed. This often results in timing differences between CCTV time and 'real' time and offsets of several minutes (or even of several hours or days) are common.

This means that an event that happens at timestamp 'A' in CCTV time may not actually have happened at the same time as an event with the same timestamp that was captured in an individual's call records.

Similarly, if data from a handset, such as events listed in a handset's call log, is aligned with the billing data for that handset, it is important to bear in mind that the date and time that are set on a handset are often set by the owner of that handset when it was being set up. As with 'CCTV time', 'handset time' could be offset from 'real time' by seconds, minutes or hours. The author has been involved in cases where handset evidence was used where the timestamps on the handset turned out to be wrong by a margin of more than 24 hours.

To summarise, just because it is possible to combine different forms of evidence with cell site evidence, it should not be assumed that all forms of evidence share a common time and date reference.

Having taken the warning given above into account, there are multiple possible examples of the use of cell site analysis to support other forms of evidence, including:

Example 1 – CCTV – Investigators have CCTV from multiple cameras that cover the route taken by the getaway car used by a gang of bank robbers following a raid. They also have the call records of one of the suspected members of the gang, which indicate that several calls were made during the period that covers the journey of the getaway car. A high-level examination of the call records showed that the cells used during that period were located close to the getaway route, so a route profile survey was ordered. The results of the route profile indicated that the cells used for those calls all served at points along the getaway route in a progression that matched the sequence in which they were used by the suspect's phone. The cell site data therefore provided additional evidence that supported the allegation that the user of the target phone was travelling in the getaway car.

Example 2 – Instant messaging – Investigators have evidence that a suspect communicated with a victim (and sent threatening messages to them) using the Internet-based instant messaging service WhatsApp and have records of the messages that were exchanged. WhatsApp logs do not include location or cell site details, but because the suspect was using the WhatsApp app on their mobile phone, their call records may be able to shed light on where the phone might have been when the messages were sent. WhatsApp and other Internet-based messaging services use a phone's PS data connection to carry their traffic and, by matching the message logs with the phone's data CDRs, it was possible to attribute an approximate location to the suspect's phone (accepting the lack of accuracy associated with PS data session records) for the times at which messages were sent.

Example 3 – Open source data – Many investigators are beginning to explore the possibilities offered by the use of 'open source' data in their cases. Open source data, as the name suggests, is data related to individuals that is made freely available via the Internet. A good example of this might the information that people choose to post to services such as Facebook or Twitter. Investigators are increasingly using open source data to test the alibis put forward by suspects and also to test claims of 'non-association' between suspects and associates in cases. 'Non-association' in this sense means that investigators have alleged that individual A and individual B are known to each other, which either or both of them have denied. An examination of open source data related to the individuals could show that they have messaged each other or that they 'follow' each other on Twitter, for example. As with the instant messaging evidence in Example 2, cellular data session records, when paired with website logs or message posting times, can be used to provide approximate location details for a suspect's mobile device at or around the time that posts, status updates or messages were uploaded.

8.2.5   Attribution

Cell site analysis typically provides evidence of where the user of a mobile phone may or may not have been when calls were made. Cell site evidence generally does not provide proof of the identity of that 'user'. A common way of describing this concept is that cell site analysis 'comments on the location of the phone and does not identify the hand holding the phone'.

It is generally recommended that cell site analysis should only be undertaken once a solid attribution for the target phone(s) relevant to a case is made, but cell site evidence can have a role to play in aiding that attribution.

There are several methods that can be used to help to attribute a target phone. The most reliable form of phone attribution is for the suspect to provide or confirm their phone number to investigators during interview or in a statement.

Another reliable method is for the billing or ownership details of the phone or its account to identify the attributed user. Most networks have a standard 'subscriber details' report, which provides details (if any are known) of the owner of a handset or an account holder. In many countries it is possible to buy 'pay as you go' prepayment handsets without needing to provide any proof of identity; in these cases the response to a subscriber details request will usually be 'no details held'.

A further source of confirmation is for a target phone to be found in the suspect's possession when arrested or to be found during a subsequent search of their vehicle, home or work address.

In circumstances in which the attribution of a target phone is disputed – where the investigators believe a phone was in the possession of a suspect but the suspect declines to confirm this – it is possible to use call records and cell site techniques to provide evidence to support attribution.

The simplest form of cell site support for attribution is a so-called 'lifestyle analysis'. This involves looking through the calls made by the disputed phone and checking for correspondences between its use and the use of handsets that can be attributed to the suspect. For example, if the majority of calls made using the disputed handset were to the suspect's girlfriend, mother and best friend then an argument for attribution can be made.

Other forms of lifestyle analysis involve looking at the cell sites used by the disputed phone: if a high proportion of the last calls made each evening and the first calls made each morning use cells that serve at the suspect's home address then that could point to the phone having been used by that suspect.

Investigators have found that some 'forensically aware' individuals use a 'clean' phone for their day to day personal calls and employ unregistered 'dirty' or 'burner' phones for criminal purposes. It is possible to compare the use of the 'clean' and 'dirty' phones to check for common numbers and also to check for 'colocation'. This occurs if the clean and dirty phones are used in the same areas for calls made in quick succession. For example, if a suspect's 'clean' phone uses Cell ID 1234 to make a call and then 2 min later the 'dirty' phone also uses the same cell, there is an argument to say that both phones are in the same area at the same time. If the same pattern of colocation keeps occurring over an extended period of hours or days then the conclusion can be drawn that either the users of the two were together or that both phones were in the possession of the same user.

8.3   Regulation of Cell Site Analysis

Cell site analysis is an intrusive investigative technique.

It allows investigators to gain access to a significant part of an individual's digital lifestyle and also sheds light on where an individual (or at least, where their attributed mobile device) may have been located at points in time.

Despite the obvious benefits that investigators can gain from this type of evidence, it is often considered desirable to limit access to this information and to tightly control both the range of agencies that are permitted to request it and the circumstances in which those requests will be honoured.

In most jurisdictions, investigators must apply for a warrant or a court order to allow them to obtain copies of an individual's call records. In many cases these court orders must be individually applied for and will be the subject of a specific decision by a judge on whether to grant the requested warrant. This individual approach can lead to delays in granting access to records and can also mean that large amounts of court time are taken up processing the requests.

In some jurisdictions, such as the United Kingdom, the process of requesting and authorising communications warrants has been streamlined by placing it within a defined legislative structure.

The United Kingdom passed the Regulation of Investigatory Powers Act (RIPA) in 2000. The main aim of this act was to put the relationship between the holders of information that is of interest to investigators (e.g. the cellular service providers) and the agencies that wish to access the information on a regulated, statutory footing. In respect of communications data, RIPA put in place a structure that identified the set of agencies that were able to request access to that data and ensured that each agency had one office – the SPoC or 'Single Point of Contact' – through which requests for that information were passed to the cellular operators. Each operator maintains an LEA (Law Enforcement Agency) liaison office that handles requests from the various SPoCs [1].

In addition to regulating the relationship between operators and LEAs, RIPA also defines the types of information that can be requested and the tests that must be applied to ensure that each request is reasonable and proportionate to the crime or incident being investigated. Responsibility for authorising RIPA requests is devolved from the courts down to nominated high-ranking individuals within each agency.

Legislative frameworks like RIPA, and similar systems that have been developed in other jurisdictions, are mainly aimed at regulating the way in which official agencies such as the police gain access to communications data. Most of these frameworks therefore work to regulate the way in which the prosecution obtain communications evidence in a case; few, if any of them, also regulate the way in which the defence are able to gain access to the same data.

In most scenarios the defence in a case gain access to copies of any communications data, such as cellular CDRs, obtained by the prosecution as part of the normal 'discovery' or 'disclosure' process – in which the prosecution are obliged to provide the defence with details of all evidence they have obtained as part of outlining the case against the defendant.

If the defence wish to gain access to additional communications data related to their client, or if they wish to gain access to data related to other individuals, then they are usually required to go through the process of requesting specific court orders to compel the CSPs to release the required information.

8.4   Components of Cell Site Analysis

A summary of the main types of input and output information related to cell site analysis is provided above. The following sections outline some of these elements in more detail and deal with the input forms of source information – call records, cell address details and forensic radio survey results – and also look at the output cell site analysis 'product' of cell site reports, call schedules and map presentations.

8.5   Call Detail Records

CDRs are produced every time a user makes or receives a call, sends or receives a text message or connects to a data service. Some records might also be produced in relation to network events such as attaches, location updates and detaches and to other events such as phone or SIM card purchases or top up transactions.

There are two different methods for capturing call details, as there are two methods for charging for those calls: post-pay (in which the user makes calls on credit and receives a monthly bill to pay for them) and prepay (in which users must deposit to or 'top up' their account before they are allowed to make calls).

For post-pay subscribers, voice/SMS CDRs are produced in the telephone exchange or 'switch' that controls each phone call or messaging event and MMS or data CDRs are produced in the packet core network elements that control PS data sessions. Post-pay CDRs are handled by an 'offline' charging system.

For prepay subscribers, CDRs are created by the 'online' charging system, which also manages the database that stores details of income (top ups) and outgoings (real time debits as calls are made) for each account.

Both the online and offline charging systems feed call records into the operator's billing system, making both forms of charging information equally available to be added to CDR files.

A new CDR record is 'opened' each time a new chargeable event starts. Once a CDR record is 'closed' (when the connection is released), it will be transmitted to the operator's billing system and stored in a centralised database. Ostensibly, CDR data is captured for billing and charging purposes but it can also be disclosed to authorised agencies such as the police. In the United Kingdom this occurs under the provisions of RIPA and may be covered by similar legislation in other jurisdictions.

Network operators provide CDRs in a wide variety of different formats and the formats employed by different operators provide a variety of information. Generally each CDR contains the following:

- Date and Time of start of call
- Duration of call
- Type of Service, for example, voice call, SMS, MMS, data and so on
- Originating Mobile Station ISDN (MS-ISDN; the 'A' number)
- Terminating MS-ISDN (the 'B' number)
- International Mobile Subscriber Identifier (IMSI) and International Mobile Equipment Identifier (IMEI) – not always provided
- Serving Cell ID and LAC (at start of call)
- Serving Cell ID and LAC (at end of call) – not always provided
- Cell site names, postcodes, GPS coordinates or map reference – not always provided.

Call data can theoretically be extracted from an operator's billing database based upon any of the above parameters. This means that it is possible to obtain a list of all calls made to/from a certain SIM (based on MS-ISDN or IMSI), but also to/from a certain phone, regardless of which SIM was used (based on IMEI).

GPRS (or data) CDRs often use a different format but provide much the same level of information and will also contain details of the IP address assigned to the phone for the data session. They may also provide details of the APN (Access Point Name) used for each session. An APN is the interconnection point between an operator's core network and an external data network, APNs typically have names such as 'Internet' (for basic Internet connections), 'Blackberry' (for links to the Blackberry messaging network) or 'MMS' (for photo messaging services).

Some operators provide details of all chargeable event types (voice, data, SMS, MMS) in the same combined CDR format, others provide separate CDR formats that show details of just some event types (e.g. a voice/SMS/MMS format and a separate GPRS data format), while others provide call data for each event type in its own completely separate format (e.g. separate voice format, SMS format, MMS format, and GPRS format CDRs).

Investigators should bear this in mind and make sure that they request the correct and full set of CDRs for each network they deal with, important evidence could be missed if an incomplete set of CDR files was requested.

The difference between 'voice' CDRs and 'data' CDRs is that a voice CDR will record each transaction (phone call, SMS, MMS) as a separate event, whereas data CDRs have traditionally provided details of an entire data connectivity session, which may have lasted for minutes or hours, but would not have provided details of individual connections (e.g. to a website) established during each session. More detailed data CDR formats may be made available in the future.

8.5.1   Voice/Text CDRs

Voice and text events typically have very distinct temporal identities, in the sense that there is a definite start time and end time for a voice call and a definite connection and delivery time for a text event.

Call records for voice calls typically capture details of the cell ID used to set the connection up (the 'start cell') and also the cell ID that was in use when the connection was released (the 'end cell') – they do not, unfortunately, provide details of any 'handover' cells used in between those times and so are less useful for long duration calls than for shorter calls. The relationship between device usage and billing timestamps is shown in Figure 8.6.

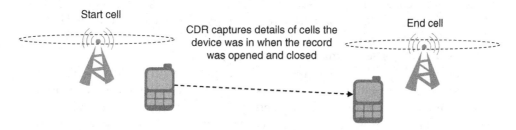

Figure 8.6   Voice/SMS CDRs

Call records for SMS events identify the cell ID that was used to transfer the text message. SMS transfer typically takes just a few milliseconds and there is no mechanism to allow a handover to take place during an SMS event, so SMS records typically detail just the 'start' cell ID.

In the case of both voice and SMS events, cell site analysts can be quite certain that a target phone was within the coverage area of the cells listed at the start time for an event and (in the case of voice records) at the end time of that event. It is therefore possible for cell site reports to conclude, in relation to a cell that serves at a significant location, that use of that cell is consistent with the possibility that the target phone was 'at or in the vicinity' of that location at either or both of those times.

8.5.2   GPRS Data CDRs

GPRS/PS data CDRs (where 'PS' stands for 'packet switched') can be far less definite in relation to their timestamps and are therefore less valuable in terms of the cell site evidence they can provide.

Unlike voice/SMS events, which record details of transactions that occurred at fixed points in time, data records relate to a more nebulous form of connection known as a 'session'.

A data session is a 'logical' connection between a mobile device and a data network such as the Internet. It provides the potential to carry data traffic, if there is any traffic to be carried, but will still be classed as active even if no data is currently flowing.

This is designed to best serve the way in which most data services operate: consider an email application, for example. The email application needs a connection to the Internet, but may only actually use that connection every 5 min or so, when it checks for new incoming mail or when the user sends an outgoing mail.

In cellular networks, a logical data session is established when a device connects to a data network, but radio resources are only assigned to that connection when there is actual traffic to carry. So a logical session may be long lasting but may only have brief periods of physical connectivity.

The data service in 2G networks is known as GPRS and is more generically known as a PS (packet switched) data service in 3G and 4G networks.

A new GPRS/PS data CDR is opened when a mobile device establishes a new logical session with a data network, such as the Internet – there is also a type of CDR that can be termed a 'follow on' or 'partial record' type, which will be described below. A PS data connection will typically carry intermittent bursts of data traffic as the user browses websites or sends email or instant messages.

PS data CDRs are 'closed' if the mobile device explicitly indicates that the session is no longer required. In many cases, however, a GPRS CDR will be closed in a more implicit fashion due to 'user inactivity'. After the successful delivery of each burst of data to or from a mobile device, the network starts a timer. If the timer expires with no further data being transmitted, the network assumes that the data

session is no longer required and releases it, triggering the closure of the associated CDR. The value of this 'user inactivity' timer is set by each network based on
their own requirements but typical values in 2G and 3G networks range between
30 minutes and 2 hours.

Mobile devices will typically be in one of two states when attached for PS services: (i) Standby (2G) or Idle (3G/4G) Mode, where the device is attached and
may have activated PS data sessions but does not have radio resources assigned
to actually carry any data; (ii) Ready (2G) or Connected (3G/4G) Mode, where
the device does have radio resources assigned and can transmit and receive data
traffic.

Devices that are in Ready/Connected Mode are required to perform a 'cell update'
or re-selection or handover when they change cells while involved in a data session,
so if a session ends directly from Ready/Connected Mode then the details of the
device's location (based on its serving cell) at the end of the session should be
accurate.

Devices that are in Standby/Idle Mode are only required to update the network
periodically or if they roam into a new Routing or Tracking Area. In between updates
the network only knows a device's location down to the current Routing or Tracking
Area, which could consist of several individual cells and the network will not necessarily know which of these cells the phone is camped on at any point in time. Sessions
that end from Standby/Idle Mode – typically as a consequence of the 'inactivity'
timer expiring – are forced to use the 'last known' serving cell ID for the session,
which was picked up the last time the device communicated with the network and
could consequently be many minutes out of date.

Although this is theoretically the case for all generations of PS data service (2G
GPRS, 3G and 4G), 4G phones are likely to establish and release a wider set of logical
data sessions than 2G or 3G phones, meaning that more detail on a device's location
can be gathered from the successive opening and closing of data CDRs.

The concepts related to PS data CDRs are illustrated in Figure 8.7 and show that
the cell ID information captured for PS data sessions can be less reliable than that
captured for voice/SMS events. Partly for this reason, networks generally do not

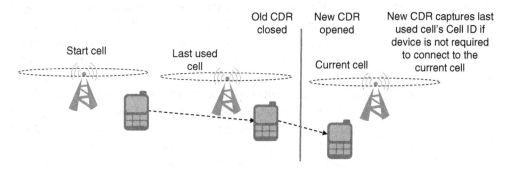

Figure 8.7   GPRS/PS data CDRs

provide 'end cell' details for GPRS/data events in CDRs, which means that location details for data records are based on the theoretically more reliable 'start cell' details. However, start cells in GPRS/PS data records can also have a high degree of uncertainty.

GPRS/PS data CDRs can be 'closed' for a variety of reasons, which include the explicit session release and 'user inactivity' causes mentioned above.

CDRs can also be closed without the associated data session ending; these cases are known as 'partial records' and are generated mid-session. Typical reasons for creating a partial record include (but are not limited to):

- Data volume limit reached – if maximum data volume for an individual record is set at, for example, 2 MB, then the current record will be closed and a new record opened when the data throughput count reaches that value.
- Time (duration) limit – if the maximum 'open time' for an individual record is set at 12 h, as an example, the current record will be closed and a new one opened when the open time reaches this value.

Networks often set the maximum data volume limit at a reasonably high value, 2–10 MB maybe, and the maximum 'open time' value at a comparatively long duration, such as 12 or 24 h, to reduce the volume of partial records being produced. The popularity of modern smartphones, with 'chatty' apps that send and receive small amounts of data almost continuously, means that many data sessions rarely trigger the 'user inactivity' process. From a forensic point, the chattiness of smartphone apps is bad news as constant connectivity means fewer sessions time out due to inactivity and consequently fewer new CDRs are opened.

If a data session remains active, its current CDR will eventually be closed due to the maximum data volume or open time values being reached. When an existing CDR is closed as a partial record, a new record will be opened immediately.

A CDR will also be closed mid-session if the mobile device roams into a new Routing Area, or hands over to a different access network type (e.g. 2G to 3G handover) or any one of several other reasons.

In the case of Routing Area change, an existing CDR will be closed and a new CDR opened when the mobile device signals a Routing Area Update to the network. The cell ID captured for the start of the new CDR will be the ID of the cell the device used to send the Update message, so the 'start cell' for the new CDR should accurately reflect the cell the device was being served by at the CDR opening time. If no further data is exchanged for a period of time after the new CDR is opened and the device continues to move, the 'start cell' will be the only cell ID captured for the duration of the device's connectivity in that RA and may quickly become out of date.

In some other 'partial record' cases (e.g. maximum time) where an RA change has not occurred, the 'start cell' listed for the new record will be the cell ID of the cell last known to have been used by the device, which may not be current information.

If the device was in Standby/Idle Mode when the partial record close/open event happened, the cell ID captured at the start of the new record could be inaccurate by the value of the expired portion of the inactivity timer period – so if a mobile device was in a car travelling at 60 km/h and had last sent data 30 min before, if the old CDR was closed due to having reached the maximum open time, the location information captured in the new CDRs 'start cell' could be inaccurate by up to 30 km.

Unlike for voice/SMS events, where cell site analysts can draw definite conclusions related to the possible location of a target device, conclusions reached in relation to GPRS data records must by necessity be less definite.

The accuracy of the start cell of CDRs for 'new' GPRS/PS data sessions in relation to the start timestamp can be relied upon, so conclusions could be drawn that state that the target phone could have been 'at or in the vicinity' of a location served by that cell at the time of the record's start timestamp. The same is true of CDRs created in relation to Routing Area change.

The accuracy of the start cell of CDRs for 'follow on' GPRS/PS data sessions in relation to the start timestamp could be relied upon if the mobile device was in Ready/Connected Mode when the partial record close/open event took place.

The accuracy of the start cell of CDRs for 'follow on' GPRS/PS data sessions in relation to the start timestamp cannot be relied upon, if the mobile device had been in Standby/Idle Mode for a period of time when the partial record close/open event took place, as they could be out of date. The connection state to the device is not capture din the CDR.

It can be difficult to distinguish these various call record closure causes from each other when examining a target phone's CDRs. Although it is relatively simple to determine that a CDR relates to a 'new' session (as there would not have been any other data CDRs recorded for a period of time before that), it is difficult, if not impossible, to tell the difference between the various scenarios for follow on records.

Cell site conclusions based on data session records are therefore forced to take a more circumspect form, such as that the target phone could have been 'within the coverage area of a serving cell sometime before or after the time of the record's start timestamp' or that the device could have been 'at or in the vicinity' of a location at which the start cell in a GPRS/PS record serves sometime before or after the record's timestamp.

It is for this reason that, in terms of best practice for cell site analysis, current guidance indicates that voice and text CDRs should take precedence over data CDRs if they are sufficient to prove the prosecution case and in all cases great care should be taken when attempting to draw conclusions based on GPRS/PS data.

Of course, if a mobile phone was used to make calls as well as connecting to data services, the voice call records when combined with the data records may provide more accuracy. So, for example, if the call records show only that a follow-on data CDR was opened at 20:20 hrs and record Cell ID 12345 as the start cell, it would only be possible to conclude that the target phone was in the serving coverage of that cell at some time before or after 20:20 hrs. However, if the data event at 20:20 hrs using Cell ID 12345 was followed at 20:21 hrs by a voice call that also used Cell ID 12345, then the conclusions reached in relation to both events can be much more definite.

8.5.3   CDR Examples

CDRs are provided in a variety of different formats. Each operator has their own preferred format and some operators even provide different types of CDR in different formats – for example one format for voice/SMS and a different format for GPRS data.

CDR formats also differ in terms of complexity: some operators employ relatively simple formats that list one entry for each call, text or data session that takes place, whilst other operators use much more complex formats that list multiple rows of data for each billable event.

As an example: the CDR format used by Vodafone UK separately lists the component charging records that detail each stage of each billable event, so for an event such as a text message transmission the CDR will list one entry showing the text leaving the source phone, another event showing the receipt of the message at the SMS messaging centre, a third event showing the forwarding of the text to the target phone and a fourth event showing the message being received at the target phone. In cases where the initial delivery of the text message fails, there could be multiple event records showing the subsequent redelivery attempts. Complex CDR formats of this kind can be extremely difficult to analyse correctly and there is a high potential for inaccurate analysis, caused by the analyst missing details of an event or by incorrectly interpreting linked records as belonging to different events.

Further complexity is caused by the different ways in which operators' CDR formats deal with call forwarding events. In standard telecoms terminology, the number making a call is known as the A number and the number receiving the call is known as the B number. In circumstances where a call from A to B is diverted – either to another phone or to a service such as voicemail – the eventual receiving number is known as the C number.

Some operators explicitly reference the numbers listed in their CDRs as A, B and C, which makes the relationship between them very clear. Other operators employ terminology such as 'Calling Number', 'Receiving Number' and 'Redirecting Number' which make the relationship between them quite difficult to determine. The typical way of interpreting this terminology is that the 'Calling Number' is the A number, the 'Receiving Number' is either the B number if no call diversions have taken place or is the C number if a redirection has taken place, and the 'Redirecting Number' is the B number in cases where call diversions have taken place. As can be seen, the terminology used by some operators can be complicated to unravel.

CDR formats also change and evolve over time, meaning that an understanding of an operator's current format does not necessarily lead to an understanding of previous or future formats.

There is such a wide variety of CDR formats employed that it would take up too much space to reproduce specific examples of real formats, so Figures 8.8 and 8.9 provide generic examples of the typical content of a simple combined voice/SMS CDR and a GPRS/PS data CDR.

Generic Voice/SMS CDR

NationalCell Network Disclosure Management System

MS-ISDN:	07700 345876
Records Requested:	Voice, SMS In & Out
Records Start Date/Time:	14/01/2014 00:00
Records End Date/Time:	14/01/2014 23:59
Request Date:	01/05/2014
Request By:	DC Davies

Date	Time	A Number	B Number	Call Type	Duration	IMEI	IMSI	Start Cell ID	Start LAC	Start Site Name	End Cell ID	End LAC	End Site Name
14/01/2014	00:15:04	447700345876	447700198019	Outgoing Voice	00:01:24	001457645213254	234950867523412	8173	124	Kingsbury West	8173	124	Kingsbury West
14/01/2014	03:34:56	447700127610	447700345876	Incoming Voice	00:00:00		234950867523412						
14/01/2014	03:57:06	447700127610	447700345876	Incoming Voice	00:00:00		234950867523412						
14/01/2014	08:17:26	447700345876	123	Outgoing Voice	00:02:13	001457645213254	234950867523412	8173	124	Kingsbury West	8173	124	Kingsbury West
14/01/2014	08:22:45	447700345876	447700127610	Outgoing Voice	00:07:16	001457645213254	234950867523412	8173	124	Kingsbury West	8173	124	Kingsbury West
14/01/2014	08:34:19	447700345876	447700198019	Outgoing Voice	00:21:29	001457645213254	234950867523412	8173	124	Kingsbury West	8173	124	Kingsbury West
14/01/2014	09:10:12	447700198019	447700345876	Incoming Voice	00:14:28	001457645213254	234950867523412	8296	124	Kingsbury Heights	8297	124	Kingsbury Heights
14/01/2014	09:16:33	447700127610	447700345876	Incoming SMS	00:00:00	001457645213254	234950867523412	8297	124	Kingsbury Heights			
14/01/2014	10:02:17	447700345876	447700067185	Outgoing SMS	00:00:00	001457645213254	234950867523412	21276	298	London Bridge			
14/01/2014	13:47:23	447700345876	447700127610	Outgoing SMS	00:00:00	001457645213254	234950867523412	6091	300	Waterloo			
14/01/2014	14:02:18	447700345876	447700127610	Outgoing Voice	00:04:11	001457645213254	234950867523412	23189	300	Globe	23190	300	Globe
14/01/2014	14:34:17	447700127610	447700345876	Incoming SMS	00:00:00	001457645213254	234950867523412	23189	300	Globe			
14/01/2014	15:04:19	447700198019	447700345876	Incoming Voice	00:00:12	001457645213254	234950867523412	21276	298	London Bridge	6091	300	Waterloo
14/01/2014	16:23:19	4477008861356	447700345876	Incoming SMS	00:00:00	001457645213254	234950867523412	6091	300	Waterloo			
14/01/2014	16:28:07	447700345876	447700127610	Outgoing Voice	00:01:38	001457645213254	234950867523412	11398	301	Oxford St	11398	301	Oxford St
14/01/2014	16:55:24	447700345876	447700019010	Outgoing Voice	00:21:16	001457645213254	234950867523412	6091	300	Waterloo	6091	300	Waterloo
14/01/2014	17:04:12	447700345876	447700345876	Incoming SMS	00:00:00	001457645213254	234950867523412	44231	217	Westminster Tower			
14/01/2014	18:26:55	447700345876	447700198019	Incoming SMS	00:00:00	001457645213254	234950867523412	3981	207	St Catherines Dock			
14/01/2014	19:01:00	447700127610	447700345876	Incoming Voice	00:07:23	001457645213254	234950867523412	6116	221	East St	28760	221	London East
14/01/2014	20:14:49	447700886135	447700345876	Incoming Voice	00:01:01	001457645213254	234950867523412	28761	221	London East	28761	221	London East
14/01/2014	20:15:59	447700345876	442079460187	Outgoing SMS	00:00:00	001457645213254	234950867523412	28761	221	London East			
14/01/2014	20:44:15	447700345876	442079460187	Outgoing SMS	00:00:00	001457645213254	234950867523412	28761	221	London East			
14/01/2014	20:07:22	447700345876	447700886135	Outgoing Voice	00:16:12	001457645213254	234950867523412	28761	221	London East	28761	221	London East
14/01/2014	21:38:11	447700019010	447700345876	Incoming SMS	00:00:00	001457645213254	234950867523412	28761	221	London East			
14/01/2014	22:07:16	447700127610	447700345876	Incoming Voice	00:00:24	001457645213254	234950867523412	31008	221	West Ham	31009	221	West Ham
14/01/2014	22:58:04	447700345876	447700019010	Outgoing SMS	00:00:00	001457645213254	234950867523412	7816	202	Royal Docks			
14/01/2014	23:09:17	447700019010	447700345876	Incoming SMS	00:00:00	001457645213254	234950867523412	7817	202	Royal Docks			
End of Data													

filename: 345876 voice in/out.csv

Figure 8.8 Generic voice/SMS CDR content

Generic GPRS PS Data CDR

Cellplus
MS-ISDN: 07700 198019
Records Requested: GPRS data
Records Start Date/Time: 14/01/2014 00:00
Records End Date/Time: 14/01/2014 23:59
Request Date: 01/05/2014
Request By: DC Davies

Date	Time	A Number	IP Address	Duration	IMEI	IMSI	Start Cell ID	Start LAC	Start Site Name	APN	Data Up	Data Down
14/01/2014	01:56:34	447700198019	10.14.107.118	02:00:00	02567629984520202	234042787261234	2009	90	Rotherhithe	Internet	2481.0kb	4698.1kb
14/01/2014	08:47:23	447700198019	10.0.176.23	00:01:34	02567629984520202	234042787261234	2009	90	Rotherhithe	Blackberry	209.4kb	598.5kb
14/01/2014	09:06:12	447700198019	10.0.176.23	00:02:10	02567629984520202	234042787261234	165	54	Green Park	Blackberry	16.0kb	187.3kb
14/01/2014	09:07:34	447700198019	10.14.107.118	02:00:00	02567629984520202	234042787261234	6577	109	Bermondsey	Internet	876.3kb	1045.5kb
14/01/2014	11:07:34	447700198019	10.14.107.118	02:00:00	02567629984520202	234042787261234	6577	109	Bermondsey	Internet	1075.9kb	22876.4kb
14/01/2014	12:12:54	447700198019	10.0.176.23	00:03:24	02567629984520202	234042787261234	165	54	Green Park	Blackberry	16.0kb	238.5kb
14/01/2014	13:07:34	447700198019	10.14.107.118	02:00:00	02567629984520202	234042787261234	2009	90	Rotherhithe	Internet	756.5kb	8756.4kb
14/01/2014	14:29:51	447700198019	10.0.176.23	00:07:12	02567629984520202	234042787261234	1627	211	Regent's Park	Blackberry	34.6kb	283.1kb
14/01/2014	15:07:34	447700198019	10.14.107.118	02:00:00	02567629984520202	234042787261234	4142	186	London Bridge Stn	Internet	146.6kb	2094.4kb
14/01/2014	17:07:34	447700198019	10.14.107.118	02:00:00	02567629984520202	234042787261234	897	203	Waterloo Underground	Internet	457.5kb	1578.3kb
14/01/2014	20:36:18	447700198019	10.0.176.23	00:11:56	02567629984520202	234042787261234	1143	176	Aston St	Blackberry	65.6kb	77.2kb
14/01/2014	21:27:44	447700198019	10.0.176.23	00:03:21	02567629984520202	234042787261234	1143	176	Aston St	Blackberry	54.9kb	165.4kb
14/01/2014	21:58:12	447700198019	10.14.107.118	02:00:00	02567629984520202	234042787261234	3029	54	Harbour Wharf	Internet	267.5kb	638.1kb
14/01/2014	22:19:34	447700198019	10.0.176.23	03:51:02	02567629984520202	234042787261234	165	54	Green Park	Blackberry	34.7kb	356.4kb
14/01/2014	23:58:12	447700198019	10.14.107.118	02:00:00	02567629984520202	234042787261234	2009	90	Rotherhithe	Internet	136.5kb	675.3kb
End of Record												

filename: 198019 GPRS.csv

Figure 8.9 Generic GPRS/PS data CDR content

Cell Address Details

Some CDR formats provide only limited address and location details for cell sites in their CDRs, with many providing only Cell ID and Site Name, while others may also include a post- or zip code for the site. Conversely, other operators add a great deal of cell address information to their CDRs, either in an 'in line' manner, where the details of the used cells are written into each call event row in the main body of the data, or in the form of a cell address table that is added to the foot of the call data extract.

In the case of CDR types that provide only limited cell address information, analysts and investigators who wish to progress a cell site enquiry are often required to request detailed cell address details from the operators. In jurisdictions with more mature cell site analysis infrastructure, this is usually managed in a 'self-service' way via the operator's online disclosure management system using facilities that allow investigators to obtain the information they need by interrogating an automated directory of cell details. In less mature systems the investigators would need to send an application form to the operators to get the required information released.

As with CDR formats, operators have a wide variety of reporting formats used to supply cell address details, typically however they will supply some or all of the details listed below:

- Cell ID or full CGI (e.g. MCC-MNC-LAC-CI)
- Site name
- Site address and post- or zip code
- Site GPS latitude/longitude or map grid reference
- Cell azimuth
- Cell technology (2G, 3G, 4G)
- Cell type – macro, micro, femto.

Investigators often request details for multiple cells as part of the same enquiry and the results are often batched into one report.

A generic example of a detailed cell address information report is shown in Figure 8.10.

CellPlus
Records Requested: Cell Locations
Request Date: 01/05/2014
Request By: DC Davies

Cell ID	LAC	Azimuth	RAT	Site Name	Address	Post Code	Site Latitude	Site Longitude
28760	221	0	2G	London East	414 White Horse Rd, London	E1 2JG	51.51569	−0.04117
28761	221	120	2G	London East	414 White Horse Rd, London	E1 2JG	51.51569	−0.04117
28762	221	240	2G	London East	414 White Horse Rd, London	E1 2JG	51.51569	−0.04117
End of Data								

filename: 345876 cell addresses.xls

Figure 8.10   Generic cell location report content

8.6   Sources of Cellular Coverage Data

Cellular coverage data can be obtained in two ways: investigators can apply to the relevant cellular operators and request general cell coverage plots; or they can commission specific forensic radio surveys.

Forensic radio survey techniques were outlined in Chapter 7 and will be recapped in Section 8.6.

Cellular operators typically provide two main types of coverage data – serving cell plots and path profiles – both of which are discussed below.

8.6.1   Serving Cell Plots

Cellular network operators employ complex radio coverage planning systems that enable them to decide where best to site base stations and give them predications as to the coverage that can be expected from each of the cells deployed to those sites. Once sites have been deployed, operators supplement the planned coverage estimates with the results of test calls and drive surveys to help them to compare predicted coverage against actual service.

Operators will typically make data from their coverage planning systems available to law enforcement agencies on request, which therefore provides a source of coverage data that can be used to aid cell site investigations.

Operators' coverage data usually takes the form of a 'cell coverage plot', which shows the predicted serving coverage of a target cell plotted on a map. Generally, there are several types of plot that can be supplied:

- Serving coverage plots, which show the areas in which the target cell is predicted to provide serving coverage;
- Best server plots, which show areas where, although one of several serving cells, the target cell is expected to be selected as serving most often;
- Single server plots, which shows areas where the target cell is expected to be the single, dominant serving cell.

There are a number of benefits and disadvantages to the use of operator-provided coverage data.

The benefits include historical availability; network operators regularly update the coverage predictions made by their planning systems, based on changed cell designs or on feedback from drive testing, and therefore may have historical versions of coverage data going back for a number of years. In scenarios in which cell site investigators are asked to examine or comment on call records that are several years old, the historical perspective provided by the relevant operator's planning data could be useful. This is especially the case if some or all of the cells in question had been reoptimised or removed in the intervening period.

Other benefits might be related to cost and availability of coverage data; it might be less expensive to request coverage data from an operator than it would be to pay a forensic surveyor to conduct a cell coverage drive survey and operator data might be available almost immediately, whereas a bespoke forensic survey might take several days to schedule and complete.

There are two main potential disadvantages related to using operator-supplied coverage data: the first one relates to the way in which the data is created, which is based on predictions made by a planning system. However sophisticated the planning system is, the coverage plots provided can be assumed to be based at least partly on educated guesswork and might not be fully representative of the actual coverage. A bespoke cell coverage survey has the benefit that it provides details of the actual coverage captured by the survey equipment on the day of the survey and so may be regarded as providing a truer picture of the coverage provided by the target cell.

The second potential disadvantage is related to the granularity of the survey data. Cellular planning systems typically break the area of the network into units known as 'pixels', with each coverage something like a 100 m^2 area. The plotted coverage is then based on the predicted average coverage within each pixel, it does not break the prediction down to the level of actual coverage at points within each pixel. This means that although a serving coverage plot might be useful for indicating the general coverage area of a cell and the high-level likelihood of coverage being provided at specific spot locations, it cannot compete with the specific data provided by a bespoke spot/location survey.

Operator serving cell plots are therefore considered by many cell site experts to be a useful source of secondary coverage information, but are not considered to be a substitute for bespoke forensic survey data. In circumstances where a bespoke survey cannot be conducted, then operator serving cell plots may provide the only source of coverage data that is available.

8.6.2   Path Profiles

A radio path profile is designed to demonstrate whether a line of sight (LOS) or 'near' LOS path exists between a cell site antenna and a spot location.

Operators' radio planning systems typically use powerful GIS (Geographical Information System) data that includes topographical details about the height of the ground in each area. If the location and height of a cell site's antenna are known, the planning system should be able to calculate whether a LOS path exists between it and any other location. This information can help to determine whether it is likely that a call could be setup between the cell and a phone, given their respective locations.

Path profile data is often used when cell site experts are attempting understand what appear to 'impossible' calls; calls that are made by target phones that are known (from other evidence) to have been in a particular location and that make calls using cells that would not be expected to serve at that location.

8.7   Forensic Radio Surveys

Forensic radio surveys provide empirical data that shows the coverage provided by different cells and networks, at the time at which the surveys were undertaken.

The impetus to discover the coverage of one or more cells is usually triggered by the analysis of suspects' mobile phone records.

8.7.1   Role of Forensic Survey Results

If the addresses of cell sites used by a target phone are known, 'high-level' cell site analysis can be attempted based on a comparison of the proximity of a cell site to a significant location – if the cell site is within 1–2 km of the significant location in a direction that is compatible with the cell's azimuth, it can be assumed that use of that cell site is potentially consistent with the target being at or near that location when the relevant calls were made. Although this is, on the face of it, a reasonable assumption to make, assumptions are not proof.

Forensic radio surveys are designed to provide solid evidence to back up the assumptions made by analysts.

Forensic radio survey results can be used to prove that particular cells serve or provide coverage at significant locations and therefore prove that it was possible for a phone using those cells to have been at or near those locations.

The only definite conclusion that can be drawn from cell site analysis is that use of a particular cell by a target phone means that the phone must have been within the serving coverage area of that cell at the time a mobile transaction was undertaken.

Forensic radio surveys, especially cell coverage surveys, can set approximate limits to the area within which the target phone must have been located. This type of evidence can be very useful when attempting to prove or disprove an alibi or other statement.

Overall, forensic radio surveys add empirical rigour to an area of investigation that would otherwise fall prey to assumptions and wishful thinking.

8.7.2   Limitations of Forensic Surveys

Like cell site analysis in general, forensic radio surveying also has limitations.

A forensic radio survey is only able to provide an indication of the radio coverage at a location at the time the survey was conducted; it cannot provide definitive details of the nature of the radio coverage at times in the past when significant calls were made.

Cell site analysts therefore have to take it somewhat on trust that the network configuration did not change significantly in the period between the calls being made and the survey being undertaken.

One way of minimising the potential for network change to have taken place is to make sure that radio surveys are conducted as close in time as possible to the events in a case – ideally within just a few hours or days. Many investigators routinely arrange for radio surveys (usually all-network profiles) to be conducted straight away after an incident to ensure that cell site evidence is secured ready for when the investigation needs it.

It is not always possible to conduct radio surveys in a timely manner, especially in the case of re-investigations, re-trials or appeals, and investigators may be forced to request radio surveys months or even years after the event. The relevance of a forensic radio survey diminishes in proportion to the length of time that has elapsed since the analysed calls were made and this is exacerbated during periods when extensive amounts of network build and optimisation work is taking place, but there are still reasonably effective options even in these circumstances.

Many police forces and cell site companies undertake large numbers of forensic radio surveys and are encouraged to keep an archive of historic survey results. If a radio survey is required that relates to historic events it may be possible to get a view of the coverage that pertained in an area by seeking out old surveys taken in the same or a nearby area. By comparing historical results with newly obtained measurements it may be possible to determine the amount of change that has taken place in the network in the intervening period. It may even be possible to prove that the coverage in an area has not experienced any significant change during that period, meaning that the more recent survey results can be considered to be accurate and relevant.

The last option for investigators is to apply to the relevant network operators and ask for copies of any coverage planning data they have from the relevant historical period or, alternatively, to ask them to provide details of any changes, upgrades or reoptimisations that may have taken place in the area. Networks are generally not incentivised to keep this type of data and are reluctant to make it available even if they have it, so this option must really be considered to be the last resort.

8.8   Cell Site Reports

8.8.1   Report Writing and Structure

The structure of a cell site report is largely down to the personal preference of the analyst or expert responsible for writing it, but most reports follow a similar pattern, the sections of which are outlined below.

Case Details

The first part of a typical cell site report confirms the case name, the names of the defendant(s) in the case, the offences that are alleged to have been committed and also lists and identifies the phone numbers attributed to each of the defendants.

As previously mentioned, cell site cases often assign a colour to each target phone number (or a colour to each individual, if they have multiple attributed phone numbers) and use a form of shorthand to identify the target phones in a case. For example,

Table 8.1 Example of a case details table from a hypothetical cell site report.

Case: R v Smith, Jones, Williams and Johnson

Allegation: Assault

Defendant	Attributed phone	Known as
John Smith	07700 345876	Red-5876
Paul Jones	07700 198019	Blue-8019
	07700 300981	Blue-0981
Peter Williams	07700 127610	Green-7610
Jim Johnson	07700 019010	Orange-9010
Victim	Attributed phone	Known as
Dave Cooper	07700 614228	Aqua-4228

a case may include a defendant known as John Smith, who is alleged to have been the user of mobile phone number 07700345876. The cell site analyst may have allocated the colour blue to phones attributed to Mr Smith, so the target phone will be referred to as 'Blue-5876'.

The case details section will usually make clear the association between the defendants and the target phones. Details of other significant phones numbers – witnesses, victims, associates and others – can also be highlighted and 'colourised' in reports, schedules and maps and would therefore also be included in a case details table. It is generally considered to be good practice to avoid the natural inclination to colour code defendants' phones as 'red' and victims' phones as 'blue', as this could be construed as attempting to unduly influence the jury by implication.

A typical case details table could be similar to the example shown in Table 8.1.

Cell Site Explanation

The intended audience for a cell site report includes the judge and lawyers in court, who might have encountered cell site evidence and the techniques used to gather it in previous cases, but it also potentially includes the members of the jury, who almost certainly will not have.

It is usually considered beneficial to include a short introduction to cell site analysis at the start of the report to give readers who have not previously encountered the discipline an opportunity to understand some of the concepts and terminology that will be used.

An important component of the explanation section of a cell site report will be one or more paragraphs that set out the limitations of cell site evidence and make it clear to readers that just because the use of a cell could be consistent with the user of a phone being located at a specific address, it is generally also consistent with the user being located anywhere else within the used cell's coverage area as well.

Cell site reports that do not make this limitation explicitly clear run the risk of misleading the courts and juries that must consider the evidence in the associated case.

Summary of Source Data

Most reports then list the set of source data files – CDRs and cell address details – that were used in the preparation of the report. This section usually also serves to provide details of the 'continuity' of the chain of evidence for each source file. Continuity in this sense means that the report gives details of the date that the analyst or expert received each source file, it describes the method by which it was transmitted (e.g. on paper, on CD, via email) and it gives details of from whom the file was received.

In theory, the continuity details for each piece of data should lead in an unbroken 'chain of evidence' all the way back to the original source of that data and, in the United Kingdom system at least, each person who received and then forwarded that data should provide a statement to the court indicating their part in that chain. Testing the continuity of the prosecution's evidence is one of the basic activities of the defence in a case; if the chain can be shown to be broken then the evidence may be deemed inadmissible by the court.

A typical case continuity table could be similar to the example shown in Table 8.2.

This indicates that the report writer received the source CDR and data files from a police officer, Detective Constable Davies, via email on 1 May 2014.

Forensic Radio Survey Results

Some report formats contain a section that provides a summary of the radio surveys that have been undertaken in relation to the case.

Such a summary would typically list the locations at which surveys were captured and also provide details of any cell coverage or route profiles that were conducted.

Some experts also like to add a summary of the results obtained at each surveyed location and include details of the location itself, sometimes including a photograph of the spot at which the survey was conducted.

Table 8.2 Example of a continuity table from a hypothetical a cell site report.

Phone	Filename	From	Method	Date
07700 345876	345876 voice in/out.csv	DC Davies	E-mail	1 May 2014
	345876 GPRS.csv	DC Davies	E-mail	1 May 2014
	345876 cell addresses. xls	DC Davies	E-mail	1 May 2014
07700 198019	198019 voice in/out.csv	DC Davies	E-mail	1 May 2014
	198019 GPRS.csv	DC Davies	E-mail	1 May 2014
	198019 cell addresses. xls	DC Davies	E-mail	1 May 2014
	198019 subscriber check.doc	DC Davies	E-mail	1 May 2014

Main Report Section

The main section of a cell site report generally provides a detailed examination of the source data in comparison to the events of the case and/or provides a point by point testing of the source data against the allegations being made by the prosecution.

The process of compiling the main section of a cell site analysis report is largely an iterative one:

- Divide the set of calls to be analysed into batches that conform to the timings of the main events in the case timeline.
- Check each batch of the defendants' calls against the allegations being made.
- Check to see if any of the calls were made at around the time of a significant event.
- If so, did those calls use cell sites near to the event location? (high-level analysis).
- If yes, do the forensic radio survey results from that location indicate that the cells used for the calls are 'serving' cells at the location? (low-level analysis).

Imagine, as an example, that a case involved an alleged attack by members of one gang against a member of a rival gang. The prosecution's case might be that the attackers were all in communication with each other by phone and arranged to meet up prior to the attack. They then made a call to the victim to lure him to a rendezvous location and when he arrived they attacked him. Following the assault, the attackers separated but continued to keep in touch by mobile phone.

The cell site report prepared for a case like this might be divided into two sections – Section 1 might deal with the timeline of events in the case and would work through the call records of the group of defendants providing details of the calls made, the cells used and the potential locations of the target phones during each of those calls; Section 2 might deal with specific questions asked or allegations made by the prosecution, such as 'were all of the target phones in contact with each other at times during the 24 h before the attack?' or 'is it possible that all of the target phones could have been at or near the same location at a specific time in the 3 h period before the attack?'

Grouping Calls

A cell site report will usually group calls into batches and each batch will cover one part of the case timeline. Each batch of calls examined in the report will usually be associated with one or more maps in the report's mapping presentation.

There are a number of strategies used to group calls for analysis or reporting, but all of them generally related back to the timeline of events or the 'period of interest' defined by the allegations being made in a case. These methods include:

- A group of calls made during a period when the target phone(s) were alleged to have been at or near a particular location would generally be discussed in the same report paragraphs and would be presented on the same map.

- A specific call or sequence of calls that are particularly significant or that took place at an important time would generally be discussed in a paragraph on their own.
- Calls made during a period when the target phones were assumed to be travelling would be described in a paragraph and would be shown on a map that details the progression of cells used. Such a 'travelling' map can be a powerful indicator of movement between significant locations, especially if a reasonable number of calls were made during the journey, as this will result in a more comprehensive view of the possible route taken during the journey.

Report Paragraphs

An example set of paragraphs from a cell site report written in relation to the hypothetical case outlined above could be as follows:

The Crown alleges that the users of Red-5876, Blue-8019, Blue-0981, Green-7610 and Orange-9010 met at Flat 16, 24 Matlock St, Limehouse at times between 20:00 and 22:00 h on 14 January 2014.

Between 20:14 and 21:38 h on 14 January 2014, five calls/texts for Red-5876 were carried via Cell ID 28761 (NationalCell network, London East site). The London East cell, Cell ID 28761 was detected as a serving cell at Flat 16, 24 Matlock St, Limehouse.

Between 20:36 and 21:54 h on 14 January 2014, four calls/texts and two data sessions for Blue-8019 and one text for Blue-0981 were carried via Cell ID 1143 (Cellplus network, Aston St site). The Aston St cell, Cell ID 1143 was detected as a serving cell at Flat 16, 24 Matlock St, Limehouse.

At 21:45 h on 14 January 2014, one call was made by Green-7610 (to the victim's phone) via Cell ID 28762 (NationalCell network, London East site). The London East cell, Cell ID 28762 was detected as a serving cell at Flat 16, 24 Matlock St, Limehouse.

Between 20:01 and 21:38 h on 14 January 2014, three calls/texts for Orange-9010 were carried via Cell IDs 28761 and 28762 (NationalCell network, London East site). As previously mentioned, both cells were detected as serving cells at Flat 16, 24 Matlock St, Limehouse.

Details of the calls made during this period are shown in Map 7.

Note that the United Kingdom *NationalCell* and *CellPlus* networks mentioned in the above paragraphs and elsewhere in this section are fictitious, as are the Cell IDs, cell names and MNCs (Mobile Network Codes) assigned to these networks.

Cell Site Report Conclusions

When considering each batch of call records for a target phone or a group of phones, a cell site report will attempt to provide conclusions to show how well the evidence provided by the combination of the call records and any forensic radio survey results agrees with the allegations being made by the prosecution.

The main indication that the analysis will check is whether the cells used by a target phone provide serving coverage at a location that the attributed user of the phone is alleged to have been at or near at the time those calls were made.

If the used cells do serve at the significant location then the report can conclude that the calls 'could have been made *at or in the vicinity* of the location' or similar such wording. Note the use of the term 'could have been made'; cell site analysis can rarely be definite about the location of a target phone when calls were made. The most that can generally be expected is that cell site analysis shows that it is possible for the phone to have been at that location but does not exclude the possibility that it could have been elsewhere in the used cell's coverage area instead.

If the used cells do not serve but do provide strong or very strong non-serving coverage at the significant location, or if they appear near the top of the table in the forensic radio survey results, the report can conclude that the calls 'could have been made *in the vicinity* of the location'. The definition of 'in the vicinity' of a location is broadly agreed by cell site experts to mean within 50–100 m of the location, but there are some differences of opinion on the topic.

If the used cells do not serve and provide moderate to weak non-serving coverage at the significant location, if they appear lower down the table in the forensic radio survey results, the report can conclude that the calls 'could have been made in *the general area* of the location'.

If the used cells were not detected at all during the forensic radio survey then the report can conclude that the calls 'are *unlikely to have been in at or in the vicinity* of the location'.

The various levels of conclusion are outlined in Figure 8.10.

In relation to the hypothetic case that has been used to provide examples in this section, as all of the cells used by the set of target phones in the period of time between 20:00 and 22:00 hrs on 14 January 2013 were cells that were detected as being serving cells during the forensic radio survey undertaken at Flat 16, 24 Matlock St, the cell site report dealing with that batch of calls could reach a conclusion similar to the following:

The use of NationalCell network cells 28761 and 28762 (London East) and Cellplus network cell 1143 (Aston St) between 20:00 and 22:00 hrs on 14 January 2014 by Red-5876, Blue-8019, Blue-0981, Green-7610 and Orange-9010 is consistent with the proposition that the users of those phones were located at or in the vicinity of Flat 16, 24 Matlock St, Limehouse at times during that period.

Summary and Declaration

The last section of a cell site report usually summarises the conclusions reached in the main section in a simpler and more abbreviated format. One way of doing this is to reproduce all of the individual conclusion paragraphs in one section.

A report typically ends with a 'declaration' made by the author which generally restates the limitations of cell site evidence, confirms that the author of the report has no connection with any of the parties associated to the case and states their willingness to review their conclusions should further evidence come to light.

Expert witnesses in most jurisdictions will operate under guidelines that have been set down to ensure that they compile their reports in an unbiased and professional manner. In the United Kingdom, as an example, the Crown Prosecution Service publishes a guide for experts called 'Disclosure: Experts' Evidence, Case Management and Unused Material' [2], which sets out the obligations and responsibilities of experts. The declaration at the end of a cell site report often acknowledges that the writer is aware of and has adhered to the relevant guidelines.

8.9   Call Schedules

Cell site reports are usually enhanced by the preparation of call tables or call schedules. These collate the relevant call records from the target phones in the case and present them in a combined and coherent document.

Each network operator provides call records in their own format, and the formats used by the operators can be very different. This makes it difficult to compare calls made by phones belonging to different networks, so cell site analysts often spend large proportions of their time processing (also known as 'normalising' or 'cleansing') call records into a common format. This process can be very time consuming if conducted manually (and is also open to inevitable human error), so many organisations have developed their own data processing macros or use a commercial data cleansing product such as Forensic Analytics CSAS (Cell Site Analysis Suite) software application to process call data automatically.

Call tables are usually prepared using Microsoft Excel and are often presented in court in A3-sized booklets.

An excerpt from a typical call table is shown in Figure 8.11, which lists some of the calls from the hypothetical cell site report that has been used to provide examples during this section.

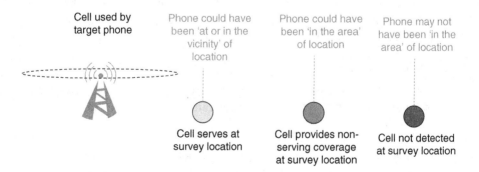

Figure 8.11   Cell site conclusions

Colours are usually assigned to each significant phone number and that colouration is employed to make the phones more identifiable in the call tables.

As cell site evidence works to identify the possible locations of a mobile device, rather than attempting to identify the user of that device, significant phone numbers are usually identified using a combination of the colour assigned to the phone and the last three or four digits of the mobile number, 'Blue-1234' or 'Red-2468' for example.

In cases that cover multiple phones, where several phones might be attributed to the same individual, analysts often assign a 'colour per person' to allow phones attributed to the same individual to be easily grouped and identified.

8.10   Maps and Graphics

Mapping and graphics presentations are often used to make the evidence presented in cell site reports simpler to understand.

Figure 8.12 provides an example of a typical cell site map and illustrates the set of calls detailed in the hypothetical case that has been used to provide examples during this section.

Many experts use a set of graphical slides to provide juries with a basic overview of cell site concepts, which usually mirrors the information presented at the start of the expert's report and is often presented to the jury at the start of the expert's evidence.

The substantive part of a mapping presentation generally begins with a slide that provides an overview of the significant locations in the case and their geographical relationship to each other.

An analyst will then usually produce a separate map to represent each batch of calls dealt with in the accompanying report. The map will be zoomed in to the general area of the cell sites used by each batch of calls and will typically have icons showing the cell locations, labels detailing the cell name and cell ID and call labels providing basic details of the calls under discussion. Cell icons and cell details labels are often coloured to match the attribution colour assigned to the target phone that used that cell – so, for example, the cell label for a cell used by the Blue-1234 phone would be coloured blue. Maps that detail the cell usage of several target phones might maintain this convention or might colour the cell labels in a neutral colour to indicate that they were used by multiple phones.

Call labels are included in maps to tie the information presented in the map to the cell site report and the call schedule. Different experts and agencies favour their own preferred label format, but the example shown in Figure 8.13 has labels that contain the following data:

- Call index number (based on call schedule numbering)
- Index number is coloured to show the target phone to which the event relates

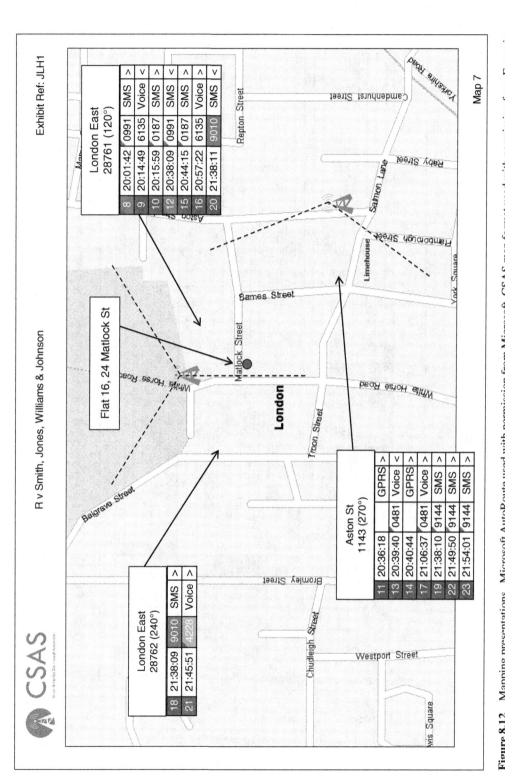

Figure 8.12 Mapping presentations. Microsoft AutoRoute used with permission from Microsoft, CSAS map format used with permission from Forensic Analytics Ltd

CSAS

Index	Date	Time	A Number	Attribution	B Number	Attribution	Call Type	Duration	Start Call ID	Azimuth	Start Site Name	End Call ID	Azimuth	End Site Name
1	14/01/2014	17:04:09	447700019010	Jim Johnson	447700345876	John Smith	Outgoing SMS	00:00:00	6115	120	East St			
2	14/01/2014	18:26:55	447700345876	John Smith	447700198019	Paul Jones	Outgoing SMS	00:00:00	3981	110	St Catherines Dock			
3	14/01/2014	18:26:55	447700345876	John Smith	447700198019	Paul Jones	Incoming SMS	00:00:00	1467	15	Aldgate West			
4	14/01/2014	19:01:00	447700127610	Peter Williams	447700345876	John Smith	Outgoing Voice	00:07:22	3981	110	St Catherines Dock	3981	110	St Catherines Dock
5	14/01/2014	19:01:00	447700127610	Peter Williams	447700345876	John Smith	Incoming Voice	00:07:23	6116	240	East St	28760	0	London East
6	14/01/2014	19:27:12	447700127610	Peter Williams	447700198019	Paul Jones	Outgoing SMS	00:00:00	3981	110	St Catherines Dock			
7	14/01/2014	19:27:12	447700127610	Peter Williams	447700198019	Paul Jones	Incoming SMS	00:00:00	1467	15	Aldgate West			
8	14/01/2014	20:01:42	447700019010	Jim Johnson	442079460991		Outgoing SMS	00:00:00	28761	120	London East			
9	14/01/2014	20:14:49	447700345876	John Smith	447700886135		Incoming Voice	00:01:01	28761	120	London East	28761	120	London East
10	14/01/2014	20:15:59	447700345876	John Smith	442079460187	APN: Blackberry	Outgoing SMS	00:00:00	28761	120	London East			
11	14/01/2014	20:36:18	447700198019	Paul Jones	APN: Blackberry		GPRS	00:11:56	1143	270	Aston St			
12	14/01/2014	20:38:09	442079460991		447700019010	Jim Johnson	Incoming SMS	00:00:00	28761	120	London East			
13	14/01/2014	20:39:40	442079460481		447700198019	Paul Jones	Incoming Voice	00:06:17	1143	270	Aston St			
14	14/01/2014	20:40:44	447700198019	Paul Jones	APN: Blackberry		GPRS	00:03:21	1143	270	Aston St			
15	14/01/2014	20:44:15	447700345876	John Smith	442079460187		Outgoing SMS	00:00:00	28761	120	London East			
16	14/01/2014	20:07:22	447700345876	John Smith	447700886135		Outgoing SMS	00:01:35	28761	120	London East	28761	120	London East
17	14/01/2014	21:06:37	447700198019	Paul Jones	442079460481		Outgoing Voice	00:11:56	1143	270	Aston St			
18	14/01/2014	21:38:09	447700345876	John Smith	447000019010	Jim Johnson	Outgoing SMS	00:00:00	28762	240	London East			
19	14/01/2014	21:38:10	447700300981	Paul Jones	447700709144		Outgoing SMS	00:00:00	1143	270	Aston St			
20	14/01/2014	21:38:11	447700345876	John Smith	447000019010	Jim Johnson	Incoming SMS	00:00:00	28761	120	London East			
21	14/01/2014	21:45:51	447700127610	Peter Williams	447700619228	Dave Cooper	Outgoing Voice	00:00:24	28762	240	London East	28762	240	London East
22	14/01/2014	21:49:50	447700198019	Paul Jones	447700709144		Outgoing SMS	00:00:00	1143	270	Aston St			
23	14/01/2014	21:54:01	447700198019	Paul Jones	447700709144		Outgoing SMS	00:00:00	1143	270	Aston St			
24	14/01/2014	22:06:31	447700127610	Peter Williams	447700345876	John Smith	Outgoing Voice	00:00:24	28761	120	London East	28761	120	London East
25	14/01/2014	22:07:16	447700127610	Peter Williams	447700345876	John Smith	Incoming Voice	00:00:24	31008	45	West Ham	31009	165	West Ham
26	14/01/2014	22:58:04	447700345876	John Smith	447700019010	Jim Johnson	Outgoing SMS	00:00:00	7816	180	Royal Docks			
27	14/01/2014	22:58:11	447700345876	John Smith	447700019010	Jim Johnson	Incoming SMS	00:00:00	7817	300	Royal Docks			

CSAS Call Schedule   Rv Smith, Jones, Willaims & Johnson

Figure 8.13 Call tables. CSAS data format, used with permission from Forensic Analytics Ltd

- Time of call start
- Abbreviated 'other party' number
- Event type – SMS, Call, GPRS
- Indication of whether the event was outgoing (>) or incoming (<)

Some label formats also include an indication of whether the cell was used for the start of the call (indicated with an 'S'), the end of the call ('E') or both ('S/E'). In the case of calls that start on one cell and end on another, there would be an 'S' call label listed under the start cell and an 'E' label under the end cell. Some cell site formats list the start time of a call on an 'S' label and the end time of the call on the corresponding 'E' label, but this is not a universally accepted practice.

Further labels could be included in mapping slides to provide details of significant events that occur during the period covered by the map.

Additional graphics are used to provide explanations for complex concepts. For example, if cell coverage or route surveys were undertaken as part of the case, then maps illustrating the results of these surveys can be included in the mapping presentation.

If the maps to be displayed in court are created using a presentational application such as Microsoft PowerPoint, the call and cell labels and icons can be animated so that they appear on each map slide in the order in which the calls were made and the cells were used. The animated progression of call and cell details can make it easier for the court and for members of the jury to grasp the relationship between the calls and the significant events in the case.

8.11   Report Checking and Peer Review

Cell site reports can develop into enormously complex collections of documents, especially if a case involves multiple handsets, and it is to be expected that the writers and compilers of these reports will make at least one mistake somewhere within them.

It is therefore absolutely vital that each report is fully proofread and fact-checked once it has been completed.

The main aspects that need to be checked include:

- Case details and continuity information are complete and correct.
- Details of survey locations and survey results are complete and correct.
- Attribution and colouration details for each target phone are correct.
- Within the body of the report, the paragraphs relating to each examined group of calls should be checked to make sure that the call times, used cell details, location details and forensic survey results mentioned are correct.
- Each conclusion should be checked back against the source data to ensure that it is sound and is supported by the evidence.

- Call schedules should be checked back against the source CDR data to make sure that no errors were introduced during the 'cleansing' process.
- Maps should be checked to ensure that the marked locations of addresses and cell sites are correct, that cell labels contain the correct cell details, that the correct set of call labels are listed for each cell and that any azimuths are correct (both in the cell labels and in the orientation of the cell icons).

Once the report writer has fully checked (and, if necessary, corrected) their work, the report should be passed to at least one equally qualified and competent peer reviewer, who should go through the whole checking process again.

Peer reviews are an essential quality assurance tool as they help to overcome any issues associated with over-familiarity with the case details. When working on a cell site case, analysts and experts become steeped in the details and often find it difficult to recognise when they have missed out important facts that would help someone less familiar with the case to understand key events.

To be truly effective, a peer reviewer should have had no involvement with the case that they are reviewing, which should ensure that any 'omissions due to familiarity' will be detected and can be corrected.

8.12   Professional and Expert Witnesses

Cell site and forensic survey evidence is typically presented in court by one of two types of witness: professional witnesses or expert witnesses. United Kingdom courts draw a distinction between a professional 'witness of fact' and an expert 'witness of opinion'.

The distinction between 'witness of fact' and 'witness of opinion' outlined below may also be one that is not recognised in all countries; again, it is included to provide an indication of the differences between types of evidence that forensic radio survey-ors would be expected to provided compared the type of evidence that a cell site expert would offer.

8.12.1   Witness of Fact

A Witness of Fact is typically regarded as being a 'professional witness' and is able to give evidence related only to things they have observed or to processes that they have undertaken.

A police officer who has been trained to undertake the tasks associated with foren-sic radio surveys, for example, would be able to give factual evidence related to sur-veys that had been undertaken and the results that had been obtained. Such a witness would be able to state that 'Cell 1234 was detected as a serving cell at Location A' but would not be permitted to offer an opinion as to, for example, why Cell 1234 served when Cell 2345 did not.

8.12.2 Witness of Opinion

A Witness of Opinion is an 'expert witness' who has been accepted by the court as an expert in their field and is therefore permitted to provide opinions related to the evidence or facts that have been put before them.

A cell site expert witness is able to review the call records and radio survey results presented in evidence and would be permitted to draw conclusions as to whether the use of certain cells is consistent with the user of a target phone being located at or near a significant address.

Forensic radio surveyors who appear in court as professional witnesses of fact would be expected to provide an account of the surveys they had undertaken and the results they had obtained, it would usually be the job of a cell site expert to offer an opinion as to whether the evidence in the case supported the prosecution's allegations or not.

8.12.3 Duties of an Expert Witness

In the United Kingdom, rules on the use of expert evidence, including guidance on the expert's duty to the court and on the content of an expert's report, are detailed in Part 33 of the Criminal Procedures Rules legislation [3]. The Crown Prosecution Service guidance booklet for experts, 'Disclosure: Experts' Evidence, Case Management and Unused Material' [2], mentioned earlier in Section 8.8 is based on the Part 33 rules. Collectively the guidance for experts is often summarized as the 3Rs – 'Record: Reveal: Retain' – as described below.

The Part 33 rules describe an expert witness as 'a person who is required to give or prepare expert evidence for the purpose of criminal proceedings'.

It goes on to describe the duties of an expert to the court (whether they are working for the prosecution or the defence). These can be summarised as:

- An expert must help the court to achieve the overriding objective by giving objective, unbiased opinion on matters within his (or her) expertise.
- This duty overrides any obligation to the person from whom he/she receives instructions or by whom he/she is paid.
- This duty includes an obligation to inform all parties and the court if the expert's opinion changes.

The Part 33 rules also make mention of the required content of an expert's report and/ or statement, which must:

- Give details of the expert's qualifications, relevant experience and accreditations.
- Give details of any literature or other information that the expert has relied upon.
- Contain a statement setting out the substance of all facts given to the expert which are material to the opinions expressed in the report.
- Make clear which of the facts stated in the report are within the expert's own knowledge.

- Say who carried out any examinations, measurements, test or experiment which the expert has used for the report.
- Where there is a range of opinion, provide a summary of those opinions and state their own opinion.
- Contain a summary of the conclusions reached.
- Contain a statement that the expert understands their duty to the court.
- Contain the same declaration of truth that is given in a witness statement.

If an expert's report can be shown not to be in accordance with the Part 33 rules then it could be excluded from the evidence in a case.

8.12.4   Defence Cell Site Reports

In most jurisdictions, a defendant has the right to appoint a defence lawyer. The process of 'disclosure' or 'discovery' typically means that the defence are entitled to see in advance of the trial all of the evidence that is to be put forward in the case, including any expert evidence such as a cell site report.

If the prosecution has had the benefit of advice from a cell site expert the defence is typically given the opportunity to appoint an expert of their own, who will be expected to produce a defence cell site report.

The role of the defence cell site report is to examine and challenge the prosecution report to ensure that it has been conducted in a technically rigorous manner and that its conclusions stand up to scrutiny. Defence reports are also often used to test the cell site evidence against alibis and alternative interpretations of the call records.

Most defence reports begin with a point by point commentary on the prosecution report, highlighting instances where the defence expert agrees or disagrees with the prosecution expert, they then move on to examine disputed or contentious aspects of the case in more detail and sometimes conclude by putting forward alternative scenarios or interpretations.

The best defence reports are those that tease out any inaccuracies or inconsistencies in the prosecution case and ensure that the prosecution expert has conducted a rigorous investigation; the worst defence reports simply put forward a blanket objection to every point made by the prosecution, whether the objection has technical merit or not, and seek by any means to undermine the evidence provided in the hope of having it excluded from the case. This second type of report could be said to be non-compliant with the requirements for expert reports outlined in the Part 33 rules.

It is often said of court cases in general that a robust and principled defence is necessary to ensure that the prosecution offers its best possible case, and this is equally true of cell site evidence. Indeed, many independent cell site experts work both sides of the industry, providing reports for the prosecution in some cases and working for the defence in others.

8.13 Court Presentations

The procedural examples provided in this chapter have all been based on United Kingdom law and court procedures, as the author is based in the United Kingdom and mainly has experience of that jurisdiction. The information provided in this chapter is not intended to reflect the processes and procedures in all jurisdictions and is included only to provide an insight into the end result of a cell site and forensic radio investigation.

8.13.1 Evidence-in-Chief

Evidence-in-Chief is the term used in United Kingdom courts to describe a witness's presentation of evidence for the 'side' that called them. For example, a witness called by the prosecution would give their evidence-in-chief under examination from a prosecution barrister and may then be subject to further cross-examination by the defence barrister. A witness called by the defence, conversely, would give their evidence-in-chief to the defence and would be cross-examined by the prosecution.

Evidence-in-Chief sessions relating to cell site evidence can be made more engaging and understandable for a jury by the addition of graphics and other presentation aides. A forensic radio surveyor or cell site expert will usually prepare some or all of the following elements for a court presentation:

- Detailed cell site report – distributed to the barristers and judge only;
- Summary cell site report – provided in the jury 'bundle' of evidence documents and to barristers, judge and defendant(s);
- Call table booklet – printed on A3 paper and provided in the jury 'bundle' and to barristers, judge and defendant(s);
- Mapping presentation booklet – printed on A4 paper and provided in the jury 'bundle' and to barristers, judge and defendant(s);
- Mapping presentation in Microsoft PowerPoint – to present via screens or a projector in court;
- Cell site overview presentation in Microsoft PowerPoint– to present via screens or projector in court.

The process of giving expert cell site evidence usually follows a fairly predictable routine:

- The witness (a forensic radio surveyor or cell site expert) is sworn in (or affirms).
- The witness's background and experience are described, which provides the jury with details of how and why they should be considered to be an expert in their field. The defence have the opportunity to challenge the expert status of a witness and if the challenge is accepted by the judge, the witness may be subjected to a 'voir dire', which

is a 'hearing within a hearing' designed to assess the witness's competency. This is not a common procedure for professional witnesses and is rare even for expert witnesses.

- A professional witness of fact, such as a forensic radio surveyor, will then usually be lead through a presentation of the surveys that they have performed and the results they have obtained. These witnesses will not be asked to provide opinions on the results they have obtained but may be required to explain and defend the methods they employed to obtain or process them.
- A cell site expert witness will usually be asked to provide a brief overview of mobile phone technologies and cell site analysis for the jury. This is often achieved by using a simple PowerPoint presentation and should not take more than 5 or 10 min.
- The prosecution barrister then usually leads the expert through their cell site report section by section, highlighting the allegations being made in respect of each relevant defendant. The format for this part of the evidence is usually:
  - Set the context for the calls, in relation to alleged events, for example 'these calls were made at the time that the Crown alleges that the user of the phone was on his way the robbery location'.
  - Describe the calls, the significant phone numbers involved, the durations of the calls and the cells that were used.
  - If the allegation relates to the calls potentially having been made at a specific location, and if a forensic radio survey was undertaken at that location, details of whether the used cells provide coverage at that location will be described, for example 'Cell ID 1235, as used by the Blue-1234 phone, served at location A'.
  - Present the cell site conclusions related to the set of calls, for example 'the use of this cell for the calls at hh:mm:ss is consistent with the phone being located at or in the vicinity of Location A at that time'.
  - If presentation aids are available (screens or a projector) it helps to show the cell map(s) associated to each set of calls whilst they are being discussed.

- Once the main cell site report has been presented, the prosecution barrister may move on to cover other matters investigated by the expert such as alternative interpretations of specific calls, cell or route coverage profiles and so on.
- Once the prosecution concludes the evidence-in-chief the defence has the opportunity to cross-examine

In general there are some things that *should* be said by a cell site expert when giving evidence and some things that *should not* be said.

Cell site experts should always ensure that a jury is made aware of the limitations of cell site evidence, that this form of evidence generally cannot pinpoint the location of a phone when a call was made and that it usually only indicates that it is possible for a phone to have been at a location.

Cell site experts should never (unless there is definitive evidence) attempt to suggest that their interpretation of calls and their estimation of the location a phone is the only possible interpretation. They should never get led, by examination or cross-examination, into a situation where they agree with a statement that they know cannot be supported by the facts, as can sometimes happen if the examining barrister is unaware of the technical limitations of cell site analysis.

8.13.2   Typical Cross-Examination Questions

The use of cell site analysis by law enforcement agencies has presented them with a powerful investigative tool, but the uncertainty involved in reaching conclusions based on cell site evidence leaves it vulnerable to challenge from the defence in a case.

Some typical avenues of attack employed by defence barristers include:

- *Attribution* – the point is often made that cell site analysis provides details of where a *phone* may have been located but does not prove in whose hand the phone was at the time. This is true and without supporting evidence or solid attribution cell site evidence cannot successfully prove that a specific individual could have been at a location.
- *Uncertainty* – another common point made is that cell site evidence only shows that it is *possible* that a phone could have been at an alleged location when specific calls were made. The phone might equally have been anywhere else within the coverage area of the cell(s) used during that period. This is also true, up to a point. If calls are analysed in isolation from each other it is not usually possible to be specific about where in a cell's coverage area the phone was located. However, in some cases when a series of calls is analysed it is possible to be more emphatic. For example, if a series of calls is made in quick succession using a set of different cells, and if the only area where all of the cells that were used provide coverage is at or near the alleged location then the defence argument falls down. In this scenario, to have used the set of cells that were used in quick succession, the target phone must be in an area where all of the used cells overlap, which potentially narrows down the area within which the phone must have been located.
- *Network Change* – the question is often asked: 'How do you know that the network had not changed between the time of the offence and the time you conducted your radio survey?' Operators undertake regular maintenance on their networks. One form of maintenance is known as 'optimisation' and is designed to improve the coverage or service offered by a site. This may involve moving (or 'reorientating') the site antennas to point in a different direction. If such a procedure takes place after significant calls were made but before a forensic radio survey is conducted then the results of the survey might not match the coverage provided when the significant calls were made. To avoid this problem it is recommended that surveys are taken as soon as possible after an offence or incident has taken place, even if no suspect has yet been identified. An All-Network Profile should cover any eventualities.

- *Network Busy* – the question often asked is: If this cell is busy, the call will be handled by a neighbouring cell which could be even further away from the alleged location, is that correct? Theoretically that is true. There is a feature called 'network directed retry' which can push calls from busy cells to ones close by which are less busy. This is an optional setting by a network and the surveyor would not know if this had been enabled or not. However, a call can only be redirected to a cell which serves or provides strong coverage at the phone's current location, so any alternative cell used is likely to appear high on the list of any forensic radio survey conducted at that location, allowing normal cell site conclusions to be reached.

In general, the cross-examination of professional and expert witnesses can be characterised into three distinct methods of attack, depending upon the defence lawyers' estimate of the strength of the expert evidence. The phases or levels of cross-examination can be summarised as the 3Ms – 'attack the material; attack the method; attack the man'.

'Attack the material' refers to standard cross-examination actions to test and scrutinise the content of the professional or expert witnesses report looking for errors, inconsistencies or contradictions.

If no significant errors can be found in the report the focus of the cross-examination often turns to examining the methods employed to compile the report. This generally focuses on the reliability of RF survey results, the accuracy of any call schedules or maps that have been produced or on any mathematical tools employed to calculate results. This phase of a cross-examination will often also examine the continuity of evidence and the strength of the handset attribution data.

If no significant problems can be found with the methods employed to compile a report, the cross-examination may focus in the 'man' (or woman), meaning that the credibility of the professional or expert witness is examined. It is comparatively rare for a cross-examination to move into this phase, as the defence would have had the entire pre-trial period to raise any objections to the expert status of any witnesses. However, the 'attack the man' phase can happen and may result in the court requiring the expert to undergo a *voir dire* hearing.

Among expert witnesses, it is often said that if the defence go straight to the 'attack the man' it should be interpreted as a good sign, as it means that they have found no significant errors or omissions in either the material or the method.

References

[1] UK Government (2014) *Regulation of Investigatory Powers Act 2000*, http://www.legislation.gov.uk/ukpga/2000/23/contents (accessed 2 June 2014).
[2] UK Crown Prosecution Service (2010) *Disclosure: Experts' Evidence, Case Management and Unused Material: 2010</fi>*, http://www.cps.gov.uk/legal/assets/uploads/files/Guidance_for_Experts_-_2010_edition.pdf (accessed 3 June 2014).
[3] UK Government (2013) *The Criminal Procedure Rules, Part 33 (7 October 2013)*, http://www.justice.gov.uk/courts/procedure-rules/criminal/docs/2012/crim-proc-rules-2013-part-33.pdf (accessed 5 June 2014).

9

Summary and Practical Activities

The preceding chapters have outlined a number of techniques that are applicable to forensic radio surveying and to cell site analysis. They have also provided details on a number of basic radio and cellular topics.

This chapter attempts to provide a simple overview of the practical activities and techniques mentioned so far to allow readers who are new to these topics to cement their understanding of them. It also restates some of the basic radio and cellular information that has previously been provided.

Each of the following sections provides a table to summarise key concepts or offers bulleted lists of the steps, actions or considerations required to perform each of the featured practical activities.

9.1 Radio and Cellular Concepts

9.1.1 Basic Radio Terminology

Basic radio terminology is summarised in Figure 9.1.

9.1.2 Decibels

The standard notation employed for base 10 dB values is as follows:

$$dB = 10 \mathrm{L\,og}_{10}(\text{value})$$

Forensic Radio Survey Techniques for Cell Site Analysis, First Edition. Joseph Hoy.
© 2015 John Wiley & Sons, Ltd. Published 2015 by John Wiley & Sons, Ltd.

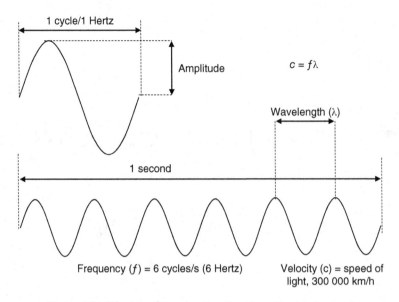

Figure 9.1 The frequency, wavelength and amplitude of a signal

Using generic values (power at transmitter = 100 mW, power at receiver = 0.000 001 mW), the benefit of using the dB scale becomes clear:

$$100\,\text{mW} = 10\text{L}\,og_{10}(100\,\text{mW}) = 10 \times 2 = 20\,\text{dBm}$$

$$0.000\,001\,\text{mW} = 10\text{L}\,og_{10}(0.000\,001\,\text{mW}) = 10 \times -6 = -60\,\text{dBm}.$$

The power loss experienced during transmission is therefore the ratio of the transmitted and received values:

$$100 / 0.000\,001\,\text{mW} = 1 \times 100\,000\,000$$

Using the law of powers with the decibel values (where exponential dB values are subtracted, as opposed to the division that would be performed on linear values):

$$20 - (-60) = 80\,\text{dB}.$$

This shows the received signal experienced a loss of 80 dB compared to the transmitted signal, which equates to it being 100 million times less powerful.

As illustrated in Table 9.1, every time a measured power level doubles, 3 dB is added and each time a power level halves, 3 dB is subtracted.

9.1.3 Decibel Milliwatts

Where dB will show the comparative difference between two values, the dBm (decibel milliwatts) scale will provide a result that can be mapped to a specific or 'absolute' milliwatt value.

dBm employs the same logarithmic scale as dB and is calibrated around a value of 1 mW, which is equal to 0 dBm. This is shown in Table 9.2.

Table 9.1 Typical decibel values.

Ratio of transmitted to received signal power	Decibels (dB)
10 000	40
1000	30
100	20
10	10
2	3
1	0
½	−3
1/10	−10
1/100	−20
1/1000	−30
1/10 000	−40

Table 9.2 Linear mW values compared to exponential dBm values.

Linear power level (mW)	Decibel milliwatts (dBm)
100 000 (100 W)	50
10 000 (10 W)	40
1 000 (1 W)	30
100	20
10	10
2	3
1	0
0.5	−3
0.1	−10
0.01	−20
0.001	−30
0.000 1	−40
0.000 01	−50

To convert mW to dBm: $dBm = 10Log_{10}(mW)$
To convert dBm to mW: $mW = 10^{(dBm/10)}$

9.1.4 Cellular Radio Bands

The set of radio bands employed to support cellular services in various regions around the world is detailed in Table 9.3.

9.1.5 Cellular Radio bands by Region

Each of the three WRC regions has its own subsets of radio bands dedicated to carrying cellular services and there are also some bands that are available in most parts of the world. Table 9.4 provides an overview of the usage patterns of the most commonly deployed cellular frequency bands around the world.

Table 9.3 Cellular radio bands.

Frequency band (MHz)	Network types	Characteristics
300	Public safety networks	Long distance, wide area cellular coverage
450	2G, 3G, 4G	
700	2G, 4G	
800	2G, 3G, 4G	
900	2G, 3G, 4G	
1500	2G, 3G	
1700	2G, 3G, 4G	Medium distance, medium area coverage
1800	2G, 3G, 4G	
1900	2G, 3G	
2000	3G	
2100	3G, 4G	
2300	3G, 4G	Short distance, local area coverage
2500	3G, 4G	
2600	4G	
3400	3G, 4G	
3500	3G, 4G	
3600	3G	

Table 9.4 Commonly used cellular frequency bands by region.

Frequency band (MHz)	Africa	Eastern Europe	Western Europe	Middle East	North America	South America	Asia Pacific
450	—	✓	—	—	—	✓	✓
700	✓	—	—	—	✓	✓	✓
800	✓	✓	✓	✓	—	—	✓
850	✓	✓	—	—	✓	✓	✓
900	✓	✓	✓	✓	—	✓	✓
1500	—	—	—	—	—	—	✓
1700	—	—	—	—	✓	✓	✓
1800	✓	✓	✓	✓	—	✓	✓
1900	✓	—	—	—	✓	✓	—
2100	✓	✓	✓	✓	✓	✓	✓
2300	✓	✓	✓	—	✓	—	✓
2500	—	✓	—	—	✓	—	✓
2600	✓	✓	✓	✓	✓	✓	✓
3500	—	✓	✓	—	—	✓	—

Source: GSM Association [1], CDMA Development Group [2] and 4G Americas [3].

9.1.6   Decimal, Binary and Hexadecimal

As shown in Table 9.5, hexadecimal (or 'hex') is a numbering system of 16 characters; 10 digits and 6 letters. It is used to condense the long strings of zeroes and ones in large binary numbers into a more manageable form. This base-16 numeric notation is frequently used to specify addresses in computer memory as

Table 9.5 Comparison of decimal, binary and hexadecimal notation.

Decimal (base 10)	Binary (base 2)	Hexadecimal (base 16)
0	0000	0
1	0001	1
2	0010	2
3	0011	3
4	0100	4
5	0101	5
6	0110	6
7	0111	7
8	1000	8
9	1001	9
10	1010	A
11	1011	B
12	1100	C
13	1101	D
14	1110	E
15	1111	F

it makes life simpler for programmers. The decimal numbers 0–9 are represented by the decimal digits 0–9 and the decimal numbers 10–15 are represented by the letters A–F.

9.2 Cellular Identifiers

9.2.1 Mobile Country Code List

The MCC (Mobile Country Code) number uniquely identifies the country in which a PLMN operates. MCC numbers are issued and controlled by the ITU (International Telecommunications Union), which is an agency of the UN that coordinates global telecoms activities. MCC assignments are listed in Table 9.6.

MNCs (Mobile Network Codes) are generally administered by each individual country's telecoms regulator and uniquely identify a network within an MCC area. The set of MNCs is too numerous and subject to change to list in this publication. There are multiple sources of information regarding current MNC assignments on the Internet.

Most countries have just one MCC, but some – such as the United States, India, United Kingdom and others – have more than one assigned to meet the demand for networks in those countries.

Some countries, mainly in the Caribbean, have networks deployed using multiple MCCs or using MCCs that are assigned to a network's parent company that operates in a different country. An example of this is Bermuda, which has two networks deployed using the Bermuda MCC (350), one network using the Jamaican MCC (338) and one using a United States MCC (310). In these circumstances, the

Table 9.6 Mobile country code list.

Alpha-3 code	Name	MCC-1	MCC-2	MCC-3	MCC-4
ABK	Abkhazia	289	—	—	—
AFG	Afghanistan	412	—	—	—
ALB	Albania	276	—	—	—
DZA	Algeria	603	—	—	—
ASM	American Samoa	544	—	—	—
AND	Andorra	213	—	—	—
AGO	Angola	631	—	—	—
AIA	Anguilla	365	—	—	—
ATG	Antigua and Barbuda	344	—	—	—
ARG	Argentina	722	—	—	—
ARM	Armenia	283	—	—	—
ABW	Aruba	363	—	—	—
AUS	Australia	505	—	—	—
AUT	Austria	232	—	—	—
AZE	Azerbaijan	400	—	—	—
BHS	Bahamas	364	—	—	—
BHR	Bahrain	426	—	—	—
BGD	Bangladesh	470	480	—	—
BRB	Barbados	342	—	—	—
BLR	Belarus	257	—	—	—
BEL	Belgium	206	—	—	—
BLZ	Belize	702	—	—	—
BEN	Benin	616	—	—	—
BMU	Bermuda	350 01	350 02	338 050	310 59
BTN	Bhutan	402	—	—	—
BOL	Bolivia	736	—	—	—
BIH	Bosnia and Herzegovina	218	—	—	—
BWA	Botswana	652	—	—	—
BRA	Brazil	724	—	—	—
BRN	Brunei Darussalam	528	—	—	—
BGR	Bulgaria	284	—	—	—
BFA	Burkina Faso	613	—	—	—
BDI	Burundi	642	—	—	—
KHM	Cambodia	456	—	—	—
CMR	Cameroon	624	—	—	—
CAN	Canada	302	—	—	—
CPV	Cape Verde	625	—	—	—
CYM	Cayman Islands	346	—	—	—
CAF	Central African Republic	623	—	—	—
TCD	Chad	622	—	—	—
CHL	Chile	730	—	—	—
CHN	China	460	—	—	—
COL	Colombia	732	—	—	—
COM	Comoros	654	—	—	—
COG	Congo	629	—	—	—
COD	Congo, DR	630	—	—	—
COK	Cook Islands	548	—	—	—

Table 9.6 (*Cont'd*)

Alpha-3 code	Name	MCC-1	MCC-2	MCC-3	MCC-4
CRI	Costa Rica	712	—	—	—
CIV	Côte d'Ivoire	612	—	—	—
HRV	Croatia	219	—	—	—
CUB	Cuba	368	—	—	—
CUW	Curaçao	362	—	—	—
CYP	Cyprus	280	—	—	—
CZE	Czech Republic	230	—	—	—
DNK	Denmark	238	—	—	—
DJI	Djibouti	638	—	—	—
DMA	Dominica	366	—	—	—
DOM	Dominican Republic	370	—	—	—
ECU	Ecuador	740	—	—	—
EGY	Egypt	602	—	—	—
SLV	El Salvador	706	—	—	—
GNQ	Equatorial Guinea	627	—	—	—
ERI	Eritrea	657	—	—	—
EST	Estonia	248	—	—	—
ETH	Ethiopia	636	—	—	—
FRO	Faroe Islands	288	—	—	—
FJI	Fiji	542	—	—	—
FIN	Finland	244	—	—	—
FRA	France	208	—	—	—
PYF	French Polynesia	547	—	—	—
GAB	Gabon	628	—	—	—
GMB	Gambia	607	—	—	—
GEO	Georgia	282	—	—	—
DEU	Germany	262	—	—	—
GHA	Ghana	620	—	—	—
GIB	Gibraltar	266	—	—	—
GRC	Greece	202	—	—	—
GRL	Greenland	290	—	—	—
GRD	Grenada	352	—	—	—
GLP	Guadeloupe	340	—	—	—
GUM	Guam	310	311	—	—
GTM	Guatemala	704	—	—	—
GGY	Guernsey	234	—	—	—
GIN	Guinea	611	—	—	—
GNB	Guinea-Bissau	632	—	—	—
GUY	Guyana	738	—	—	—
HTI	Haiti	372	—	—	—
VAT	Vatican City	225	—	—	—
HND	Honduras	708	—	—	—
HKG	Hong Kong	454	—	—	—
HUN	Hungary	216	—	—	—
ISL	Iceland	274	—	—	—
IND	India	404	405	—	—

(*Continued*)

Table 9.6 (*Cont'd*)

Alpha-3 code	Name	MCC-1	MCC-2	MCC-3	MCC-4
IDN	Indonesia	510	—	—	—
IRN	Iran	432	—	—	—
IRQ	Iraq	418	—	—	—
IRL	Ireland	272	—	—	—
IMN	Isle of Man	234	—	—	—
ISR	Israel	425	—	—	—
ITA	Italy	222	—	—	—
JAM	Jamaica	338	—	—	—
JPN	Japan	440	—	—	—
JEY	Jersey	234	—	—	—
JOR	Jordan	416	—	—	—
KAZ	Kazakhstan	401	—	—	—
KEN	Kenya	639	—	—	—
KIR	Kiribati	545	—	—	—
PRK	North Korea	467	—	—	—
KOR	South Korea	450	—	—	—
UNK	Kosovo	212 01	293 41	212 01	—
KWT	Kuwait	419	—	—	—
KGZ	Kyrgyzstan	437	—	—	—
LAO	Laos	457	—	—	—
LVA	Latvia	247	—	—	—
LBN	Lebanon	415	—	—	—
LSO	Lesotho	651	—	—	—
LBR	Liberia	618	—	—	—
LBY	Libya	606	—	—	—
LIE	Liechtenstein	295	—	—	—
LTU	Lithuania	246	—	—	—
LUX	Luxembourg	270	—	—	—
MAC	Macao	455	—	—	—
MKD	Macedonia	294	—	—	—
MDG	Madagascar	646	—	—	—
MWI	Malawi	650	—	—	—
MYS	Malaysia	502	—	—	—
MDV	Maldives	472	—	—	—
MLI	Mali	610	—	—	—
MLT	Malta	278	—	—	—
MHL	Marshall Islands	551	—	—	—
MTQ	Martinique	340	—	—	—
MRT	Mauritania	609	—	—	—
MUS	Mauritius	617	—	—	—
MEX	Mexico	334	—	—	—
FSM	Micronesia,	550	—	—	—
MDA	Moldova	259	—	—	—
MCO	Monaco	212 01	—	—	—
MNG	Mongolia	428	—	—	—
MNE	Montenegro	297	—	—	—
MSR	Montserrat	354	—	—	—

Table 9.6 *(Cont'd)*

Alpha-3 code	Name	MCC-1	MCC-2	MCC-3	MCC-4
MAR	Morocco	604	—	—	—
MOZ	Mozambique	643	—	—	—
MMR	Myanmar	414	—	—	—
NAM	Namibia	649	—	—	—
NRU	Nauru	536	—	—	—
NPL	Nepal	429	—	—	—
NLD	Netherlands	204	—	—	—
NCL	New Caledonia	546	—	—	—
NZL	New Zealand	530	—	—	—
NIC	Nicaragua	710	—	—	—
NER	Niger	614	—	—	—
NGA	Nigeria	621	—	—	—
NIU	Niue	555	—	—	—
NFK	Norfolk Island	505	—	—	—
NOR	Norway	242	—	—	—
OMN	Oman	422	—	—	—
PAK	Pakistan	410	—	—	—
PLW	Palau	552	—	—	—
PSE	Palestine	425	—	—	—
PAN	Panama	714	—	—	—
PNG	Papua New Guinea	537	—	—	—
PRY	Paraguay	744	—	—	—
PER	Peru	716	—	—	—
PHL	Philippines	515	—	—	—
POL	Poland	260	—	—	—
PRT	Portugal	268	—	—	—
PRI	Puerto Rico	330	—	—	—
QAT	Qatar	427	—	—	—
REU	Réunion	647	—	—	—
ROU	Romania	226	—	—	—
RUS	Russian Federation	250	—	—	—
RWA	Rwanda	635	—	—	—
KNA	Saint Kitts and Nevis	356	—	—	—
LCA	Saint Lucia	358	—	—	—
SPM	Saint Pierre and Miquelon	308	—	—	—
VCT	Saint Vincent and the Grenadines	360	—	—	—
WSM	Samoa	549	—	—	—
SMR	San Marino	292	—	—	—
STP	Sao Tome and Principe	626	—	—	—
SAU	Saudi Arabia	420	—	—	—
SEN	Senegal	608	—	—	—
SRB	Serbia	220	—	—	—
SYC	Seychelles	633	—	—	—
SLE	Sierra Leone	619	—	—	—
SGP	Singapore	525	—	—	—

(Continued)

Table 9.6   *(Cont'd)*

Alpha-3 code	Name	MCC-1	MCC-2	MCC-3	MCC-4
SXM	Sint Maarten	362	—	—	—
SVK	Slovakia	231	—	—	—
SVN	Slovenia	293	—	—	—
SLB	Solomon Islands	540	—	—	—
SOM	Somalia	637	—	—	—
ZAF	South Africa	655	—	—	—
SSD	South Sudan	659	—	—	—
ESP	Spain	214	—	—	—
LKA	Sri Lanka	413	—	—	—
SDN	Sudan	634	—	—	—
SUR	Suriname	746	—	—	—
SWZ	Swaziland	653	—	—	—
SWE	Sweden	240	—	—	—
CHE	Switzerland	228	—	—	—
SYR	Syrian Arab Republic	417	—	—	—
TWN	Taiwan	466	—	—	—
TJK	Tajikistan	436	—	—	—
TZA	Tanzania	640	—	—	—
THA	Thailand	520	—	—	—
TLS	Timor-Leste	514	—	—	—
TGO	Togo	615	—	—	—
TON	Tonga	539	—	—	—
TTO	Trinidad and Tobago	374	—	—	—
TUN	Tunisia	605	—	—	—
TUR	Turkey	286	—	—	—
TKM	Turkmenistan	438	—	—	—
TCA	Turks and Caicos Islands	338	—	—	—
TUV	Tuvalu	553	—	—	—
UGA	Uganda	641	—	—	—
UKR	Ukraine	255	—	—	—
ARE	United Arab Emirates	424	—	—	—
GBR	United Kingdom	234	235	—	—
USA	United States of America	310	311	313	316
URY	Uruguay	748	—	—	—
UZB	Uzbekistan	434	—	—	—
VUT	Vanuatu	541	—	—	—
VEN	Venezuela	734	—	—	—
VNM	Viet Nam	452	—	—	—
VGB	Virgin Islands, British	348	—	—	—
YEM	Yemen	421	—	—	—
ZMB	Zambia	645	—	—	—
ZWE	Zimbabwe	648	—	—	—
	International	901	—	—	—
	Test	001	—	—	—

Source: From ITU Report No. 1005 [4], used with permission from ITU.

individual networks have been identified in Table 9.6 using a combination of their MCC and MNC.

Alpha-3 codes are assigned by the ISO (International Standards Organisation) and provide a common three-letter abbreviation for countries and territories.

9.3 Cellular Network Types

9.3.1 2G GSM Networks

The basic characteristics of 2G GSM networks are as shown in Table 9.7.

Assigned Radio Bands

The radio bands and channel numbering employed by 2G GSM are as shown in Table 9.8.

9.3.2 3G UMTS Networks

The basic characteristics of 3G UMTS networks are as shown in Table 9.9.

Assigned Radio Bands

The radio bands and channel numbering employed by 3G UMTS are as shown in Table 9.10.

Table 9.7 Basic characteristics of 2G GSM.

Sub-generations/variants	2G GSM – basic voice, SMS and dial – up data
	2.5G GPRS – packet switched data
	2.75G EDGE – faster PS data
	2.75G EDGE Evolution – faster PS data
Air interface method	TDMA
Channel size	Nominally 200 kHz wide (actually 270 kHz)
	EDGE Evolution can aggregate two carriers
Duplexing options	FDD only
Frequency reuse	Yes – no single frequency network option
Channel numbering	ARFCNs – see Table 9.8
Physical layer ID	BSIC – six-bit identifier (64 BSICs in total)
Cell discrimination	ARFCN + BSIC
Cell ID format	CGI – MCC-MNC-LAC-Cell ID
Key measurements	RXLev in dBm
	Measurements taken of BCCH (idle), TCH (connected)
Typical values	Very strong > –84 dBm
	Very weak < –100 dBm

Table 9.8   2G GSM radio bands and channel numbering.

Band name	Uplink range (MHz)	Downlink range (MHz)	DL UARFCN		Deployment
			Low	High	
GSM450	450.4–457.6	460.4–467.6	259	293	Not used
GSM480	478.8–486.0	488.8–496.0	306	340	Not used
GSM700	698.0–716.0	728.0–746.0	Dynamic		Not used
GSM700	747.0–763.0	777.0–793.0	438	511	Not used
GSM850	824.0–849.0	869.0–894.0	128	251	Americas
E-GSM900	880.0–890.0	925.0–935.0	0, 955	1023	Global
P-GSM900	890.0–915.0	935.0–960.0	1	124	Global
GSM-R	873.0–890.0	918.0–935.0	940	1023	Europe
GSM-R	890.0–915.0	935.0–960.0	0	124	Europe
GSM1800	1710.0–1785.0	1805.0–1880.0	512	885	Global
GSM1900	1850.0–1910.0	1930.0–1990.0	512	810	Americas

Source: Based on 3GPP TS 45.005:2 [5], used with permission from 3GPP.

Table 9.9   Basic characteristics of 3G UMTS.

Sub-generations/variants	3G UMTS – voice, SMS, medium fast PS data
	3.5G HSPA – voice, SMS, faster PS data
	3.5G HSPA + – voice, SMS, very fast PS data
	UMTS-FDD
	UMTS-TDD$_{HCR}$
	UMTS-TDD$_{LCR}$ (TD-SCDMA)
Air interface method	WCDMA
Channel size	Typical 5 MHz
	1.6 and 10 MHz options also exist for TDD
	HSPA + can aggregate up to eight carriers
Duplexing options	FDD and TDD versions
Frequency reuse	No, deployed as multiple single frequency network layers
Channel numbering	UARFCN – see Table 9.10
Physical layer ID	PSC (Primary Scrambling Code) – 512 available
Cell discrimination	UARFCN + PSC
Cell ID format	CGI – MCC-MNC-RNC ID-Cell ID
Key measurements	RSCP – wanted signal in dBm
	RSSI – channel noise in dBm
	Ec/No – signal to noise ratio in dB
	Measurements taken of cell CPICH
Typical values	Ec/No – very strong > −5 dB
	Ec/No – very weak < −20 dB

Table 9.10   3G UMTS radio bands and channel numbering.

Band	Uplink range (MHz)	Downlink range (MHz)	DL UARFCN low	DL UARFCN high	Band name	Deployment
I	1920–1980	2110–2170	10 562	10 838	UMTS2100	Global
II	1850–1910	1930–1990	9662	9938	UMTS1900	Americas
III	1710–1785	1805–1880	1162	1513	UMTS1800	
IV	1710–1755	2110–2155	1537	1738	UMTS1700	Americas
V	824–849	869–894	4357	4458	UMTS850	Global
VI	830–840	875–885	4387	4413	UMTS800	Japan
VII	2500–2570	2620–2690	2237	2563	Used by LTE	Global
VIII	880–915	925–960	2937	3088	UMTS900	Global
IX	1749.9–1784.9	1844.9–1879.9	9237	9387	UMTS1700	Japan
X	1710–1770	2110–2170	3112	3388	UMTS1700	Americas
XI	1427.9–1447.9	1475.9–1495.9	3712	3787	UMTS1500	Japan
XII	699–716	729–746	3842	3903	UMTS700	Americas
XIII	777–787	746–756	4017	4043	UMTS700	Americas
XIV	788–798	758–768	4117	4143	UMTS700	Americas
XIX	830–845	875–890	5112	5413	UMTS800	Japan
XX	832–862	791–821	5762	5913	Used by LTE	Europe
XXI	1447.9–1462.9	1495.9–1510.9	10 562	10 838	Used by LTE	Japan
XXII	3410–3490	3510–3590	9662	9938	Not yet used	
XXV	1850–1915	1930–1995	1162	1513	Not yet used	
XXVI	814–849	859–894	1537	1738	Not yet used	

Source: Based on 3GPP TS 25.104-v12.2.0:5.2 [6], reproduced with permission from 3GPP.

9.3.3   2G cdmaOne and 3G CDMA2000 Networks

The basic characteristics of 2G cdmaOne and 3G CDMA2000 networks are as shown in Table 9.11.

Assigned Radio Bands

The band classes and channel numbering employed by 2G cdmaOne and 3G CDMA2000 are as shown in Table 9.12.

9.3.4   4G LTE Networks

The basic characteristics of 4G LTE networks are as shown in Table 9.13.

Assigned Radio Bands

The radio bands and channel numbering employed by 4G LTE are as shown in Table 9.14.

Table 9.11 Basic characteristics of 2G cdmaOne and 3G CDMA2000.

Sub-generations/variants	2G cdmaOne IS95/IS95A/IS95B
	3G CDMA2000 1xRTT
	3.5G EVDO Rev0/Rev A/Rev B
Air interface method	CDMA
Channel size	1.25 or 3.75 MHz (deployed in a 5 MHz carrier)
	EVDO can aggregate 3 × 1.25 MHz carriers
Duplexing options	FDD only
Frequency reuse	No, deployed as multiple single frequency network layers
Channel numbering	Band Classes have their own set of Channel Numbers – see Table 9.12
Physical layer ID	PN Offset – 512 available
Cell discrimination	Band Class + Channel Number + PN Offset
Cell ID format	MCC-MNC-SID-NID-BSID
Key measurements	RSCP – wanted signal in dBm
	RSSI – channel noise in dBm
	Ec/Io – signal to noise ratio in dB
	Measurements taken of Pilot Channel (F-PICH)
Typical values	Ec/Io – very strong, > −6 dB
	Ec/Io – very weak, < −28 dB

Table 9.12 2G cdmaOne and 3G CDMA2000 band classes and channel numbering.

Band class	Forward link frequencies	Reverse link frequencies	Channel numbers	Description
0	860–894	815–849	1–1323	800 MHz band
1	1930–1990	1850–1910	0–1199	1.8–2.0 GHz PCS band
2	917–960	872–915	0–2108	872–960 MHz TACS band
3	832–870	887–925	1–1600	832–925 MHz JTACS band
4	1840–1870	1750–1780	0–599	1.75–1.87 Korean TACS band
5	420–493	410–483	1–2108	450 MHz NMT band
6	2110–2170	1920–1980	0–1199	2 GHz IMT2000 band
7	746–758	776–788	0–240	Upper 700 MHz band
8	1805–1880	1710–1785	0–1499	1800 MHz band
9	925–960	880–915	0–699	900 MHz band
10	851–940	806–901	0–919	Secondary 800 MHz band
11	420–493	410–483	0–2016	400 MHz European PAMR band
12	915–921	870–876	0–239	800 MHz PAMR band
13	2620–2690	2500–2570	0–1399	2.5 GHz IMT2000 band
14	1930–1995	1850–1925	0–1299	US 1.9 GHz PCS band
15	2110–2155	1710–1755	0–899	AWS band
16	2624–2690	2502–2568	140–1459	US 2.5 GHz band
17	Not specified			US 2.5GHz Forward link only band
18	757–769	787–799	0–240	700 MHz public safety band
19	728–746	698–716	0–360	Lower 700 MHz band
20	1525–1559	1626.5–1660.5	0–680	L-band
21	2180–2200	2000–2020	0–399	S-Band

Source: Based on 3GPP2 Specification C.S0057-E [7], used with permission from 3GPP2.

Table 9.13 Basic characteristics of 4G LTE.

Sub-generations/variants	4G LTE – very fast PS data
	4G – LTE-Advanced – very fast PS data
	LTE-FDD/FD-LTE
	LTE-TDD/TD-LTE
Air interface method	OFDMA (downlink), SC-FDMA (uplink)
Channel size	Scalable from 1.4 to 20 MHz
	LTE-A can aggregate up to five carriers
Duplexing options	FDD and TDD options exist
Frequency reuse	Yes and can also be deployed as multiple single frequency network layers
Channel numbering	EARFCNs – see Table 9.14
Physical layer ID	PCI (Physical-layer Cell ID) – 504 available
Cell discrimination	EARFCN + PCI
Cell ID format	eCGI – MCC-MNC-eNB ID-Cell ID
Key measurements	RSRP – wanted signal in dBm
	RSSI – channel noise in dBm
	RSRQ – signal to noise ratio in dB
	Measurements taken of reference signals
Typical values	RSRQ – very strong > −5 dB
	RSRQ – very weak < −25 dB

Table 9.14 4G LTE radio bands and channel numbering.

Band	Uplink range	Downlink range	Duplex mode	DL Channel Numbers
1	1920–1980	2110–2170	FDD	0–599
2	1850–1910	1930–1990	FDD	600–1199
3	1710–1785	1805–1880	FDD	1200–1949
4	1710–1755	2110–2155	FDD	1950–2399
5	824–849	869–894	FDD	2400–2649
6	830–840	875–885	FDD	2650–2749
7	2500–2570	2620–2690	FDD	2750–3449
8	880–915	925–960	FDD	3450–3799
9	1749.9–1784.9	1844.9–1879.9	FDD	3800–4149
10	1710–1770	2110–2170	FDD	4150–4749
11	1427.9–1447.9	1475.9–1495.9	FDD	4750–4949
12	699–716	729–746	FDD	5010–5179
13	777–787	746–756	FDD	5180–5279
14	788–798	758–768	FDD	5280–5379
15	Reserved			
16	Reserved			
17	704–716	734–746	FDD	5730–5849
18	815–830	860–875	FDD	5850–5999
19	830–845	875–890	FDD	6000–6149
20	832–862	791–821	FDD	6150–6449
21	1447.9–1462.9	1495.9–1510.9	FDD	6450–6599

(Continued)

Table 9.14 *(Cont'd)*

Band	Uplink range	Downlink range	Duplex mode	DL Channel Numbers
22	3410–3490	3510–3590	FDD	6600–7399
23	2000–2020	2180–2200	FDD	7500–7699
24	1626.5–1660.5	1525–1559	FDD	7700–8039
25	1850–1915	1930–1995	FDD	8040–8689
26	814–849	859–894	FDD	8690–9039
27	807–824	852–869	FDD	9040–9209
28	703–748	758–803	FDD	9210–9659
29	No uplink	717–728	FDD	9660–9769
30	2305–2315	2350–2360	FDD	9770–9869
31	452.5–457.5	462.5–467.5	FDD	9870–9919
32	Not used			
33	1900–1920		TDD	36000–36199
34	2010–2025		TDD	36200–36349
35	1850–1910		TDD	36350–36949
36	1930–1990		TDD	36950–37559
37	1910–1930		TDD	37550–37749
38	2570–2620		TDD	37750–38249
39	1880–1920		TDD	38250–38649
40	2300–2400		TDD	38650–39649
41	2496–2690		TDD	39650–41589
42	3400–3600		TDD	41590–43589
43	3600–3800		TDD	43590–45589
44	703 – 803		TDD	45590–46589

Source: Based on 3GPP TS 36.104:5.2 [8], reproduced with permission from 3GPP.

9.4 Forensic Radio Surveys

A recap of forensic radio survey techniques follows.

9.4.1 Spot/Location Surveys

'Spot' or location surveys are designed to capture details of the set of serving and non-serving cells that provide coverage at or near a given location. Generally, the locality chosen for the survey is the address where an incident has occurred or where a person of interest in an investigation lives, works or has specified as an alibi address. The basic concepts related to spot and location surveys are illustrated in Figure 9.2.

Spot/Location Survey Actions

The set of individual surveys to be conducted at a location will vary depending upon the networks/technologies to be surveyed, which is dictated by the networks and technologies used by the case's target phones as detailed in their call records. The geography of the location and the circumstances of the case will also influence the survey

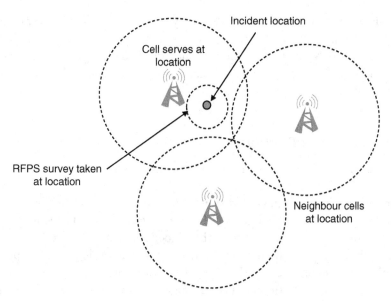

Figure 9.2   Spot/location survey

list, but the full set of captures that could be undertaken for spot or location surveys
includes the following:
- Per required network:
  - ○ Free-running 2G survey, with survey device technology locked to 2G;
  - ○ Free-running 3G survey, with survey device technology locked to 3G;
  - ○ Additional 3G surveys with channel or band locks in place if the network employs
    Idle Mode behaviour settings that make surveys of HSPA 'data' channels
    difficult;
  - ○ Lock file to capture details of 3G neighbours if the survey device does not
    automatically capture neighbour cell IDs;
  - ○ Free-running 4G survey, with survey device technology locked to 4G;
  - ○ Additional 4G surveys with channel or band locks in place if the network employs
    Idle Mode behaviour settings that make surveys of some channels difficult;
  - ○ Lock file to capture details of 4G neighbours if the survey device does not
    automatically capture neighbour cell IDs;
  - ○ Potentially, also capture WiFi surveys at each location, if required.

Multiple 3G and 4G surveys are suggested due to the possibility (indeed, the strong
likelihood) that operators will employ some form of Idle Mode behaviour control to
limit mobile devices to camping on a specific frequency layer when idle. Channel or
band lock surveys therefore allow the surveyor to guarantee that they capture cell
coverage details of the non-camping bands.

The set of surveys suggested above would be conducted in Idle Mode but would
typically also each include a set of Connected Mode test calls.

Obviously, such an extended set of surveys could take a considerable amount of time and surveyors may be able to reduce the intensity of surveys by conducting just 'free running' Idle Mode surveys on each network/technology first and then supplementing these with 'channel lock' or 'band lock' surveys and Connected Mode test calls if required.

Spot/Location Survey Procedures

The suggested set of procedures for a spot/location survey includes the following:

- Ensure that each survey device or sub-module in the case of multi-receiver devices has the appropriate network's SIM inserted, the appropriate technology lock in place and that any further channel or band locks are set if required.
- Conduct the survey as a non-static location survey rather than a static spot survey if possible, safe and practical (unless a spot survey is specified).
- Spend at least 5 min in the immediate vicinity of the surveyed address but extend the survey up to 50 or 100 m in all directions from the target address.
- Conduct the surveys in free-running Idle Mode first and perform additional locked surveys or Connected Mode test calls (or test data sessions) only if the necessary information was not captured in Idle Mode.
- If possible, capture more than one simultaneous survey, using different devices or even different types of device to ensure a more broadly representative set of survey results – results from different devices can be combined into one overall set of results during post-processing.
- For 3G surveys, make each test call last for an extended period – 1 min, for example – to ensure that a representative range of soft handover serving cells are captured.
- For 3G and 4G surveys, if the survey equipment does not automatically capture neighbour cell IDs, consider recording a 'lock file' (if cell locks are supported by the survey equipment) to ensure that Cell IDs for neighbour cells are captured.
- Make sure that GPS fixes of the locations(s) of the survey are captured as evidence that the surveys were conducted in the reported location, possibly also take a photo of the target address, if this can be done safely and without causing distress to victims or witnesses.

9.4.2   All-Network Profiles

Spot/location surveys are typically undertaken to gather evidence related to a specific set of target phones and are therefore often conducted on just a subset of networks or technologies at a time.

All-network profiles are usually undertaken on all networks and all technologies at a spot or location and can really be thought of as a linked set of location surveys. An example of an All-network Profile is shown in Table 9.15.

Table 9.15 Example of an all-network profile.

Network 1				Network 2				Network 3			
	Channel/PCI	Cell ID (status)	Ave. signal		Channel/PCI	Cell ID (status)	Ave. Signal		Channel/PCI	Cell ID (status)	Ave. Signal
2G	45/27	20 456 (s)	−82.34	2G	23 12	234 (s)	−75.12	2G	787/34	45 398 (s)	−82.04
	47/29	30 456 (s)	−87.23		67 13	345 (n)	−98.12		821/42	2123 (n)	−87.38
	105/23	10 456 (n)	−93.87		97 16	12 765 (n)	−100.67		698/31	54 901 (s)	−88.13
	51/23	38 765 (n)	−99.56						670/32	19 801 (n)	−90.87
									701/36	20 700 (n)	−91.45
3G	10 637/145	10 987 (s)	−4.56	3G	10 712 87	32 154 (s)	−7.38	3G	10 836/198	56 901 (n)	−5.78
	10 637/146	20 987 (n)	−14.56						10 811/198	55 901 (s)	−6.87
									10 811/199	55 902 (s)	−9.21
									10 836/198	56903 (n)	−5.78
									10 836/199	56 902 (n)	−12.72
									10 761/14	8790 (n)	−16.87
4G	6400/31	10 976 542 (s)	−8.32	4G	6300 121	435 213 (s)	−7.02	4G	1617/341	62541 (s)	−8.3
	6400/32	10 976 543 (n)	−14.34		6300 122	435 212 (s)	−11.59				
					2850 21	143 256 764 (n)	−18.34				

9.4.3   Cell Coverage Surveys

Cell coverage surveys are intended to determine the extent of serving coverage of a particular cell in a way that allows the approximate 'footprint' of the cell to be mapped. This type of survey is generally performed as a drive survey and the results provide a snapshot of cell coverage at the time the survey was taken.

Coverage Survey Procedures

The recommended procedures to be undertaken during a coverage survey include:

- Ensure that each survey device or sub-module in the case of multi-receiver devices has the appropriate network's SIM inserted and the appropriate technology, band or channel lock in place.
- Cell locks or Connected Mode surveying for cell coverage surveys are not recommended, as the purpose of the survey is to map the natural boundary of the cell's serving coverage. Setting a cell lock for the target cell would force the survey device to continue to treat it as the serving cell way beyond the point where a free-running device would have reselected to a neighbour cell. A cell lock can therefore distort the results of the survey.
- Ensure that the device's GPS receiver is online and that GPS fixes are being captured. Coverage survey data is valueless without accurate location information.
- Follow the drive survey safety recommendations; drive surveys should always be a two-person job.
- Work out the route to be driven in advance and record the route as it is being driven (some survey devices provide a map with the driven route overlaid on it to aid surveyors), this avoids unnecessary duplication and can shorten the survey time.
- If the survey is intended to capture details of only specific sectors of a site rather than all coverage, then it might be beneficial to confirm the azimuth of the required sectors first to avoid driving in unnecessary directions.
- Consider using multiple survey devices simultaneously, especially if the survey instructions require the surveyor to capture details of different cells on the same site that share the same sector azimuth, as would be the case in a 3G or 4G multi-layer single frequency network. In this case each survey device would need to have the appropriate band or channel lock in place to ensure that it captured details of the correct frequency layer.

In some cases the surveyor may be required to undertake surveys of cells belonging to different networks that happen to be broadcast from base stations that are sharing the same site. The site sharing agreements between operators in many countries mean that this scenario is becoming more common. In these cases it is possible to capture details of multiple cells during just one drive survey if multiple survey devices are used.

One final recommendation relates to the post-processing of survey data to create coverage maps. In a large majority of cases, if cell coverage surveys are required, the cell site analyst will request survey data for multiple cells in the same area. Surveyors may complete the whole survey set by conducting individual drive surveys for each cell or sector on their list and some or all of these surveys may include overlapping areas.

Given the potential that exists for cells to provide non-contiguous patches of coverage beyond the main contiguous coverage area – for there to be small 'islands' of coverage separate from the main area of coverage – it is considered sensible to 'pool' the results of all of the surveys before extracting details of each individual cell to be mapped.

9.4.4   Route Profiles

A Route Profile survey employs similar methods to a cell coverage survey but, whereas a coverage survey seeks to determine the area served by a single cell, a route profile attempts to represent the progression of cells that serve along a given route.

Route Profile Actions

The usual methodology for a route profile survey is really no more complicated than that outlined in the previous paragraph:

- Ensure that each survey device or sub-module in the case of multi-receiver devices has the appropriate network's SIM inserted and the appropriate technology lock in place, and that any further channel or band locks are set if required.
- Ensure that the device's GPS receiver is online and that GPS fixes are being captured. Route profile survey data is valueless without accurate location information.
- Follow the drive survey safety recommendations; drive surveys should always be a two-person job.
- Work out the route to be driven in advance and stick to it; unscheduled deviations from the prescribed route could cause inaccurate results.
- Try to drive the route at the slowest speed that is practicable and safe, this should ensure that a greater amount of detail is captured at each point along the route. Survey accuracy can also be improved by stopping at regular intervals, for 1–2 min at a time, to allow the survey device to capture long term variations in coverage.
- Alternatively, consider driving the route multiple times or use multiple survey devices simultaneously to broaden the depth of information captured.
- If target phones were connected to different networks or used different technologies, consider using multiple devices, each set to capture a different network/technology to minimise the number of repeat drives along the route that need to be taken.

Typically, when compiling route profile data, only details of the serving cell at each location are required; details of neighbour cells are usually only examined if the serving cell data fails to detect an expected cell or a cell that was used by a target phone for a significant call.

9.4.5   Survey Specifications

Before undertaking a forensic radio survey it is considered good practice for the surveyor to have an indication of the cells that are expected to be found at each location.

A cell site analyst will use the call records for each target phone to draw up a list of the cells used by those phones at times when the users of those phones were suspected of being at locations of interest. They will use this information to draw up a target cell list of each location, area or route to be surveyed.

An example target cell list might include some of the following information:

- Location/cell coverage area/route to be surveyed.
- Networks and technologies to be surveyed.
- List of cells of interest (based on cell ID) that are expected to be detected on each network/technology at that location.
- Additional information, such as 'cell X was only used for data sessions', 'end cell only' or 'cell Y was used extensively' can aid the survey engineer's understanding of the objectives of the survey and the relative importance of each cell.
- An indication of whether an attempt should be made to arrange for the survey to be conducted inside an address. This would be an issue in cases where the suspected user of a target phone was assumed to have been indoors when calls were made and is especially relevant to locations such as flats and apartments, where an outdoor survey would have difficulty replicating the coverage provided at altitude.
- Any potential dangers or sensitivities that should be taken into consideration by the surveyor.
- Cell address/location details for expected cells, including cell azimuth details.
- Cell and survey location map, which would usefully include details of the cell locations and azimuths in relation to the survey location.
- Providing extracts of the relevant CDR (call records) that show the calls made using the cells to be surveyed can be useful, this can add useful context for a surveyor by indicating, for example, the sequence in which particular cells were used.

An example Target Cell List is shown in Figure 9.3 and the accompanying Cell Location Map is shown in Figure 9.4.

9.4.6   Preparing Survey Devices

Each type of forensic radio survey has its own generic actions and activities, but the specific actions to be performed for each individual survey are typically dictated by the circumstances of the case, by the location at which the survey is to be based and by the surveyor's personal preferences.

It is therefore difficult to draw up a set of suggested activities and guidelines that are relevant in all scenarios and that match with every surveyor's preferences. The

Case Name:	R v Moriarty		Case Ref:		CSAS-1-2014			

link to map

Location Surveys		Expected Cells						
Location	C - 24 Lower St, Anytown	2G		3G		4G		
Significance	Robbery address	Cell ID	Comment	Cell ID	Comment	Cell ID	Comment	
Vodafone	Surveyor Comments	6776	voice & data use	4144	data use only	not required		
		244	used once					
		245	end cell only					
O2	Surveyor Comments	20234	only used once	10298	data use only	not required		
		30234	end cell only	30298	end cell only			
		13423	used extensively					

Coverage Surveys

Survey 1

Network	O2	Technology	2G	Cell ID	13423	link to map
Site Name	Bubbenhall	Azimuth	20	Site Location	Bubbenhall Rd, Ryton - 52.45672, -1.09765	

Route Surveys

Route ID	1	Description	Marston to Kenilworth via A423, A45, B4115					
			Expected Cells					
Comments:		2G		3G		4G		
		Cell ID	Comment	Cell ID	Comment	Cell ID	Comment	
O2	Surveyor Comments	free		free		not required		

Surveyor	Jim Smith	Organisation	RF Group Ltd	Comments:
Survey Date		Equipment		
Requested By	Jill Jones	Date	02/02/2014 15:12	

Figure 9.3 Target cell list. Source: CSAS Target Cell List reproduced with permission from Forensic Analytics Ltd

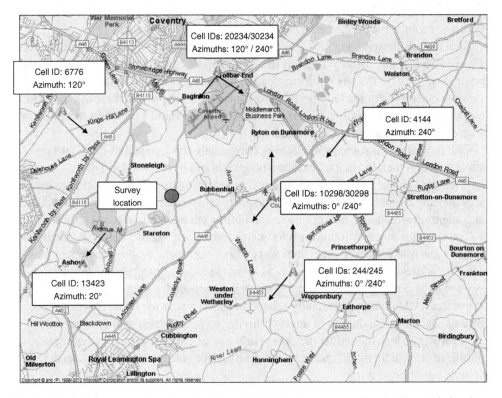

Figure 9.4 Cell location map. Source: Microsoft AutoRoute map reproduced with permission from Microsoft, CSAS Map format used with permission from Forensic Analytics Ltd

following suggested actions and activities should therefore be seen as at least partly subjective, as they are based on the author's experiences and preferences.

Before commencing each new survey, it is recommended that surveyors check the following:

- Make sure the survey device has sufficient charge for the predicted survey duration (if not on external power).
- Make sure the survey device has sufficient free memory or enough spare data cards to store the expected survey data file(s).
- Make sure any previously applied technology, channel or cell locks have been reset if not required for the new survey.
- Make sure the required technology, channel or cell locks for the new survey are set.
- Make sure that the 'save as' filename for the survey is correct and reflects the survey being undertaken – this could include making sure that the correct location, network and technology are listed in the filename if a specific file naming convention is being employed.

Failure to perform these simple checks can lead to surveys failing to capture the required information, meaning that they may have to be rerun at extra cost.

9.4.7 Survey Safety

Forensic radio surveys can be hazardous for a number of reasons.

Spot and all-network surveys, for example, are generally undertaken outside, often close to roads and usually require the surveyor to concentrate on the information being displayed on the test equipment. Location surveys are often undertaken as walk surveys but also require the surveyor to devote at least part of their attention to the output displayed on the survey device's screen. Care should be taken to maintain an awareness of the surveyor's surroundings.

Cell coverage and route profile surveys are usually undertaken as drive surveys. It is strongly recommended that at least two people should be involved in a drive survey, one to drive the vehicle and one to operate the survey equipment. Lone working surveyors are exposing themselves and other road users to danger due to driving without due care and attention whilst also attempting to monitor or operate their survey equipment.

The circumstances surrounding a forensic radio survey should also be borne in mind: surveys are often undertaken at locations where traumatic events have taken place or near the addresses of witnesses, victims or suspects in a case or their families. A surveyor may unwittingly cause further distress if the reason for their presence at a location is guessed. Additionally, especially related to surveys near the addresses of suspects or defendants, there is a danger of attack from the suspect or their family or friends.

A further risk is related to the surveyor's use of expensive survey equipment in a public area, which might make them a target for mugging or might lead to their vehicle being broken into.

The reason for a surveyor's presence at a location is often guessed due to the nature of the survey equipment they are carrying and the amount of time it is necessary to spend at or around a location in order to capture spot or location survey data. Experience has shown that the risk of being discovered increases greatly if the surveyor elects to take a photo of the survey location to act as proof that they surveyed at the correct location. If a photo is required then it usually a good idea to take it after the survey has been completed.

Some surveyors decide to provide themselves with an 'alibi' for being at a location, sometimes using false identification credentials so that they can claim to be from a mobile phone company or a public utility. Some surveyors have gone to the length of acquiring false ID cards or even branded work clothing (such as a high visibility jacket) from the organisation they wish to claim to be working for.

To ensure the surveyor's safety during a survey it is recommended that they:

- Always let colleagues know where they are planning to survey and when.
- Consult the investigators in the case to determine if there are any specific risks or sensitivities related to any survey locations.
- Arrange a police escort for particularly dangerous or sensitive locations.
- Conduct a local coverage survey (where the surveyor moves around), which is less likely to attract attention than a static spot survey (where they stay still in one place).
- Keep survey equipment out of sight if at all possible.
- Only take site photos if they are sure it is safe.
- Maintain awareness of their location, especially if near a road or other hazards.

  Use a driver when conducting drive surveys.

9.5 Survey Results: Checking and Confirmation

9.5.1 Confirming the Expected Results

When a survey is underway, most types of survey equipment provide an on-screen list of the cells currently being detected and measured. The forensic survey engineer will use this output to check the progress of a survey against their target cell list.

If all of the cells listed in the target cell list are detected, the survey can be deemed a success.

9.5.2 Expected Results Not Found

If one or more of the cells on the target cell list fails to be detected, the forensic survey engineer has a number of options:

- If the survey device supports channel or cell locking, they could lock to the channel or cell ID and see if the 'missing' cell is detectable that way, although this would be done purely to determine whether the 'missing' cell was on air or not and would not be used as the basis for radio measurements of that cell – measurements taken with a cell lock in place cannot be regarded as being representative of the cells normal operation.

- They could try making test calls to see if the 'missing' cell is used in Connected Mode – this is particularly relevant in 3G mode, where some cells may only show up during soft handover, and so it is important that test calls are considered.
- If the survey equipment supports 'band scan' function, which scans through all channels in a radio band and identifies the cells that are detected, the surveyor could run this test to see if the missing Cell ID appears.
- If the survey equipment provides only details of the cells that are currently being detected, it might be that the 'missing' cell is appearing in the data only intermittently and too quickly for the equipment to register onscreen. If the surveyor is able to review or process the captured survey data after the survey period has ended the 'missing' cell ID might appear in a summary of the collated data.
- If cell address details have been provided the surveyor could walk or drive towards the cell site to see if a signal can be detected as they move closer (noting the location at which a signal is eventually detected) bearing in mind the drive survey safety recommendations mentioned above.
- They could visit the cell address to check that the site is still there and still 'on air'.
- They could perform an 'orbit' test of the site in an attempt to determine the current orientation (azimuth) of the cell sectors to see if they agree with the information provided (making a note of any revised azimuth estimations they calculate).
- If they are working for a law enforcement agency, they could ask their service provider liaison department to check with the network operator to determine why the cell is not being detected. This would include asking for details of if/when a site had been reoptimised, reorientated, relocated or retired.

In all cases, if an expected cell is not detected the surveyor should make a note of the steps taken to try to detect it and if the cell is still not discovered a note to this effect should be made in the post-survey report.

9.6   Survey Notes and Progress Maps

Forensic radio surveyors are under the same obligations to make contemporaneous notes during the course of their investigations as any other forensic investigator. These notes must be retained as they could be requested by the court when the case comes to trial.

Contemporaneous notes for cell site surveys usually take the form of a series of notes indicating:

- The location or address being surveyed, possibly including a GPS fix;
- A note of the specific spot at which the survey was taken (e.g. balcony outside Flat 7) for a static spot survey or a basic description of the route followed (e.g. walked around perimeter of building) during a non-static location survey;
- The time and date the survey started;

- An indication of the weather conditions and any other significant factors;
- The network and technology being surveyed;
- Details of any channel or cell locks applied before or during the survey;
- A list of each detected cell ID, with an indication of whether it served, was a first neighbour or was simply detected;
- The list of cell IDs could be added to during the survey with details of the changes in signal strength;
- An indication of whether additional survey types were captured (e.g. lock files) or whether multiple devices were used simultaneously;
- If any specific conclusions are reached during the survey then these should be noted for future reference;
- Details of any expected but missing cells should be noted, as should the steps taken by the surveyor to determine whether the missing cell was actually on air or not.

It can be argued that if the surveyor is using a forensic survey device that automatically captures and records survey details then it may not be necessary to manually record some of the detail listed above. It may also be difficult to manually capture all of the details listed above if the surveyor is undertaking concurrent surveys using multiple devices.

Information such as a description of the survey location, the weather conditions and the types of survey undertaken should always be recorded in the surveyor's contemporaneous notes, as should details of any specific conclusions and actions taken to 'find' a 'missing' cell.

9.7 Survey Results

9.7.1 Spot/Location Survey Results

Several types of survey device produce their output in a tabulated format that shows, for each measurement event, the serving and neighbour cells, their identities and radio signal strengths, along with a timestamp and a GPS fix.

Although the tabulated form of results data has its uses, it does not help to provide an immediate understanding of which cells, in areas of non-dominance, serve most often or offer the most consistently strong signals. A summarised table is much more useful for this kind of understanding and analysis.

An example of a set of processed and summarised spot/location survey results is shown in Figure 9.5.

A typical summarised survey results table will provide some or all of the following types of information:

- Survey location, start date/time of survey, surveyor name
- Duration of survey
- GPS fix (or average GPS fix) of survey location or locality
- Network and technology surveyed
- List of detected cells, ranked in order of signal strength, showing:

CSAS					<Case name>			Spot/Location Survey Summary	
	Location	A - 14 High St							
	Ave GPS	Lat: 52.123		Long: -1.2345		Height: 110m			
	Network	Vodafone UK		Technology	2G	# of Test Calls	3		
Survey Results									
Cell ID	LAC/TAC	Channel/PCI	Ave. Strength	Description	Detected (Serving)	Serving Percentage	Azimuth	Site Name	Served in Call
12345	345	654 12	-84.12	Strong	123 (123)	100%	120	Town Hall	Yes
2468	345	101 14	-94.32	Moderate	123 (0)	0%	240	Town Hall	No

source filename - C:/surveys/operation csr/ddp/Feb 2014/sodisodisodi.csv

Surveyor	Jim Smith	Organisation	RF Group Ltd	Comments: survey conducted 15m from front door,
Survey Date	01/02/2014 13:45-13:55	Equipment	Csurv MuNST	with walk survey around perimeter of building
Processed By	Jill Jones	Date	02/02/2014 15:12	after 5 minutes

Figure 9.5   Spot/location survey results. Source: CSAS data format, reproduced with permission from Forensic Analytics Ltd

- ○ Cell ID
- ○ Cell site name details (if known)
- ○ Channel number and physical layer cell identifier (BSIC/PSC/PCI/PN Offset)
- ○ Average signal strength (RXLev, Ec/No, RSRQ, Ec/Io)
- ○ Number of times the cell was detected in total and the number of times it was selected as serving
- ○ Indication of whether the cell served in Idle Mode
- ○ Indication of whether the cell served during a test call

- Overall number of test calls made
- Surveyors comments.

This type of summarised table is often the forensic radio survey 'product' that is passed on to the case investigators or cell site analysts and is therefore used to enable them to form conclusion on the potential locations of target phones during significant calls.

9.7.2   All-Network Profile Results

An all-network profile is essentially a 'summary of summaries' for each surveyed spot or location and is designed to draw together the summary data for each separate network/technology survey into one overall coverage summary table.

An example of an all-network profile report – taken in a hypothetical country that has three network operators who offer 2G, 3G and 4G services, is shown in Table 9.16.

Table 9.16 Example of all-network profile report.

	Network 1			Network 2			Network 3		
	Channel/PCI	Cell ID (status)	Average signal	Channel/PCI	Cell ID (status)	Average signal	Channel/PCI	Cell ID (status)	Average signal
2G	45 27	20 456 (s)	−82.34	23 12	234 (s)	−75.12	787 34	45 398 (s)	−82.04
	47 29	30 456 (s)	−87.23	67 13	345 (n)	−98.12	821 42	2123 (n)	−87.38
	105 23	10 456 (n)	−93.87	97 16	12 765 (n)	−100.67	698 31	54 901 (s)	−88.13
	51 23	38 765 (n)	−99.56				670 32	19 801 (n)	−90.87
							701 36	20 700 (n)	−91.45
3G	10 637 145	10 987 (s)	−4.56	10 712 87	32 154 (s)	−7.38	10 836 198	56 901 (n)	−5.78
	10 637 146	20 987 (n)	−14.56				10 811 198	55 901 (s)	−6.87
							10 811 199	55 902 (s)	−9.21
							10 836 198	56903 (n)	−5.78
							10 836 199	56 902 (n)	−12.72
							10 761 14	8790 (n)	−16.87
4G	6400 31	10 976 542 (s)	−8.32	6300 121	435 213 (s)	−7.02	1617 341	62 541 (s)	−8.3
	6400 32	10 976 543 (n)	−14.34	6300 122	435 212 (s)	−11.59			
				2850 21	143 256	−18.34			
					764 (n)				

A typical all-network profile presents an abbreviated set of details for each network and technology that was surveyed and will typically show, for each network and each technology, the set of detected cells ranked in order of average signal strength along with these details for each cell:

- Channel number and physical layer ID (BSIC, PSC, PCI or PN Offset)
- Cell ID (or Base ID) and status (serving or neighbour)
- Average signal strength (RXLev, Ec/No, RSR or Ec/Io)

All-network profile reports are often undertaken to allow investigators to capture a 'snapshot' of network coverage at a crime scene or other significant location in the immediate aftermath of an event. This allows them to preserve evidence of the cellular coverage at the site as it was at the time of the event and guards against changes in network configuration affecting the accuracy of any cell site reports or conclusions that are reached later in the investigation.

9.7.3   Coverage Survey Results

The output of a coverage survey will usually be a map showing the route driven and indicating measurement locations that selected the target cell as serving (and optionally another set of markers showing where the target cell was first neighbour).

The raw material for such a map is a set of GPS coordinates followed by details of the serving cell (and optionally the first neighbour cell) from the measurement events captured at each surveyed point along the driven route.

An example of the raw results data table for a coverage survey is shown in Table 9.17.

Table 9.17   Example of coverage survey results raw data, showing details of only the 'target' cell.

Longitude	Latitude	Cell ID	Status
−1.463 342	52.521 378	27 165	0 – serving
−1.463 342	52.521 378	27 165	0 – serving
−1.463 257	52.521 317	27 165	0 – serving
−1.463 257	52.521 317	27 165	0 – serving
−1.463 257	52.521 317	27 165	0 – serving
−1.463 257	52.521 317	27 165	0 – serving
−1.463 257	52.521 317	27 165	0 – serving
−1.463 170	52.521 255	27 165	0 – serving
−1.463 170	52.521 255	27 165	0 – serving
−1.463 170	52.521 255	27 165	1 – 1st neighbour
−1.463 170	52.521 255	27 165	1 – 1st neighbour
−1.463 170	52.521 255	27 165	1 – 1st neighbour
−1.463 085	52.521 198	27 165	1 – 1st neighbour
−1.462 998	52.521 141	27 165	0 – serving

Such a data set could then be manually filtered, so that only events that showed the target cell as serving or first neighbour are retained, and the filtered results could be imported into a mapping tool such as Microsoft AutoRoute.

Depending upon the level of detail required in the coverage map, it may be necessary to import three separate data sets in to each cell coverage map:

- The first data set shows the entire route driven during the survey and consists of just the GPS latitude/longitude data for the survey. The route would usually be represented in the map using small pushpins in a neutral colour.
- The second data set shows the locations of survey measurements where the target cell was detected as the first neighbour cell. These locations would usually be shown on the map using a larger pushpin and a different colour to the route pins.
- The last data set shows the locations of survey measurements where the target cell was detected as the serving cell. These locations would use the same size pushpins as for first neighbour locations but would be in a third colour.
- The map should contain a legend indicating the significance of each type of pushpin shown.

An example of a completed cell coverage survey map, created using Microsoft AutoRoute and where all three levels of data have been imported, is shown in Figure 9.6.

This type of map is usually the form in which forensic radio survey product for cell coverage surveys is passed to the investigators or cell site analysts in a case and allows them to draw conclusions as to the area within which a target phone could have been located when making significant calls via that cell.

9.7.4   Route Profile Results

Like coverage survey results, the processed output of a route survey is usually a data set containing GPS fixes and serving cell details, which can then be turned into a route map that indicates the progression of serving cells detected along a route.

Unlike coverage survey data, which details the locations at which a specific target cell was selected as serving, route profile data usually lists numerous serving cells, depending upon the length of the surveyed route and the progression of coverage encountered along that route. Table 9.18 provides a generic example of a route profile raw data table.

The raw data contained in such a table can then be converted into a more visual representation of serving coverage along a route by importing it into a mapping application such as Microsoft AutoRoute.

Depending upon the level of detail required in the route profile map, it may be necessary to import two separate data sets into each map:

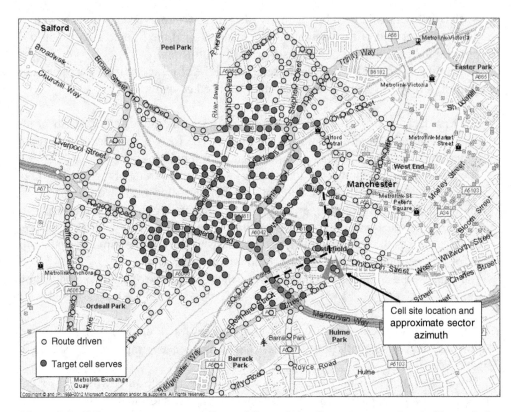

Figure 9.6 Cell coverage survey map. Source: Microsoft AutoRoute map reproduced with permission from Microsoft

Table 9.18 Example of route survey results raw data showing a succession of serving cells along a surveyed route.

Longitude	Latitude	Cell ID	Status (serving)
−1.462 905	52.521 084	27 165	0
−1.462 897	52.521 064	27 165	0
−1.462 894	52.521 057	6785	0
−1.462 871	52.521 062	6785	0
−1.462 862	52.521 067	6785	0
−1.462 859	52.521 076	6784	0
−1.462 848	52.521 071	6784	0
−1.462 834	52.521 068	6785	0
−1.462 832	52.521 055	6784	0
−1.462 832	52.521 051	6784	0
−1.462 825	52.521 043	6784	0
−1.462 817	52.521 039	33 145	0
−1.462 811	52.521 032	33 145	0
−1.462 802	52.521 028	33 145	0

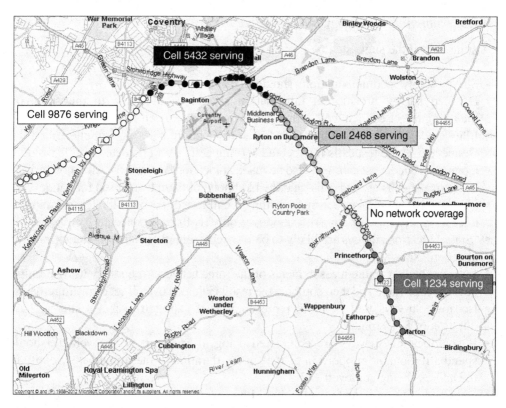

Figure 9.7   Route profile survey map. Source: Microsoft AutoRoute map reproduced with permission from Microsoft

1. The first data set shows the entire route driven during the survey and consists of just the GPS latitude/longitude data for the survey. The route would usually be represented in the map using small pushpins in a neutral colour.
2. The second data set shows the locations of survey measurements where cellular coverage was detected and should indicate the details of serving cell at each point. These locations would usually be shown on the map using a larger pushpin and a different colour to the route pins.

An example of a completed route profile survey map, created using Microsoft AutoRoute and where both levels of data have been imported, is shown in Figure 9.7.

This type of map is usually the form in which forensic radio survey product for route profile surveys is passed to the investigators or cell site analysts in a case and allows them to draw conclusions as to the parts of a route within which a target phone could have been located when making particular significant calls.

9.7.5 Summary of Survey Best Practice

A summary of the suggested best practice for forensic radio surveys is as follows:

- Survey safety recommendations should be followed at all times.
- Detailed survey preparation information should be available to the surveyor, including target cell list, cell addresses, azimuths and a cell locations map.
- Idle Mode should be employed for spot/location surveys, supplemented by Connected Mode test calls if expected serving cells aren't detected in Idle Mode.
- Connected Mode test calls can also be made at locations to prove specific points, such as the use of cells as 'end cells' and the likelihood of them being used for handovers.
- 3G test calls should last for up to 1 min (if call length is controllable) to provide a decent opportunity for any 'soft handover only' cells to be selected.
- 'Static' location surveys are likely to be less representative than non-static 'location coverage' surveys.
- Surveys should capture measurements of the same network/technology using multiple devices, if possible and then combine the results into one overall set of measurements.
- Location surveys will be more representative if captured over a relatively long duration of 5–10 min.
- If spot/location surveys are conducted using a drive survey, the vehicle should spend a reasonable proportion of the survey duration in the immediate vicinity of the address being surveyed.
- Expected but non-detected cells should be investigated, by visiting the cell site if necessary.
- Lock-on surveys could be undertaken if required (and if supported by the survey device) to capture Cell ID details of 3G/4G neighbours.
- Cell coverage surveys should be undertaken in Idle Mode only.
- Route Profile surveys should be undertaken in Idle Mode, supplemented by Connected Mode test calls in specific circumstances, such as to prove the use of 'end cells'.
- Contemporaneous notes should be taken and post-survey reports compiled.

9.8 Cell Site Analysis

Cell Site Analysis (or cell tower tracking as it is sometimes known in the United States) attempts to provide evidence of where a mobile phone may have been located when certain calls were made. It is a useful tool for investigators and is based on a combination of network-provided CDR (Call Detail Record) data and forensic radio survey results.

9.8.1 Limitations of Cell Site Analysis

Cell site evidence works best as supporting evidence.

On its own, cell site evidence is generally considered to be too open to interpretation to be used as the sole or the primary evidence in a case. There have been cases where the cell site evidence was so strong that it could be used as the primary evidence, but there are dangers inherent in using cell site evidence in this way.

9.8.2   Components of Cell Site Analysis

A summary of the main types of input and output information related to cell site analysis follows and deals with the input forms of source information – call records, cell address details and forensic radio survey results – and also looks at the output cell site analysis 'product' of cell site reports, call schedules and map presentations.

Call Detail Records

CDRs are produced every time a user makes or receives a call, sends or receives a text message or connects to a data service. Some network records might also be produced in relation to events such as attaches, location updates and detaches.

Network operators provide CDRs in a wide variety of different formats and the formats employed by different operators provide a variety of information. Generally each CDR contains the following:

- Date and Time of start of call
- Duration of call
- Type of Service, for example voice call, SMS, MMS, data, and so on
- Originating MSISDN (A number)
- Terminating MSISDN (B number)
- IMSI and IMEI of target phone
- Serving Cell ID and LAC (at Start of call)
- Serving Cell ID and LAC (at End of call) – not always provided
- Cell site names, postcodes, GPS coordinates or map reference – not always provided.

GPRS (or data) CDRs often use a different format but provide much the same level of information as voice CDRs and will also contain details of the IP address assigned to the phone for the data session.

Cell Address Details

As with CDR formats, operators have a wide variety of reporting formats used to supply cell address details, typically however they will supply some or all of the details listed below:

- Cell ID or full CGI (MCC, MNC, LAC, CI)
- Site name

- Site address and post/zip code
- Site GPS latitude/longitude or map grid reference
- Cell azimuth
- Cell technology (2G, 3G, 4G)
- Cell type – macro, micro, femto

Investigators often request details for multiple cells as part of the same enquiry and the results are often batched into one report or spreadsheet.

Cell Site Reports

The structure of a cell site report is largely down to the personal preference of the analyst or expert responsible for writing it, but most reports follow a similar pattern, the sections of which are outlined below:

- Case Details
- Cell Site Explanation
- Summary of Source Data
- Continuity Statement
- Significant Locations
- Forensic Survey Details
- Main Report Section
- Conclusions
- Summary
- Declaration.

Call Schedules

Cell site reports are usually enhanced by the preparation of call tables. These collate the relevant call records from the target phones in the case and present them in a combined and coherent document.

Each network operator provides call records in their own format, and the formats used by the operators can be very different. This makes it difficult to compare calls made by phones belonging to different networks, so cell site analysts often spend large proportions of their time processing (also known as 'normalising' or 'cleansing') call records into a common format. This process can be very time consuming if conducted manually (and is also open to inevitable human error), so many organisations have developed their own data processing macros or use a commercial data cleansing product such as Forensic Analytics CSAS (Cell Site Analysis Suite) software application to process call data automatically.

Call tables are usually prepared using Microsoft Excel and are often presented in court in A3-sized booklets.

Colours are usually assigned to each significant phone number and that colouration is employed to make the phones more identifiable in the call tables.

As cell site evidence works to identify the possible locations of a mobile device, rather than attempting to identify the user of that device, significant phone numbers are usually identified using a combination of the colour assigned to the phone and the last three or four digits of the mobile number, 'Blue-1234' or 'Red-2468', for example.

In cases that cover multiple phones, where several phones might be attributed to the same individual, analysts often assign a 'colour per person' to allow phones attributed to the same individual to be easily grouped and identified.

Maps and Graphics

Mapping and graphics presentations are often used to make the evidence presented in cell site reports simpler to understand.

Many experts use a set of graphical slides to provide juries with a basic overview of cell site concepts, which usually mirrors the information presented at the start of the expert's report and is often presented to the jury at the start of the expert's evidence.

The substantive part of a mapping presentation often begins with a slide that provides an overview of the significant locations in the case and their geographical relationship to each other.

An analyst will then usually produce a separate map to represent each batch of calls dealt with in the accompanying report. The map will be zoomed in to the general area of the cell sites used by each batch of calls and will typically have icons showing the cell locations, labels detailing the cell name and cell ID and call labels providing basic details of the calls under discussion. Cell icons and cell details labels are often coloured to match the attribution colour assigned to the target phone that used that cell – so, for example, the cell label for a cell used by the Blue-1234 phone would be coloured blue. Maps that detail the cell usage of several target phones might maintain this convention or might colour the cell labels in a neutral colour to indicate that they were used by multiple phones.

Call labels are often included in maps to tie the information presented in the map to the cell site report and the call schedule. Different experts and agencies favour their own preferred label format, but a representative label format might contain the following data:

- Call index number (based on call schedule numbering)
- Index number is coloured to show the target phone to which the event relates
- Time of call start
- Abbreviated 'other party' number
- Event type – SMS, Call, GPRS
- Indication of whether the event was outgoing (>) or incoming (<)

Some labelling formats also include details of whether the associated cell was the Start cell ('S') or the End cell ('E') for a call or whether the call started and ended ('S/E') on the same cell.

Further labels could be included to provide details of significant events that occur during the period covered by the map.

Additional graphics are used to provide explanations for complex concepts. For example, if cell coverage or route surveys were undertaken as part of the case, then maps illustrating the results of these surveys can also be included in the mapping presentation.

As mapping graphics to be displayed in court are often created using a presentational application such as Microsoft PowerPoint, the call and cell labels and icons are often animated so that they appear on the slide in the order in which the calls were made and the cells were used. The animated progression of call and cell details can make it easier for the court and for members of the jury to grasp the relationship between the calls and the significant events in the case.

9.8.3   Report Checking and Peer Review

Cell site reports can develop into enormously complex collections of documents, especially if a case involves multiple handsets, and it is to be expected that the writers and compilers of these reports will make at least one mistake somewhere within them.

It is therefore absolutely vital that each report is fully proofread and fact checked once it has been completed.

The main aspects that need to be checked include:

Case details and continuity information are complete and correct.

- Details of survey locations and survey results are complete and correct.
- Attribution and colouration details for each target phone are correct.
- Within the body of the report, the paragraphs relating to each examined group of calls should be checked to make sure that the call times, used cell details, location details and forensic survey results mentioned are correct.
- Each conclusion should be checked back against the source data to ensure that it is sound and is supported by the evidence.
- Call schedules should be checked back against the source CDR data to make sure that no errors were introduced during the 'cleansing' process.
- Maps should be checked to ensure that the marked locations of addresses and cell sites are correct, that cell labels contain the correct cell details, that the correct set of call labels are listed for each cell and that any azimuths are correct (both in the cell labels and in the orientation of the cell icons).

Once the report writer has fully checked (and, if necessary, corrected) their work the report should be passed to at least one equally qualified and competent peer reviewer, who should go through the whole checking process again.

9.9   End-to-End Process

The preceding sections of this chapter have recapped the basic concepts and techniques of cell site analysis in general and forensic radio surveying in particular. The following sections provide an example of the end-to-end process of compiling a cell site analysis report, followed by a detailed overview of the practical steps related to some of those activities.

The specific steps taken whilst compiling a cell site analysis report will be determined by the circumstances of the case and also by the analyst's own personal preferences, so the following should be taken as a generic guide rather than as a rigid specification. This list has also been compiled to reflect the activities related to producing a 'prosecution' report; the activities related to producing 'defence' reports are generally similar and will not be outlined here.

The practical activities highlighted in *italics* will be discussed in more detail in following sections:

1. Receive case details, descriptions of events, specific allegations and general instructions from case investigators.
2. Receive source CDRs and cell address details. Record details of each and capture file continuity information.
3. *Create a master list of significant locations and events.*
4. *Create a 'source files and attributions' list that links target phones to attributed individuals and shows the source data files received for each phone.*
5. *Normalise the source CDRs into a standard call schedule format.*
6. Assign identifying colours to each target phone number and add these to the call schedule.
7. Compare the call schedule to the case events timeline and, if necessary, cut the data down into POIs (Periods of Interest) and identify batches of calls within each POI that correspond to the timing of events and allegations in the case.
8. *Create a list of the cells used in the case, or in each POI.*
9. If necessary, request additional cell address details for any significant cells for which full details are not held.
10. Integrate returned cell address information into the call schedule and record continuity details of any received files.
11. *Create a case master map showing all locations and cell sites – typically, cell sites will be coloured to identify the target phone that used them.*
12. Perform an initial high-level cell site analysis to see how many allegations and alibis can be dealt with from a simple comparison of the used cell site to the significant location or alibi location.
13. If further low-level analysis is required, determine which locations, cells or routes require forensic radio surveys to be undertaken and compile the required survey specifications.
14. Undertake forensic radio surveys.
15. *Process survey results, compile survey summary tables and maps.*
16. Check that the survey results seem accurate and that all required information was captured – rerun surveys if information is missing.

17. Store or backup the raw survey results data in case it is requested later.
18. Combine all cell coverage survey data into one 'pool' before processing to ensure the best chance that any non-contiguous areas of coverage are captured.
19. Create an outline cell site report based on high level and low level analysis techniques, check to see that all case events/allegations can be discussed. Commission further surveys if required.
20. Begin to create final cell site report.
21. Finalise call schedule, making sure that schedule has been cut down into POIs, that all cell address details and phone colouration have been added and that call numbering (i.e. the index number assigned to each call within the schedule) is complete and contiguous.
22. *Create call and cell labels from the call schedule to be used in PowerPoint map presentations.*
23. *Create cell site mapping presentation – each map typically details one batch of calls showing the potential location of target phone(s) at a significant address, showing travelling between locations or showing a general pattern of calls within a period of time.*
24. Produce final version of cell site report that references the relevant call index numbers and map slides for each batch of calls.
25. Proof read the report and check the conclusions.
26. Have report peer reviewed.
27. Final corrections and amendments.
28. Sign report declaration and send to investigators.
29. Save all case material, even unused material, plus any notes, survey results, draft copies of reports and so on and keep ready to be produced as part of case discovery.

9.10   Master List of Events/Locations

Each case will have one or more significant events that occur at one or more significant locations. The main role of cell site analysis is to attempt to determine whether any of the target phones in a case could have been at or near those locations at around the time of the events.

It is therefore beneficial to have a coherent summary of events and locations to work from, as illustrated in Table 9.19.

9.11   Source Files Attribution List

A significant proportion of cell site cases involve multiple target phones and analysts may be provided with a large number of source data files. To ensure that file continuity information is correctly and coherently captured, it is recommended that report writers compile a list that makes clear the association between each target phone, the attributed user and the source data files the writer has relied upon when drawing conclusions related to each phone.

An example of a typical 'phones/attributions' table is shown in Table 9.20.

Table 9.19   Example of locations/events list.

Locations

Ref.	Significance	Address	Map Ref./GPS
A	Smith H/A	14 Preston Hill, London	51.5842/ −0.2859
B	Jones H/A	292 Chapter Rd, London	51.5509/ −0.2401
C	Meeting	Flat 16, 24 Matlock St	51.5159/ −0.0403
D	Offence	Car Park, Red Lion pub	51.5166/ −0.0470

Events

Ref.	Times	Details	Location
1	08:30	Smith leaves home address	A
2	11:45	Jones leaves home address	B
3	20:00–22:00	Meeting at Flat 16, 24 Matlock St	C
4	21:45	Call from Williams to Cooper (Victim)	C
5	22:30	Assault at Red Lion pub car park	D

N.B. 'H/A' in Table 9.19 is an abbreviation of 'Home Address'.

Table 9.20   Example 'phones/attributions' table.

Attribution	Phone	Filename
John Smith	07700 345876	345876 voice in/out.csv
		345876 GPRS.csv
		345876 cell addresses.xls
Paul Jones	07700 198019	198019 voice in/out.csv
		198019 GPRS.csv
		198019 cell addresses.xls
		198019 subscriber check.doc
	07700 300981	300981 voice in/out.csv
Peter Williams	07700 127610	127610 GPRS.csv
		127610 MMDS.csv
		127610 voice in/out.csv
Jim Johnson	07700 019010	019010 voice in.csv
		019010 voice out.csv
		019010 subscriber check.xls
		019010 cell addresses.xls

9.12   Normalise Call Data into a Standard Format

The key source data for a cell site investigation is contained in CDR files.

Historically, networks have compiled and retained CDR data for two main purposes: to provide proof of network usage if a customer disputes the charges that have been applied to their account; to provide an indication of the cell sites used to carry calls in case customers make complaints about poor quality or dropped calls.

CDRs were not originally designed to act as a source of investigative data for cell site analysts; they were designed for internal administrative use within each operator. Networks are therefore generally free to employ whichever format best suits their purposes and most employ formats that differ from those used by other operators.

If an investigator or cell site analyst is working on a case where all of the target phones belong to the same network, then all of the CDRs can be expected to be in the same format and it is comparatively simple to compare the usage of the phones against each other.

In cases where phones belong to different networks and where those networks employ different CDR formats, it can be very difficult to perform 'cross-phone' analysis, as the data for the different phones will be presented in different ways.

It is therefore generally necessary to process or 'normalise' CDR data into a common format before analysis can take place. This process is also known as 'cleansing' CDRs and there are a variety of ways of achieving it.

Larger organizations often develop macros or applications to automatically normalise CDR data into a common format, there are also commercial products available, such as the Forensic Analytics CSAS tool, that perform this function. For the majority of analysts, however, this process must be achieved manually by using Microsoft Excel to manipulate the CDR data.

An example of just such a manual 'normalisation' process is outlined below.

9.12.1   CDR Normalisation

This example will show the normalisation of two CDRs, one provided by the NationalCell network and the other by the MetroCell network (both are fictitious).

The NationalCell CDR format shows the following data columns:

- Date of event (in dd/mm/yyyy format, e.g. 02/04/2014)
- Time of event start (in hh:mm:ss format, e.g. 17:34:12)
- A and B Numbers (in international format, e.g. 447700 123456)
- Event type (event descriptions – incoming/outgoing voice, incoming/outgoing text, GPRS)
- Duration (in hh:mm:ss format, e.g. 00:01:39)
- IMEI and IMSI of target phone
- Start and End cell details (Cell ID, LAC, Site Name).

An example of this format is shown in Figure 9.8.

The MetroCell CDR format shows the following data columns:

- Event type (descriptions incoming/outgoing call, incoming/outgoing SMS, data)
- Event date (in dd-mmm-yy format, e.g. 02-Apr-14)
- Event time (in h:mm:ss AM/PM format, e.g. 5:34:12 PM)
- A, B and C numbers (in domestic format, e.g. 07700 123456)

NationalCell Network Disclosure Management System

MS-ISDN: 07700 345876
Records Requested: Voice, SMS In & Out
Records Start Date/Time: 14/01/2014 00:00
Records End Date/Time: 14/01/2014 23:59
Request Date: 01/05/2014
Request By: DC Davies

Date	Time	A Number	B Number	Call Type	Duration	IMEI	IMSI	Start Cell ID	Start LAC	Start Site Name	End Cell ID	End LAC	End Site Name
14/01/2014	00:15:04	447700345876	447700198019	Outgoing Voice	00:01:24	00145764521 3254	234950867523412	8173	124	Kingsbury West	8173	124	Kingsbury West
14/01/2014	03:34:56	447700127610	447700345876	Incoming Voice	00:00:00		234950867523412						
14/01/2014	03:57:06	447700345876	447700345876	Incoming Voice	00:00:00		234950867523412						
14/01/2014	08:17:26	447700345876	123	Outgoing Voice	00:02:13	0014576452 13254	234950867523412	8173	124	Kingsbury West	8173	124	Kingsbury West
14/01/2014	08:22:45	447700345876	447700127610	Outgoing Voice	00:07:16	0014576452 13254	234950867523412	8173	124	Kingsbury West	8173	124	Kingsbury West
14/01/2014	08:34:19	447700345876	447700198019	Outgoing Voice	00:21:29	0014576452 13254	234950867523412	8173	124	Kingsbury West	8173	124	Kingsbury West
14/01/2014	09:10:12	447700198019	447700345876	Incoming Voice	00:14:28	0014576452 13254	234950867523412	8296	124	Kingsbury Heights	8297	124	Kingsbury Heights
14/01/2014	09:16:33	447700127610	447700345876	Incoming SMS	00:00:00	0014576452 13254	234950867523412	8297	124	Kingsbury Heights			
14/01/2014	10:02:17	447700345876	447700067185	Outgoing SMS	00:00:00	0014576452 13254	234950867523412	21276	298	London Bridge			
14/01/2014	13:47:23	447700127610	447700127610	Outgoing SMS	00:00:00	0014576452 13254	234950867523412	6091	300	Waterloo			
14/01/2014	14:02:18	447700345876	447700127610	Outgoing Voice	00:04:11	0014576452 13254	234950867523412	23189	300	Globe	23190	300	Globe
14/01/2014	14:34:17	447700127610	447700345876	Incoming SMS	00:00:00	0014576452 13254	234950867523412	23189	300	Globe			
14/01/2014	15:04:19	447700198019	447700345876	Incoming Voice	00:00:12	0014576452 13254	234950867523412	21276	298	London Bridge	6091	300	Waterloo
14/01/2014	16:23:19	447700886135 6	447700345876	Incoming SMS	00:00:00	0014576452 13254	234950867523412	6091	300	Waterloo			
14/01/2014	16:28:07	447700345876	447700127610	Outgoing Voice	00:01:38	0014576452 13254	234950867523412	11398	301	Oxford St	11398	301	Oxford St
14/01/2014	16:55:24	447700345876	447700019010	Outgoing Voice	00:21:16	0014576452 13254	234950867523412	6091	300	Waterloo	6091	300	Waterloo
14/01/2014	17:04:12	447700019010	447700345876	Incoming SMS	00:00:00	0014576452 13254	234950867523412	44231	217	Westminster Tower			
14/01/2014	18:26:55	447700345876	447700198019	Outgoing SMS	00:00:00	0014576452 13254	234950867523412	3981	207	St Catherines Dock			
14/01/2014	19:01:00	447700127610	447700345876	Incoming Voice	00:07:23	0014576452 13254	234950867523412	6116	221	East St	28760	221	London East
14/01/2014	20:14:49	447700886135	447700345876	Incoming Voice	00:01:01	0014576452 13254	234950867523412	28761	221	London East	28761	221	London East
14/01/2014	20:15:59	447700345876	442079460187	Outgoing SMS	00:00:00	0014576452 13254	234950867523412	28761	221	London East			
14/01/2014	20:44:15	447700345876	442079460187	Outgoing SMS	00:00:00	0014576452 13254	234950867523412	28761	221	London East			
14/01/2014	20:07:22	447700886135	447700886135	Outgoing Voice	00:16:12	0014576452 13254	234950867523412	28761	221	London East	28761	221	London East
14/01/2014	21:38:11	447700019010	447700345876	Incoming SMS	00:00:00	0014576452 13254	234950867523412	28761	221	London East			
14/01/2014	22:07:16	447700127610	447700127610	Incoming Voice	00:00:24	0014576452 13254	234950867523412	31008	221	West Ham	31009	221	West Ham
14/01/2014	22:58:04	447700345876	447700019010	Outgoing SMS	00:00:00	0014576452 13254	234950867523412	7816	202	Royal Docks			
14/01/2014	23:09:17	447700019010	447700345876	Incoming SMS	00:00:00	0014576452 13254	234950867523412	7817	202	Royal Docks			
End of Data													

filename: 345876 voice in/out.csv

Figure 9.8 NationalCell CDR format

METROCELL
MS-ISDN: 07700 228671
Records Requested: Voice In/Out
Records Start Date/Time: 02/41/2014 00:00
Records End Date/Time: 14/01/2014 23:59
Request Date: 01/05/2014
Request By: DC Davies

Event Type	Date	Time	A Number	B Number	C Number	Duration	Start Cell ID	Start Site Name	Postcode	Map Ref	End Cell ID	End Site Name	Postcode	Map Ref
Outgoing Call	02-Apr-14	2:12:45 AM	07700228671	02079460164		14	122-2009	Rotherhithe	SE4 1RT	412341 651762	122-2009	Rotherhithe	SE4 1RT	412341 651762
Outgoing SMS	02-Apr-14	8:17:10 AM	07700228671	02079460164		00	122-2009	Rotherhithe	SE4 1RT	412341 651762	122-2009	Rotherhithe	SE4 1RT	412341 651762
Outgoing Call	02-Apr-14	8:22:45 AM	07700228671	07700228671	1417700228671	184	176-165	Green Park	W1B 6YR	426716 612846	176-165	Green Park	W1B 6YR	426716 612846
Incoming Call	02-Apr-14	9:01:01 AM	07700285617	07700228671		12	220-6577	Bermondsey	SE5 9SG	412467 652987	220-6577	Bermondsey	SE5 9SG	412467 652987
Outgoing Call	02-Apr-14	9:02:56 AM	07700228671	07700285617		03	220-6577	Bermondsey	SE5 9SG	412467 652987	220-6577	Bermondsey	SE5 9SG	412467 652987
Incoming SMS	02-Apr-14	9:25:51 AM	02079460164	07700228671		00	176-165	Green Park	W1B 6YR	426716 612846	176-165	Green Park	W1B 6YR	426716 612846
Outgoing Call	02-Apr-14	10:43:12 AM	07700228671	07700285617		276	122-2009	Rotherhithe	SE4 1RT	412341 651762	122-2009	Rotherhithe	SE4 1RT	412341 651762
Outgoing Call	02-Apr-14	11:13:34 AM	07700228671	02079460123		872	176-1627	Regent's Park	W1C 3JH	426789 612587	176-1627	Regent's Park	W1C 3JH	426789 612587
Outgoing Call	02-Apr-14	11:58:19 AM	07700228671	07700285617		19	314-4142	London Bridge Stn	SE1A 7TY	412587 653198	314-4142	London Bridge Stn	SE1A 7TY	412587 653198
Voicemail/Divert	02-Apr-14	12:00:04 PM	02079460164	07700228671	1417700228671	24	314-897	Waterloo Underground	SE1B 9DG	412501 653456	314-897	Waterloo Underground	SE1B 9DG	412501 653456
Incoming SMS	02-Apr-14	1:16:27 PM	07700285617	07700228671		00	66-1143	Aston St	E3 4HB	414578 667819	66-1143	Aston St	E3 4HB	414578 667819
Outgoing Call	02-Apr-14	2:18:32 PM	07700228671	02079460123		45	66-1143	Aston St	E3 4HB	414578 667819	66-1143	Aston St	E3 4HB	414578 667819
Outgoing Call	02-Apr-14	5:55:37 PM	07700228671	02079460164		310	66-3029	Harbour Wharf	E2 7DS	415699 667001	66-3029	Harbour Wharf	E2 7DS	415699 667001
Voicemail/Divert	02-Apr-14	8:38:01 PM	07700285617	07700228671	1417700228671	47	66-165	Green Park	W1B 6YR	426716 612846	66-165	Green Park	W1B 6YR	426716 612846
Incoming SMS	02-Apr-14	8:45:43 PM	02079460164	07700228671		00	122-2009	Rotherhithe	SE4 1RT	412341 651762	122-2009	Rotherhithe	SE4 1RT	412341 651762
End of Record														

Figure 9.9 MetroCell CDR format

- Duration (in seconds, e.g. 99)
- Start/End cell (in LAC-Cell ID format)
- Start/End site name
- Start/End site postcode
- Start/End site map coordinates.

An example of this format is shown in Figure 9.9.

In order to normalise these different data formats it is necessary to select a preferred 'standard' format that includes both column layout and data format choices.

The normalised format for this example will be:

- Call index number
- Date (dd/mm/yyyy)
- Time (hh:mm:ss)
- Event type (descriptions incoming/outgoing call, incoming/outgoing SMS, data)
- A/B/C numbers (in international format)
- Duration (in hh:mm:ss)
- Start/End LAC-Cell ID
- Start/End Site Name
- Start/End Site Postcode
- Start/End Site Map Reference.

An example of this format is shown in Figure 9.10.

In this example, the NationalCell data will require the least amount of work to normalise it, whereas the MetroCell data will need to be substantially reformatted.

Normalisation Activities – NationalCell

The following steps will be taken to normalise the NationalCell data:

- Column A – Call index number
- Column B – Date – no change to existing format
- Column C – Time – no change to existing format
- Column D – Event Type – change descriptions to match standard descriptions (e.g. change 'incoming voice' to 'incoming call')
- Column E – A number – no change to existing format
- Column F – B number – no change to existing format
- Column G – C number – leave blank (NationalCell CDRs do not provide C number details)
- Column H – Duration – no change to existing format
- Column I – Start Cell LAC and Cell ID – combine separate LAC and Cell ID columns together

CSAS — Call Schedule Blue-8671

Index	Date	Time	Event Type	A Number	B Number	C Number	Duration	Start Cell ID	Start Site Name	Postcode	Map Ref	End Cell ID	End Site Name	Postcode	Map Ref
1	02/04/2014	02:12:45	Outgoing Call	447700228671	442079460164		14	00:00:14	122-2009	Rotherhithe	SE4 1RT	412341 651762	122-2009	Rotherhithe	SE4 1RT	412341 651762
2	02/04/2014	08:17:10	Outgoing SMS	447700228671	442079460164		00	00:00:00	122-2009	Rotherhithe	SE4 1RT	412341 651762	122-2009	Rotherhithe	SE4 1RT	412341 651762
3	02/04/2014	08:22:45	Outgoing Call	447700228671	447700228671	44141770028671	184	00:03:04	176-165	Green Park	W1B 6YR	426716 612846	176-165	Green Park	W1B 6YR	426716 612846
4	02/04/2014	09:01:01	Incoming Call	447700255617	447700228671		12	00:00:12	220-4577	Bermondsey	SE5 9SG	412467 652987	220-4577	Bermondsey	SE5 9SG	412467 652987
5	02/04/2014	09:02:56	Outgoing Call	447700228671	447700285617		03	00:00:03	220-4577	Bermondsey	SE5 9SG	412467 652987	220-4577	Bermondsey	SE5 9SG	412467 652987
6	02/04/2014	09:25:51	Incoming SMS	442079460164	447700228671		00	00:00:00	176-165	Green Park	W1B 6YR	426716 612846	176-165	Green Park	W1B 6YR	426716 612846
7	02/04/2014	10:43:12	Outgoing Call	447700228671	447700285617		276	00:04:37	122-2009	Rotherhithe	SE4 1RT	412341 651762	122-2009	Rotherhithe	SE4 1RT	412341 651762
8	02/04/2014	11:13:34	Outgoing Call	447700228671	442079460123		872	00:14:32	176-1627	Regent's Park	W1C 3JH	426789 612587	176-1627	Regent's Park	W1C 3JH	426789 612587
9	02/04/2014	11:58:19	Outgoing Call	447700228671	447700285617		19	00:00:19	314-4142	London Bridge Stn	SE1A 7TY	412587 653198	314-4142	London Bridge Stn	SE1A 7TY	412587 653198
10	02/04/2014	12:00:04	Voicemail/Divert	442079460164	447700228671	44141770028671	24	00:00:24	314-897	Waterloo Underground	SE1B 9DG	412501 653456	314-897	Waterloo Underground	SE1B 9DG	412501 653456
11	02/04/2014	13:16:27	Incoming SMS	447700285617	447700228671		00	00:00:00	66-1143	Aston St	E3 4HB	414578 667819	66-1143	Aston St	E3 4HB	414578 667819
12	02/04/2014	14:18:32	Outgoing Call	447700228671	442079460123		45	00:00:45	66-1143	Aston St	E3 4HB	414578 667819	66-1143	Aston St	E3 4HB	414578 667819
13	02/04/2014	17:55:37	Outgoing Call	447700228671	447700228671		310	00:05:10	66-3029	Harbour Wharf	E2 7DS	415699 667001	66-3029	Harbour Wharf	E2 7DS	415699 667001
14	02/04/2014	20:38:01	Voicemail/Divert	447700285617	447700228671	44141770028671	47	00:00:47	66-165	Green Park	W1B 6YR	426716 612846	66-165	Green Park	W1B 6YR	426716 612846
15	02/04/2014	20:45:43	Incoming SMS	442079460164	447700228671		00	00:00:00	122-2009	Rotherhithe	SE4 1RT	412341 651762	122-2009	Rotherhithe	SE4 1RT	412341 651762

CSAS Call Schedule

Figure 9.10 Normalised format. Source: CSAS call schedule reproduced with permission from Forensic Analytics Ltd

- Column J – Start Site Name – no change to existing format
- Column K – Start Site Postcode – leave blank, will need to be requested from network operator with additional cell address data
- Column L – Start Site Map reference – leave blank, will need to be requested from network operator with additional cell address data
- Column M – End Cell LAC and Cell ID – combine separate LAC and Cell ID columns together
- Column N – End Site Name – no change to existing format
- Column O – End Site Postcode – leave blank, will need to be requested from network operator with additional cell address data
- Column P – End Site Map reference – leave blank, will need to be requested from network operator with additional cell address data.

Normalisation Activities – MetroCell

The following steps will be taken to normalise the MetroCell data:

- Column A – Call index number
- Column B – Date – change from dd-mmm-yy to dd/mm/yyyy format
- Column C – Time – change from h:mm:ss AM/PM to hh:mm:ss format
- Column D – Event Type – no change to existing format
- Column E – A number – change to international format
- Column F – B number – change to international format
- Column G – C number – change to international format
- Column H – Duration – convert from seconds to hh:mm:ss (in Excel the way to do this is to multiple the raw seconds value by 0.0000115740740740741, then format the column as Time = hh:mm:ss)
- Column I – Start Cell LAC and Cell ID – no change to existing format
- Column J – Start Site Name – no change to existing format
- Column K – Start Site Postcode – no change to existing format
- Column L – Start Site Map reference – no change to existing format
- Column M – End Cell LAC and Cell ID – no change to existing format
- Column N – End Site Name – no change to existing format
- Column O – End Site Postcode – no change to existing format
- Column P – End Site Map reference – no change to existing format.

Combined and Normalised Data

Once call data has been normalised into a common format it becomes much simpler to query the data.

It is common for analysts to normalise each CDR file separately, so that data for each phone is available in a standardised form, but then to also combine all of the individual files together to create an 'all phones' combined call schedule. Combined sheet are usually sorted by Date and then by Time to get all calls into the correct chronological order.

Each individual call data file will have its own set of index numbers for calls and these are preserved when the all-phones table is put together, but each call in the 'all-phones' table then also has a 'combined' call index number applied to it as well. This allows for consistent call numbering throughout a report, but also allows calls to be cross-referenced back to their source CDR file if necessary.

The call index number column is often coloured to match the assigned colour of the target phone that each call relates to.

9.13   Create a Master Cell List

Another benefit of having normalised call data is that it puts all of the various cell ID and cell address details used by different operators into a common format too.

Once the call data has been normalised it is possible to extract details related to the cells that are associated with the case and compile them into a master cell list, which can then be used to create accurate maps showing the cell locations.

It is common to create separate master cell lists for each target phone – this allows the data for each phone to imported into a map individually, so that cell locations for each phone can be shown with the appropriate colour of dot (that matches the attribution colour assigned to the associated phone). For example, if a case has 'red', 'blue' and 'green' target phones, then the analyst would import the cell list for the red phone and colour the resulting pushpins red, then do the same for the blue phone/blue pushpins and green phone/green pushpins.

An example of a master cell list is shown in Table 9.21.

Table 9.21   Master cell list.

Cell ID	Site name	Azimuth	Map Reference
11 876	Barchester East	0	412 654 – 345 241
11 877	Barchester East	120	412 654 – 345 241
11 878	Barchester East	240	412 654 – 345 241
19 087	Wimborne	110	408 176 – 309 876
20 016	South St Haswell	120	309 816 – 009 876
20 017	South St Haswell	225	156 981 – 300 914
23 956	Norton	90	561 781 – 890 145
28 431	Wells Insurance Building	20	418 091 – 510 098
30 645	Camelford	0	217 689 – 900 816
31 221	Manchester Piccadilly	120	513 241 – 819 372
54 098	Repton	330	614 360 – 104 682
55 100	Derby West 3	300	519 352 – 228 292

The format of the 'map ref' shown in Table 9.21 is an example of the OS (Ordnance Survey) grid reference employed in the United Kingdom, although the specific map references shown in the table are fictitious. Other countries and regions will employ different map reference standards. Some operators might substitute GPS latitude/longitude details or post/zip codes instead of map references. As long as the geographical method employed is capable of identifying the location of the site on a map to a reasonable level of accuracy then it doesn't really matter which reference scheme is employed.

9.14   Creating a Case Master Map

A master map showing the relationship between all of the significant locations in a case and the cells that were used by the target phones provides a powerful tool for analysts.

Not only does it provide the basis for an initial high-level analysis of the details in a case but it also serves as the basis for any detailed, low-level analysis maps that might need to be created when compiling the cell site analysis report.

There are multiple computer-based mapping applications available as well as numerous Internet-based mapping sites, but among the most commonly used (at least in regions that are covered by versions of the product) are Microsoft's AutoRoute (which is also known as Streets and Trips in the United States) and MapPoint products.

The following mapping examples will be based on the use of Microsoft AutoRoute 2013, but similar activities will be required for whichever mapping application is used.

This example will be based on a case that has five target phones and four attributed users; the details of these users is given in Table 9.22.

The case also has an Events/Locations table, as detailed in Table 9.23, that has been created as a Microsoft Excel file. The case events details could be stored in a sheet called 'Events' and locations in a sheet called 'Locations'.

The steps involved in creating a master cell map (based on using Microsoft AutoRoute 2013) are as follows:

- Open a new Microsoft AutoRoute map and save it as 'Master Cell Map v1'.
- Go to Data > Import Data Wizard.

Table 9.22   Phone attribution table for mapping example.

Defendant	Attributed phone	Known as	Network
John Smith	07700 345876	Red-5876	NationalCell
Paul Jones	07700 198019	Blue-8019	CellPlus
	07700 300981	Blue-0981	CellPlus
Peter Williams	07700 127610	Green-7610	NationalCell
Jim Johnson	07700 019010	Orange-9010	NationalCell

Table 9.23   Example case locations file.

Ref.	Significance	Address	Latitude	Longitude
A	Smith H/A	14 Preston Hill, London	51.584 2	−0.285 9
B	Jones H/A	292 Chapter Rd, London	51.550 9	−0.240 1
C	Meeting	Flat 16, 24 Matlock St	51.515 9	−0.040 3
D	Offence	Car Park, Red Lion pub	51.516 6	−0.047 0

- Select the Excel file that contains case 'Events/Locations' list (e.g. 'events and locations.xls'), then identify the worksheet within the file that contains the required list (e.g. 'locations').
- The 'import data wizard' will provide a list of the data found on the selected sheet. The column headers from the file must be matched up with the associated data type – for example, the Ref column would be matched to the 'Name' data type, Significance to 'Name 2', Address to 'Address 1' and Latitude and Longitude to the corresponding descriptions. This instructs AutoRoute as to how to interpret each column of source data.
- When the data type mapping is complete, hit 'Finish' to start the data import process.
- The case locations will be plotted on the map using the default pushpin type. AutoRoute supports alphabetical pushpins, so it is possible to swap the generic pushpin for each location to one that matches the location's reference letter. To do this, find the icon for Location A, click on the icon to activate it and then right click to bring up the context menu. Select properties and then click the dropdown arrow next to the 'Symbol' box. Select the required icon (in this case 'A') from the icon list, then click 'OK'. Repeat this process with the appropriate letters for the remaining location icons.
- With the case locations mapped, it is time to import the case cell locations on a phone by phone basis.
- Repeat the 'Import Data Wizard' process but this time select the file and sheet that contain the master cell list for the Red phone.
- Set the Cell ID column as 'Name', the Site name column as 'Name 2' and the cell location details as 'OS Grid Reference' or latitude/longitude, whichever method is employed.
- Once the data is imported use the 'Edit Pushpins' method to select a red dot icon to represent the cell icons for the Red phone. Change the pushpin properties name to 'Red-5876' to clearly identify the set of icons that relate to that target phone.
- Repeat for the Blue, Green and Orange phones.

At the end of this process an analyst will have an enormously useful, case-specific reference tool available, which will form the basis of most of the analytical activities undertaken as part of the case.

9.15   Compile Radio Survey Summary Tables

The raw results of a forensic radio survey are typically contained in a data table that: lists the set of measurement events that were captured during the survey; shows the GPS fix for each event; and provides details of the serving and neighbour cells observed at that time.

An example of a raw data table (taken from the FMS CSU-4 L device) is shown in Figure 9.11.

In the example in Figure 9.11, each event row displays details of each detected cell in a cluster of data types, including – Status (serving or neighbour), MCC, MNC, TAC (as this is a set of 4G LTE measurements), Cell ID, PCI, EARFCN, RSRP and RSRQ.

Each event row will contain as many clusters as there were detected cells, so there could conceivably be 10 or more clusters per row in urban areas. The number of event rows captured will be proportional to the length of the survey and the interval between measurements events; if the survey equipment captures on event per second and the survey lasts for 10 min then there will be around 600 rows of data. So a typical spot/location survey could contain 6000 separate clusters of cell readings.

To produce a summary table it is necessary to calculate average values for each cell ID that was detected. There are applications available that will automatically process raw survey data and produce summary tables, the Forensic Analytics CSAS tool for example, and many organisations have developed their own macros to perform the same function.

The following sections present details of how to process a raw results table into usable data for spot/location, cell coverage and route profile surveys. Each section provides an overview of the manual process that could be followed to compile a set of survey summary results and is based on the assumption that the raw data to be summarised matches the format of the data in Figure 9.11. It is further assumed that the raw results data is contained within a Microsoft Excel workbook.

Spot/Location Survey Data

- The raw results data will be contained in a worksheet, which is referred to below as the 'tabulated values' worksheet.
- Create a new sheet in the survey spreadsheet file and paste in a copy of the survey summary template to be used; name the new sheet 'Survey Summary'. An example of a survey summary template is shown in Figure 9.12 below.
- From the tabulated values worksheet, copy only the GPS latitude/longitude data columns (columns C and D in the example in Figure 9.11) to a new tab, name the tab 'GPS'.
- Sort the GPS data by latitude and delete any rows that have '0' values for latitude, then repeat the process for the longitude column. This will remove any failed location measurements from the data set.

Date	Time	Latitude	Longitude	Test	RAT	Type0	MCC0	MNC0	TAC0	Cell-Id0	PCI	DlEARFCN0	RSRP0	RSRQ0	Type1	MCC1	MNC1	TAC1	Cell-Id1	PCI	DlEARFCN1	RSRP1	RSRQ1
22.04.14	13:53:44	52.4876	-2.0727	idle	LTE	S	234	30	11772	2613506	147	1617	-105	-11	N	234	30			293	1667	-105	-12
22.04.14	13:53:45	52.4876	-2.0726	idle	LTE	S	234	30	11772	2613506	147	1667	-105	-11	N	234	30			293	1617	-105	-13
22.04.14	13:53:46	52.4876	-2.0724	idle	LTE	S	234	30	11772	2613506	147	1667	-105	-11	N	234	30			293	1617	-105	-12
22.04.14	13:53:47	52.4875	-2.0723	idle	LTE	S	234	30	11772	2613506	147	1667	-105	-10	N	234	30			293	1617	-105	-11
22.04.14	13:53:48	52.4875	-2.0721	idle	LTE	S	234	30	11772	2613506	147	1667	-98	-9	N	234	30			293	1617	-105	-12
22.04.14	13:53:49	52.4875	-2.0720	idle	LTE	S	234	30	11772	2613506	147	1667	-98	-9	N	234	30			293	1617	-105	-11
22.04.14	13:53:50	52.4874	-2.0718	idle	LTE	S	234	30	11772	2613506	147	1667	-98	-10	N	234	30			293	1617	-105	-12
22.04.14	13:53:51	52.4874	-2.0716	idle	LTE	S	234	30	11772	2613506	147	1667	-98	-9	N	234	30			293	1617	-105	-11
22.04.14	13:53:52	52.4874	-2.0714	idle	LTE	S	234	30	11772	2613506	147	1667	-98	-8	N	234	30			293	1617	-105	-11
22.04.14	13:53:53	52.4874	-2.0713	idle	LTE	S	234	30	11772	2613506	147	1667	-98	-9	N	234	30			293	1617	-105	-12
22.04.14	13:53:54	52.4873	-2.0711	idle	LTE	S	234	30	11772	2613506	147	1667	-98	-9	N	234	30			293	1617	-105	-12
22.04.14	13:53:55	52.4873	-2.0709	idle	LTE	S	234	30	11772	2613506	147	1667	-98	-10	N	234	30			293	1617	-105	-11
22.04.14	13:53:56	52.4873	-2.0707	idle	LTE	S	234	30	11772	2613506	147	1667	-98	-8	N	234	30			293	1617	-105	-12
22.04.14	13:53:57	52.4872	-2.0705	idle	LTE	S	234	30	11772	2613506	147	1667	-98	-9	N	234	30			293	1617	-105	-11
22.04.14	13:53:58	52.4872	-2.0703	idle	LTE	S	234	30	11772	2613506	147	1667	-98	-10	N	234	30			293	1617	-105	-11
22.04.14	13:53:59	52.4872	-2.0701	idle	LTE	S	234	30	11772	2613506	147	1667	-98	-9	N	234	30			293	1617	-105	-10
22.04.14	13:54:00	52.4871	-2.0699	idle	LTE	S	234	30	11772	2613506	147	1667	-98	-10									
22.04.14	13:54:01	52.4871	-2.0697	idle	LTE	S	234	30	11772	2613506	147	1667	-98	-9									
22.04.14	13:54:02	52.4871	-2.0695	idle	LTE	S	234	30	11772	2613506	147	1667	-98	-9									
22.04.14	13:54:03	52.4870	-2.0694	idle	LTE	S	234	30	11772	2613506	147	1667	-98										

Figure 9.11 Radio survey results table. Source: FMS CSU-4 L raw results table reproduced with permission from FMS Ltd

CSAS

<Case name>

Spot/Location Survey Summary

Location	A - 14 High St		
Ave GPS	Lat: 52.123	Long: -1.2345	Height: 110m

Network	Technology	# of Test Calls
Vodafone UK	2G	3

Survey Results

Cell ID	LAC/TAC	Channel/PCI	Ave. Strength	Description	Detected (Serving)	Serving Percentage	Azimuth	Site Name	Served in Call
12345	345	654 12	-84.12	Strong	123 (123)	100%	120	Town Hall	Yes
2468	345	101 14	-94.32	Moderate	123 (0)	0%	240	Town Hall	. No

source filename - C:/surveys/operation cassidy/feb 2014/acdfacdfacdf.csv

Surveyor	Jim Smith	Organisation	RF Group Ltd
Survey Date	01/02/2014 13:45-13:55	Equipment	Csurv MuNST
Processed By	Jill Jones	Date	02/02/2014 15:12

Comments: survey conducted 15m from front door, with walk survey around perimeter of building after 5 minutes

Figure 9.12 Survey summary table. Source: CSAS survey summary table reproduced with permission from Forensic Analytics Ltd

- Sum the remaining values in the Latitude column and then divide by the number of rows of GPS data, this provides an 'average latitude' value. Repeat the process of the longitude column to obtain an 'average longitude'. These values can then be copied to the appropriate boxes in the survey summary template
- From the tabulated values sheet, copy the cell readings clusters (but not the GPS or event header data) to a new tab, name the tab 'cell readings working data'.
- The next step can be quite time consuming but must be performed accurately. In its raw form, the tabulated data is held in a table that is X number of clusters wide by Y number of rows high. It needs to be reformatted into a table that is one cluster wide and $X \times Y$ rows high, so in other words it is necessary to move all of the clusters into one set of cluster data columns.
- Once this has been achieved, the data could be sorted by Cell ID to group the clusters that relate to the same cell. However, there may be instances where cell measurements were captured but a cell ID was not (especially in the case of 3G and 4G neighbour cells with some types of survey equipment), so grouping by cell ID might result in lost measurements. It is instead more effective to sort the data first by Channel Number and then by physical layer ID, so in the case of 4G data this would be sorted by EARFCN and then by PCI (or ARFCN/BSIC for 2G, UARFCN/ PSC for 3G). As each cell is associated with one channel/PCI pair this approach allows measurements related to the same cell to be grouped even if a cell ID was not captured.
- Once the data has been grouped by channel/PCI it should be possible to copy and paste each group of measurement clusters to their own worksheet. Each new worksheet should be named for the channel and PCI (e.g. 1617-147) for which it contains data.
- Within each cluster's worksheet it should then be possible to calculate average values for signal strength readings (in the 4G case, this would be average values for RSRP and RSRQ).
- First, a count of measurement rows for each cell should be calculated. This will provide the basis for the 'number of times detected' count for each cell. This figure can then be split (using the Status field) to show the number of times the cell was detected as serving and the number of times as neighbour. These three values (total detected, serving detected and neighbour detected) should be retained to be used on the summary sheet entry for that cell.
- As with the GPS averaging, any empty values for the signal strength measurements need to be removed before the data columns can be summed. Once the empty values have been removed, the average signal strengths for each cell can be calculated by summing each signal strength value's column and then dividing each by the number of remaining valid measurement rows (which will be different from the 'number of times detected' value if rows containing empty measurements have been removed).
- The resulting average values can be transferred to the summary sheet entry for that cell.
- This process must be repeated for each detected cell.

- Once all cells have been processed and the details have been copied to the survey summary sheet, it will be necessary to sort the rows on the survey summary that contain cell results. If the sort is conducted using the main signal strength measurement column (RXLev for 2G, Ec/No for 3G or RSRQ for 4G), the resulting table will rank the detected cells in order of average received signal strength.
- Other elements in the summary sheet can then be calculated or added – serving percentage, signal strength descriptions, cell site names and details – as can details of whether each cell was detected as serving during a test call.
- In the example raw data table in Figure 9.11, the 'Test' column indicates the state of the survey equipment when each measurement event was captured; 'idle' indicates that the device was in Idle Mode, 'connected' indicates that it was in Connected Mode. It should therefore be a simple matter to scan through the table to determine which cells had a Status of 'S' during each test call period.

Once the compilation of the survey summary is complete the results of the survey will be available to investigators and analysts and will provide an indication of the serving coverage at the surveyed spot or location.

An example of a completed survey summary table is shown in Figure 9.12.

Cell Coverage Surveys

The process of converting a raw survey results table into a dataset that can be used to create a cell coverage map shares much of the stages of the spot/location survey method outlined above but is somewhat simpler.

It is assumed that the raw survey data is in a Microsoft Excel spreadsheet at the start of this process and that coverage maps will be created using Microsoft AutoRoute. It is also assumed that analysts will be using the 'pooled' results method to ensure the best chance of capturing details of non-contiguous areas of cell coverage.

- Create a 'pool' data workbook with tabs named 'route', 'serving' and 'first neighbour'.
- Begin with the 'single survey' raw results workbook for the first cell coverage survey.
- Copy just the GPS latitude and longitude columns to a new worksheet within the single survey workbook named 'Route' and ensure that the columns have the correct headers, for example 'Latitude' and 'Longitude'. Remove any rows where latitude and/or longitude as '0' as these will distort the resulting map. This data is then available to import directly into a mapping application to show the route driven during the survey.
- The second stage of the process is to copy the entire contents of the raw data table to a new worksheet within the 'single survey' workbook named '<cell ID > serving', where < cell ID > identifies the cell whose coverage is being mapped.

- In the new sheet, using the table in Figure 9.11 as an example, delete the date, time, test and RAT columns, leaving only the GPS latitude/longitude columns. Then delete all columns after the first cluster, so only retain details of the serving cell cluster in each row.
- Once the data has been cut down to show only the serving cell cluster, sort the data by Cell ID.
- Delete any rows that do not show the target cell ID as the serving cell.
- Repeat the process by copying the raw data table to a sheet named '<cell ID > first neighbour', then delete the event header columns apart from the GPS latitude/longitude columns, then delete all columns after the first neighbour cluster and finally delete the columns relating to the serving cell clusters.
- What remains should be details of the first neighbour in each event row.
- Sort the data by Cell ID and delete any rows that do not show the target cell as being first neighbour.
- Save the spreadsheet.
- The contents of the 'route', 'serving' and 'first neighbour' worksheets in the 'single survey' workbook should then be copied to the 'pooled data' workbook. If data has already been copied to the pool workbook then the new data should follow on below the existing data.
- The above process should then be repeated for all remaining single survey files, at the end of which the 'pooled data' workbook should contain details from all of the pooled surveys.
- It is now necessary to copy the details related to each detected cell to individual worksheets within the 'pooled data' workbook.
- Sort the contents of the pooled 'serving' worksheet by cell ID. Copy the set of measurements that relate to each detected cell ID to a separate worksheet each, named '<cell ID > serving'.
- Sort the contents of the pooled 'first neighbour' worksheet by cell ID. Copy the set of measurements that relate to each detected cell ID to a separate worksheet each, named '<cell ID > first neighbour'.
- The pooled data workbook now contains separate worksheets for each detected cell and that data is available for mapping.
- The mapping process below would then be repeated for each cell ID that is to be mapped.
- In Microsoft AutoRoute, open a new map and save it as '<cell ID > coverage'.
- Use the Import Data Wizard to:
  - Import data from the 'Route' tab first, change the pushpin icon to a small yellow circle and change the pushpin name to 'Route'.
  - Import data from the '<cell ID > first neighbour tab next, change the pushpin icon to a blue dot, for example, and change the pushpin name to '<cell ID > first neighbour'.
  - Import data from the '<cell ID > serving' tab last, change the pushpin icon to a red circle, for example, and change the pushpin name to '<cell ID > serving'.

- Save the map.
- After resizing the resulting map to show the entire extent of serving and first neigh-bour coverage for the cell, the map will be available to be copied into a Microsoft PowerPoint mapping presentation.
- Repeat for all remaining detected cells. It is acceptable to skip the mapping process for cells that were only detected as first neighbour cells and which never served.

An example of a complete cell coverage survey map is shown in Figure 9.13.

Route Profile Surveys

The process of converting a raw survey results table into a dataset that can be used to create a route profile map shares many of the stages of the cell coverage survey method outlined above but with important differences.

It is assumed that the raw survey data is in a Microsoft Excel spreadsheet at the start of this process and that route profile maps will be created using Microsoft AutoRoute.

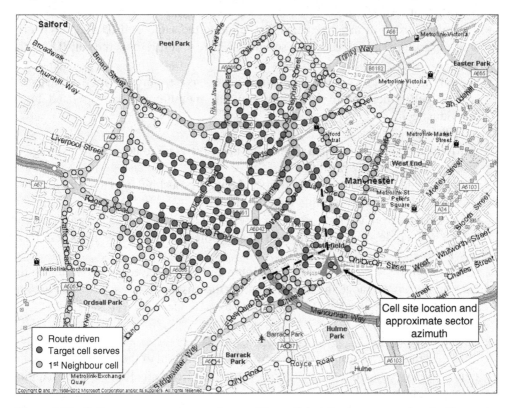

Figure 9.13 Cell coverage survey map. Source: Microsoft AutoRoute map reproduced with permission from Microsoft

- To begin with, copy just the GPS latitude and longitude columns to a new worksheet named 'Route' and ensure that the columns have the correct headers, for example 'Latitude' and 'Longitude'. Remove any rows that show latitude or longitude as '0' as these will distort the route map. This data is then available to import directly into a mapping application to show the route driven during the survey.
- The second stage of the process is to copy the entire contents of the raw data table to a new spreadsheet named 'all serving cells'.
- In the new sheet, using the table in Figure 9.11 as an example, delete the date, time, test and RAT columns, leaving only the GPS latitude/longitude columns. Then delete all columns after the first cluster, so only retain details of the serving cell cluster in each row.
- Once the data has been cut down to show only the serving cell cluster, sort the data by Cell ID.
- Delete any rows that do not show a Cell ID for the serving cell.
- It will now be necessary to create separate worksheets for each cell that was detected as serving along the surveyed route. This step is required to ensure that the coverage of each cell can be assigned a different pushpin colour when separately imported into AutoRoute.
- Copy each group of measurements that relate to the same Cell ID to a new worksheet, name the sheet '<cell ID > serving', where < cell ID > is the cell ID for the group of measurements.
- Repeat the above process until a separate sheet has been created for each detected serving cell, where each sheet only contains measurement rows related to an individual cell.
- Save the spreadsheet.
- In Microsoft AutoRoute, open a new map and save it as 'route serving coverage'.
- Use the Import Data Wizard to:
  - Import data from the 'Route' tab first, change the pushpin icon to a small yellow circle and change the pushpin name to 'Route'.
  - Import data from the first '<cell ID > serving' tab last, change the pushpin icon to a red circle, for example, and change the pushpin name to '<cell ID > serving'.
  - Repeat until the contents of all remaining cell sheets have been imported in the same map, but change the colour used for the pushpins for each cell so that different cell coverage along the route can be easily identified.

- Save the map.
- After resizing the resulting map to show the entire extent of the surveyed route, the map will be available to be copied into a Microsoft PowerPoint mapping presentation.

An example of a complete route profile survey map is shown in Figure 9.14.

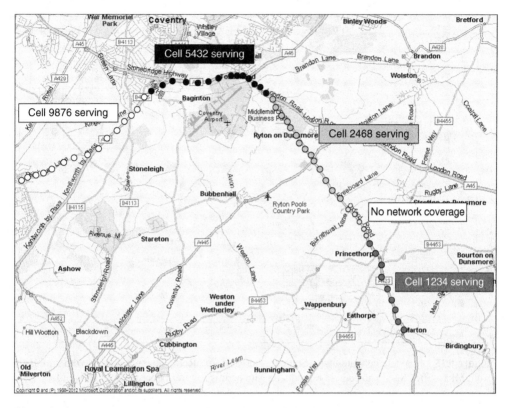

Figure 9.14   Route profile survey map. Source: Microsoft AutoRoute map reproduced with permission from Microsoft

9.16   Creating Call and Cell Labels

The mapping presentations that accompany cell site reports can be made more relevant and interactive by including labels that detail each cell being used during the period covered by the map and each call that takes place during that period.

An example of a cell site map, complete with cell and call labels, is shown in Figure 9.15.

The source for the information contained in the cell and call labels is the case call schedule. It is important that the call schedule is finalised before any call or cell labels are created from it; if details such as call index numbers are changed within the call schedule (or if calls are added to or taken away from the schedule) then it may mean that any labels created from that data will also need to be changed.

A number of methods can be used to create call and cell labels. Larger organisations may have developed macros or applications that allow them to be created automatically and commercial applications such as the Forensic Analytics CSAS tool also automate the production of these elements.

Many analysts are required to produce labels manually, however, so this section provides an overview of ways in which that can be achieved.

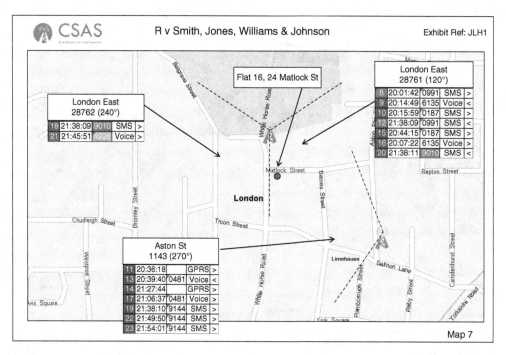

Figure 9.15   Example cell site map.   Source: Microsoft AutoRoute map reproduced with permission from Microsoft, CSAS map content reproduced with permission from Forensic Analytics Ltd

The assumption in this section is that the call schedule is stored in a Microsoft Excel spreadsheet and that the call and cell labels are to be transferred to a slide in a Microsoft PowerPoint presentation.

Cell labels are typically created in PowerPoint by drawing a rectangular object and then copying the cell name, cell ID and azimuth details into it. If the label is to be employed to highlight a cell that was only used by one target phone then the analyst might set the background colour of the label to match the attribution colour of the target phone. If the cell was used by multiple phones then a more neutral background colour might be used instead.

The content of a cell label might differ depending on the type of map slide that it is being created for. In the case of maps that show details of target phones when they are alleged to have been located at a particular address and which show details of only the cell(s) that serve at that address, cell labels are generally very detailed and might, for example, show cell name, postcode, cell ID and azimuth. This is illustrated in the left hand label in Figure 9.16.

Labels for maps designed to show the potential movement of target phone(s) around an area that contain details of several cell sites might use the more abbreviated format shown in the example on the right hand side of Figure 9.16.

In all cases the cell label is there to provide context for the calls and to indicate the location of the cell sites used by the target phones. It should be noted that the

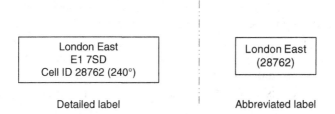

Figure 9.16 Cell detail labels. Source: CSAS cell labels reproduced with permission from Forensic Analytics Ltd

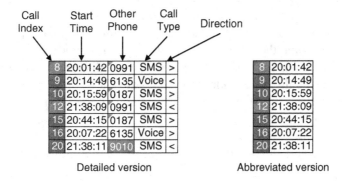

Figure 9.17 Call detail labels. Source: CSAS call labels reproduced with permission from Forensic Analytics Ltd

information shown on cell site maps almost always shows the location of the *cell sites used by target phones,* they do not usually indicate the location of the target phone itself when those calls were made.

Call labels are designed to provide details of the calls that take place during the period covered by a map slide. As shown in Figure 9.17, analysts can choose between a variety of designs and styles of label, which offer more or less information depending upon the scale of the map they are intended to be displayed on.

Typically, the minimum amount of information contained in a call label is the call index number and the time. The identity of the target phone to which the call relates is usually inferred from the colour applied to the call index number, which will match the colour assigned to that phone.

Less abbreviated formats might also include details of the type of event (call, SMS, GPRS), the other phone number involved in the call and the direction of the call (using '>' to signify and outgoing call and '<' to signify incoming).

Call labels are typically either copied directly from the call schedule and pasted into PowerPoint, or the details are copied from the spreadsheet into a 'table' object in PowerPoint. The first of these methods is quicker and simpler but results in a 'picture' object that cannot be edited, whereas the second method takes longer but results in a text object that can be edited.

If the 'copy and paste' method is to be used then it is recommended that the background colour of any non-coloured data cells is set to 'white' rather than the default 'transparent'. This should ensure that details of the map background do not show through the labels after they have been pasted to the PowerPoint slide.

The creation of cell and call labels is often the most time consuming aspect of cell site report writing and is an area that greatly benefits from some form of automation.

9.17   Cell Site Mapping Presentations

A cell site mapping presentation is a valuable tool for analysts and experts as it allows them to visualise the events and calls associated with a case in a more engaging and understandable way. If map presentations are created using an application such as Microsoft PowerPoint, then further enhancements can be made in the presentation by making use of the slide animation features. Cell and call labels can be set to appear in the order in which calls were made if the slides are shown in 'presentation' mode.

The typical process involved in producing a cell site report is to cut the call data into periods of interest and then into batches of calls that cover each stage of the events relevant to the case.

For example, a cell site report might group calls in the following way:

- Call 1 – target phone at or near Location A
- Calls 2–8 – target phones travelling in London
- Calls 8–23 – target phones at or near Location C
- Calls 23–27 – target phone travelling from Location C to Location D.

It is common for the maps created to support a cell site report to each concentrate on one batch of calls and the type of map created will depend upon the scale of the analysis being performed for that batch.

To use the list of calls provided above as an example, the set of maps an analyst would produce to details these calls might be something like this:

- Map 1– call 1 – small-scale map showing Location A and the serving cell(s) for that location. Detailed cell call labels would be used.
- Map 2 – calls 2–8 – very large-scale travelling map showing all Locations. No cell labels would be used, just abbreviated call labels with arrows pointing to the location of the associated cell site.
- Map 3 – calls 8–23 – small-scale map showing Location C and the serving cell(s) for that location. Detailed cell and call labels would be used.
- Map 4 – calls 23–27 – large-scale travelling map showing all cell sites used during the alleged journey from Location C to Location D.

An example of a small-scale, 'location detail' map is shown in Figure 9.18.

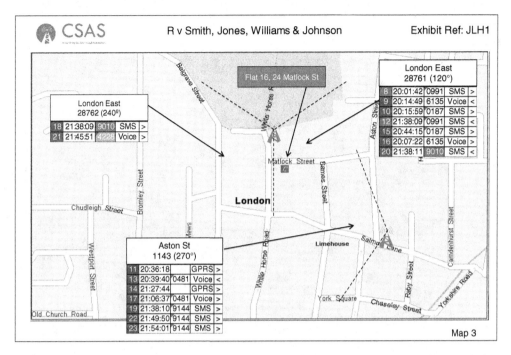

Figure 9.18 Location detail map. Source: Microsoft AutoRoute map reproduced with permission from Microsoft, CSAS map content reproduced with permission from Forensic Analytics Ltd

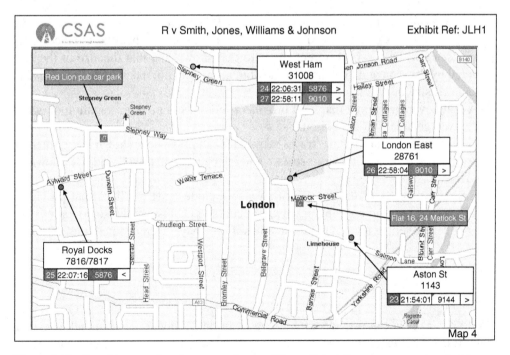

Figure 9.19 Travel between locations map. Source: Microsoft AutoRoute map reproduced with permission from Microsoft, CSAS map content reproduced with permission from Forensic Analytics Ltd

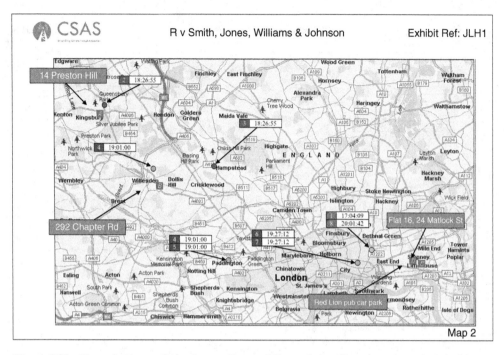

Figure 9.20   General pattern of travel map. Source: Microsoft AutoRoute map reproduced with permission from Microsoft, CSAS map content reproduced with permission from Forensic Analytics Ltd

The cell site 'sector arms' shown in Figure 9.18 are intended to be indicative of the cell's azimuth. It is usually considered important to point out that the actual cell coverage area will be a very different shape to that shown and that the size of the area covered by the icon and the length of the dotted lines are not meant to be indicative of actual cell size or coverage area.

A cell coverage survey map will provide a more accurate representation of actual cell coverage area if it is required.

An example of a large-scale 'travel between locations' map is shown in Figure 9.19.

An example of a very large-scale 'general pattern of travel' map is shown in Figure 9.20.

Cell site mapping presentations often also include a set of slides that provide a basic overview of the concepts involved in cell site analysis and also contain examples of cell coverage and route profile maps if they have been deemed relevant to the case.

9.18   Summary

Cell site analysis can provide compelling additional evidence to criminal investigators. It can highlight the possible locations of target phones during significant phone calls and can be used to support or cast doubt on alibis and witness accounts.

Call record data on its own is generally too imprecise to be used for anything more than a high-level analysis of a case – for example, it can answer broad questions such as 'was this phone in the east of London during this call?'

For cell site analysis to be able to answer low-level questions – such as 'could this phone have been at this specific address?' – investigators need to also employ forensic radio survey techniques.

Forensic radio surveys add detail and depth to the investigator's understanding of where target phones may have been located. The data provided by radio surveys adds forensic rigour to the discipline and ensures that the conclusions drawn by investigators can be backed up with testable evidence.

Cell site analysis has a number of limitations in the accuracy and certainty it can provide in its conclusions, but those limitations are rendered less severe by the addition of forensic radio survey evidence to a case.

It should be borne in mind that cell site analysis is a technique that can be as useful to the defence as it is to the prosecution and that it has as much power to support a person's alibi as it does to indicate their involvement in a crime.

References

[1] GSM Association: GSMA Intelligence (2014) *Home Page*, https://gsmaintelligence.com (accessed 30 May 2014).

[2] CDMA Development Group (2014) *Worldwide Deployments*, https://www.cdg.org/worldwide/index.asp (accessed on 30 May 2014).

[3] 4G Americas (2104) *3G/4G Deployment Status*, http://www.4gamericas.org/index.cfm?fuseaction=pageandpageid=939 (accessed 30 May 2014).

[4] International Telecommunications Union (2014) *Operational Bulletin No.1005 (1.VI.2012) and Annexed List: List of Mobile Country or Geographical Area Codes (Complement to ITU-T Recommendation E.212 (05/2008)) (Position on 1 June 2012)*, http://www.itu.int/pub/T-SP-OB.1005-2012 (accessed 2 June 2014).

[5] 3GPP Technical Specification (2013) *Radio Transmission and Reception*, TS 45.005 v11.3.0 Section 2, www.3gpp.org (accessed 29 July 2014).

[6] 3GPP Technical Specification (2013) *Base Station (BS) Radio Transmission and Reception (FDD)*, TS 25.104 v11.7.0 Section 5.2, www.3gpp.org (accessed 29 July 2014).

[7] 3GPP2 Technical Specification (2010) *Band Class Specification for cdma2000 Spread Spectrum Systems Revision E*, C.S0057-E v1.0, www.3gpp2.com (accessed 29 July 2014).

[8] 3GPP Technical Specification (2013) *Evolved Universal Radio Access (E-UTRA); Base Station (BS) Radio Transmission and Reception*, TS 36.104 v11.6.0 Section 5.2, www.3gpp.org (accessed 29 July 2014).

Index